HANDBOOK OF IMAGE PROCESSING OPERATORS

HANDBOOK OF IMAGE PROCESSING OPERATORS

Reinhard Klette
Berlin Technical University
Germany

Piero Zamperoni
Braunschweig Technical University
Germany

JOHN WILEY & SONS
Chichester • New York • Brisbane • Toronto • Singapore

Published 1996 by John Wiley & Sons Ltd.
Baffins Lane, Chichester,
West Sussex PO19 1UD, England

National 01243 779777
International (+44) 1243 779777

Originally published in German language by Friedr. Vieweg & Sohn Verlagsgesellschaft mbH, D-65183 Wiesbaden, Germany, under the title "Handbuch der Operatoren für die Bildbearbeitung. 2. Auflage (2nd Edition)".

Other Wiley Editorial Offices

John Wiley & Sons, Inc., 605 Third Avenue,
New York, NY 10158-0012, USA

Jacaranda Wiley Ltd, 33 Park Road, Milton,
Queensland 4064, Australia

John Wiley & Sons (Canada) Ltd, 22 Worcester Road,
Rexdale, Ontario M9W 1L1, Canada

John Wiley & Sons (Asia) Pte Ltd, 2 Clementi Loop #02-01,
Jin Xing Distripark, Singapore 0512

Library of Congress Cataloging in Publication Data

Klette, Reinhard.
[Handbuch der Operatoren für die Bildbearbeitung. English]
Handbook of image processing operators / Reinhard Klette, Piero Zamperoni.
p. cm.
Includes bibliographical references and index.
ISBN 0 471 95642 2
1. Image processing – Handbooks, manuals, etc. 1. Zamperoni, Piero. 11. Title.
TA 1632. K57 1995
621.3' 67' 028551 – dc20 95 20540
CIP

British Library Cataloguing in Publication Data

A catalogue record for this book is available from the British Library

ISBN 0 471 95642 2

Produced from camera-ready copy supplied by the author using Ms Word for Macintosh.
Printed and bound in Great Britain by Bookcraft (Bath) Ltd.
This book is printed on acid-free paper responsibly manufactured from sustainable forestation, for which at least two trees are planted for each one used for paper production.

CONTENTS

Bilder, die man aufhängt umgekehrt,
mit dem Kopf nach unten, Fuß nach oben,
ändern oft verwunderlich den Wert,
weil ins Reich der Phantasie erhoben.

Christian Morgenstern in *Die Galgenlieder*, 1905

PREFACE

Image processing is that domain of computer vision dealing with mappings of images onto images. These mappings can aim, e.g., at image improvement, image restoration, or at object labeling. Desk-top publishing systems represent, for example, a recent application area of image processing. In general, computer vision comprises the computer-based processing, analysis, classification, or interpretation of pictorial information. Image processing, the topic of this book, covers a first step of computer vision for "an esthetic enhancement" of images for visual analysis, or for enhancement of features for a further automatic image analysis. This image processing can be satisfying for interactive image evaluations if analytical processes are performed by a human. Image analysis is a further subfield of computer vision having as its subject the progressive replacement of human by computer-based methods for carrying out image analysis processes. This book is not focusing on analytical topics of computer vision. They are mentioned here only as motivation for performing mappings of raw image data onto digitized images.

The perspective of this book on image processing is centered on the fast access to applicable methods or algorithms for a fast development of user solutions. The book is intended to support the user's own design of image processing solutions. Our intention is that it may also be used as a reference book for obtaining methodical or algorithmic hints or suggestions for desired mappings of images onto "new images". With a few exceptions, the book deals only with the processing of gray value images.

For manuscript correction help we thank especially *Kjell Oppermann* and *Wolfram Schimke*. Both students have developed a floppy disk with C-source code (cp. mailing card enclosed to this book) and also some figures for this book. Careful proof-reading was also performed by *Gisela Klette*. Some figures were provided by *Wolfgang Schwanke*. For the English translation we thank *Ashutosh Malaviya* for his very valuable collaboration. We thank also *Gunter Bellaire,* Dr. *Andreas Koschan, Karsten Schlüns* and Prof. Dr. *Horst Völz* for some comments. For careful typewriting during the iterated manuscript modifications we thank *Sabine Hagedorn* .

Reinhard Klette **Piero Zamperoni**

INSTRUCTIONS TO THE READER

THE ROLE OF THE INTRODUCTORY CHAPTERS

This book was written as a Handbook for implementing image processing operators. Chapters 1 to 3 are fundamentals and general concepts. Specific operators are described in Chapters 4 to 7. The Glossary at the end of the book provides references to operators.

The material in Chapters 1 to 3 may be used to support the understanding of specific operators. A fast glance through these Chapters at the beginning should be sufficient to recognize what can be found here if some questions arise during the implementation of a specific operator.

STANDARD OPERATOR PRESENTATION SCHEME

The description of each operator in Chapters 4 to 7 is structured into five paragraphs numbered **(1)** to **(5)**.

(1) Characterization: A brief description of the operator, describing its general effects, is given here. Under the subsection "Attributes", some properties are listed:

- type of input images for this operator,
- type of operator, and
- (in case of a local operator) characterization of the operator kernel,

cp. synopsis in Subsections 1.1.3, 1.3.1 and 1.4. Under subsection "Inputs", are listed some parameters which have to be entered by the user when running the program, as, e.g., window size or desired variant of operator.

(2) Mathematical Definition: Here, a formal mathematical description of the relations between gray values in input images and gray values in output images is provided.

(3) Comments: Some remarks on the effects of the operator, its applications and relationships between effect and performed operations can be found here. In some cases, algorithmic aspects and ways for improving the run-time efficiency are described. For some operators the methodical background is sketched.

(4) Algorithm: Algorithms are described by using a (simplified) pseudo-programming language which is derived from Pascal. For brevity, informal descriptions are used if the interpretation is straightforward. Many operators can be implemented just by defining a specific operator kernel procedure using one of a few control structures (see Subsection 3.3.3). In these cases, only the kernel realization is specified and in the text layout is graphically emphasized by enclosing it in a box.

For the implementation of point or local operators the reader can start with selected control structures as given in Subsection 3.3.3.

The given algorithms are based on a row- or column-wise access to pictorial data. Recent computers allow random access to full images during processing (random access has been assumed for the programs on the floppy disk which can be ordered by using the reader-reply card at the back of the book), but in specific applications memory restrictions still have to be taken account of.

(5) Bibliographic References: Examples of relevant literature are given with or without comments.

OFTEN USED VARIABLES

IMAGES

f	input image, default: gray value image
h	output image, default: gray value image
F	window in input image with spiral indexing $F(z)$, cp. Fig. 3.5, or with ij-window coordinates $F(i, j)$

IMAGE PARAMETERS AND INDICES

A	image size, $A = M \cdot N$
M	number of image columns
N	number of image rows
p, q	image points, e.g. $p = (x, y)$
x, y	column index $1 \leq x \leq M$ and row index $1 \leq y \leq N$
G	number of gray values, default: $G = 256$
u	image value (in input image f), default: gray value between 0 and $G-1$
v	image value (in output image h), default: gray value between 0 and $G-1$

WINDOW PARAMETERS AND INDICES

a	window size, $a = m \cdot n$
m, n	number of columns or number of rows of a window, default: $m = n$ odd
k	for $n = m$ it holds $k =$ **integer**$(\frac{n}{2})$
i, j	column and row index, for $n = m$: $-k \leq i, j \leq k$ (n odd), $-k \leq i, j \leq k-1$ (n even) or $1 \leq i, j \leq n$

1 IMAGES, WINDOWS AND OPERATORS

This Chapter contains basic definitions. Fundamental notations of digital images are explained, e.g. connected sets of image points, directional encoding, gray value gradient, or color channels. Important functions on image data are introduced such as gray value histograms, median, variance, or modifications of color models. *Images* are the objects to be processed. *Operators* are the tools for performing such tasks. *Windows* moving on images characterize a typical way of computing operators. For images, windows, window functions (operator kernels), and operators some attributes are summarized. These attributes are used in Chapters 4 to 7 for the brief characterization of operators.

Specifying an image processing operator means that several decisions have to be taken, e.g., selection of a coordinate system, choice of parameters fitting the operator to the specific type of images which have to be processed. These decisions require some standard definitions which are also briefly discussed in this Chapter.

1.1 IMAGES

An image is a mapping of *spatial coordinates* (x, y) into a certain set of image values, often identified with a set of gray values. Formally, an image is a function f defined on a set of image points. An image f designates at each *image point* $p = (x, y)$ a univocal function value as its *image value* $f(p) = f(x, y)$ [1]. In general, such an image value can be assumed to be a numerical *gray value* u characterizing some gray shade. Formally, $f(x, y) = u$. A triple $(x, y, f(x, y)) = (x, y, u)$ is a *pixel* (abbreviation of picture element).

---

[1] Formally correct, for $p = (x, y)$ it follows that $f(p) = f((x, y))$. Considering the conventions on the variable names, it will be clear that $f(x, y)$ and $f(p, t)$ where p denotes a point and t the time, represent different cases.

Gray values are numerical representations of certain *gray shades*. Such an interpretation of gray values is selected by defining a *look-up table*[2] for representing images on a screen, and it can be modified by changing brightness or contrast of the screen. The perception of gray shades is also subjective for humans. In general, a human can visually discriminate about 30 different gray shades - between White and Black - in a single image. The discrimination between gray shades can be influenced by the nature of the specific images. For example, in general a small square with gray value 187 can visually be recognized on a background of constant gray value 188. However physiological image perception is not considered in this book. Images are considered to be numerical objects.

It is assumed that images can be displayed on a screen or inside a window within the scope of image processing. In discussing the spatial placement of image points, an *iconic image representation* is assumed. This means that image data are not encoded in some way but are in "natural two-dimensional" array arrangement. Notations such as "square sub-image" or "isolated image point" have their intuitive meaning for iconic representations.

1.1.1 Discrete Image Coordinates and Digital Geometry

In this book an *xy-coordinate system* is taken as a basis for image coordinates, as shown in Figure 1.1. The origin is located at the bottom left corner. This system has a mathematical positive orientation. For both coordinate axes, an identical scaling is assumed. The programs in this book are based on the *xy*-coordinate system as shown in Figure 1.1. This is especially important for the geometrical transformations of Chapter 4.

Attention: If the reader wishes to extend a computer vision system for which image procedures have already been implemented, it is recommended to leave the coordinate system unchanged. To avoide errors, a fixed coordinate system should be used within a computer vision system. The specifications of index definition intervals for coordinates x, y and gray values u are of even greater importance. Incorrect indexing can lead to serious errors in the programs.

Optical systems generate *analog images* where spatial coordinates (x, y) and gray values u can be considered as continuous quantities. In computer-based image processing, images are data which can be stored in a memory, and manipulated by a processor. For such data, only a finite number of spatial coordinates, and gray values are possible. Here, images are *digital images* or *discrete images*.

---

[2] See Glossary.

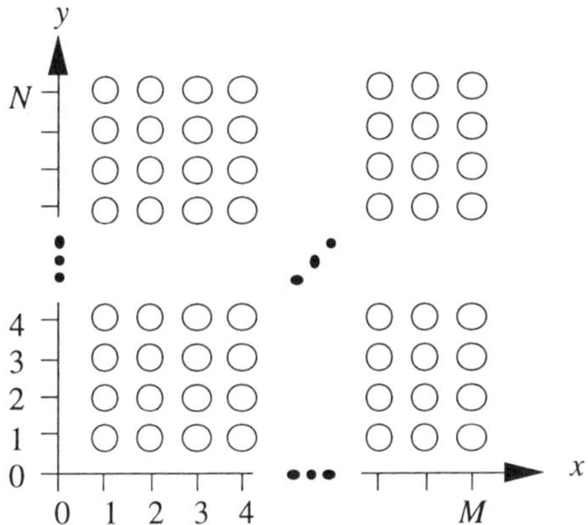

Figure 1.1: Orientation of a Cartesian xy-coordinate system as used in this book, for $A = M \cdot N$ image points. For image point $p = (x, y)$, x is its column index and y its row index. Columns are numbered from left to right, and rows are numbered from bottom to top.

For image points (x, y) of digital images, x and y are assumed to be integers (*discrete grid*) with index intervals $1 \leq x \leq M$ and $1 \leq y \leq N$. The values M and N specify the *image resolution*. The value $A = M \cdot N$ characterizes the *image size*. A digital image f consists of $A = M \cdot N$ function values $f(x, y)$ in a uniform spatial distribution on a $M \times N$ image *raster*[3]

$$\mathbf{R} = \{(x, y) \colon 1 \leq x \leq M \wedge 1 \leq y \leq N\}.$$

Instead of a grid with *unity grid step* other grid steps could be assumed such as grid step 0.1 and x- or y- values 0.1, 0.2, 0.3, ... Also, x can run between 0 and $M - 1$, and y between 0 and $N - 1$. As discussed for the coordinate system, a univocal convention is suggested within a computer vision system. Quite often for images it is assumed that $N = M$ (*square images*), and more specifically that N is a power of two. In general, programs for image processing should work with any values M and N in acceptable limits.

Attention: Typically a single image represents a very large amount of data. Operations on images are often time-consuming. To save time during program development, small test images should be used.

---

[3] By $A = M \cdot N$ (only) a number A is given. $M \times N$ also denotes a two-dimensional array constituting of M columns and N rows.

Image points (x, y) of the image raster are *grid points* of an equidistant orthogonal grid with identical scaling in the x- and y-direction. Especially for older displays, the screen geometry can be different from this ideal assumption. Then, for image visualization certain scaling factors should be used, in order to reproduce true circles and to eliminate the skewing error which represents circles as ellipses.

The consideration of the image raster as a set of points represents a mathematical abstraction. Geometrically, a point is of dimension 0, i.e. can be seen to be "arbitrarily small". Image values $f(x, y)$ are considered to be values in isolated grid points. Compared with the physical process of sampling for the generation of digital images, this is of course an ideal assumption. Alternatively, the image raster could be defined as a set of $M \times N$ square *grid cells* (x, y) of unity side length, and a value $f(x, y)$ could be assumed to be constant in the grid cell (x, y). Such an area-oriented definition of the image raster and of image values would correspond to the graphical representations of images on a screen. Here, each elementary screen dot (circles, squares or geometrically distorted figures) is "colored" by a pixel value. A VGA screen has 800×600 elementary screen dots. However in this book, picture elements (pixels) are gray values in image points, instead of image cells. This perspective reflects the fact that, in practice, gray values are stored into a digital memory.

The values of a digital image f on the image raster are isolated events at first glance. For defining relations between these events, certain neighborhood relations have to be specified. It is quite straightforward to consider the eight nearest image points

$(x-1, y+1)$	$(x, y+1)$	$(x+1, y+1)$
$(x-1, y)$	$\mathbf{(x, y)}$	$(x+1, y)$
$(x-1, y-1)$	$(x, y-1)$	$(x+1, y-1)$

as neighborhood. The 8-*neighborhood* of an image point $p = (x, y)$ is the set

$$\begin{aligned}\mathbf{N_8}(\mathrm{p}) &= \{(x, y-1), (x, y+1), (x-1, y), (x+1, y), (x-1, y-1), \\ &\qquad (x-1, y+1), (x+1, y-1)\ (x+1, y+1)\} \\ &= \{\ (i, j) : \ \mathbf{max}\ \{\ |i-x|, |j-y|\ \} = 1\ \}.\end{aligned}$$

From a different perspective, the 4-*neighborhood*

$$\begin{aligned}\mathbf{N_4}(\mathrm{p}) &= \{(x, y-1), (x, y+1), (x-1, y), (x+1, y)\} \\ &= \{\ (i, j) : \ |i-x| + |j-y| = 1\ \}\end{aligned}$$

of a point $p = (x, y)$ can be chosen. The neighborhood choice has direct consequences upon algorithms running on the image raster.

An image processing task can consider the representation of "connected sets of image points" in the pictorial objects, e.g. of "object regions". Such a subset **Q** of the image raster **R** can be represented as a *binary image f*, where $f(x,y) = 1$, if (x,y) is in the set **Q**, and $f(x,y) = 0$ otherwise.

Two image points p and q of the image raster **R** are denoted to be 4-*neighbors*, if p is contained in the 4-neighborhood $\mathbf{N_4}(q)$ of q, i.e. if q is contained in $\mathbf{N_4}(p)$. A 4-*path* is a finite sequence $p_1, p_2, ..., p_n$ of image points where p_i and p_{i+1} are 4-neighbors, for $i = 1, 2, ..., n-1$. A set **G** of image points is 4-*connected*, if for each pair of points p, q in **G** there is at least one 4-path from p to q which contains only points belonging to **G**. An arbitrary set **Q** of image points can consist of several maximum 4-connected subsets called 4-*components* of **Q**. The notations 8-*neighbors*, 8-*path*, 8-*connected*, and 8-*components* are defined in analogous way if the 8-neighborhood is assumed. The recognition of connected components is considered in Section 7.1.1.

In Figure 1.2, on the left an analog binary image is shown consisting of four squares, two of which are white ($f(x, y) = 1$) and two are black ($f(x, y) = 0$). On the right, in the digital binary image the pair of black as well as the pair of white squares form an 8-connected set each. This can be seen as a contradiction to our perceptual experience with the analog image, i.e. that a connection also implies a separation. Under the assumption of 4-neighborhood, neither the black squares nor the white squares are 4-connected. Here, a separation implies no connection. The analysis of connected sets of image points constitutes an important topic in the field of image analysis.

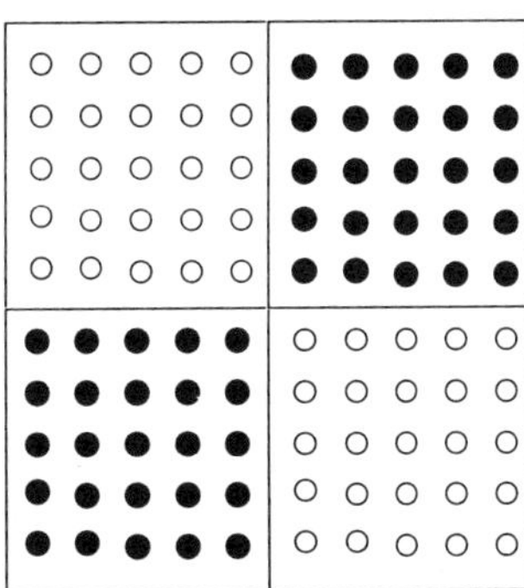

Figure 1.2: Is there a topological connection between the black squares and/or between the white squares? In the analog image one should consider the point in the middle which does not exist in the digital image.

For discrete binary images, a useful approach can be to assume 4-neighborhood for all the white image points, and 8-neighborhood for all the black image points, or vice-versa. Then, in Figure 1.2 a pair of digital squares is connected, and the other pair is not. Such *topological problems* have to be considered for some operators given in Chapter 7 (Sections 7.1 and 7.2).

For the analysis of sets of image points, basic concepts of the Euclidean geometry (angle, length, congruence, area etc.) should be reconsidered for the corresponding digital approaches. This is not a straightforward process.

For the distance definition for image points, assume points $p_i = (x_i, y_i)$ for $i = 1, 2, 3, \ldots$ The *Manhattan* or *city-block-metric*, also called l_1-*metric* in mathematical analysis,

$$d_1(p_1, p_2) = |x_1 - x_2| + |y_1 - y_2| ,$$

corresponds to distance measurements by measuring the length of shortest connecting 4-paths (4-way-stepping). The *Euclidean metric*, or l_2-*metric*,

$$d_2(p_1, p_2) = \sqrt{(x_1 - x_2)^2 + (y_1 - y_2)^2} ,$$

can also be used, of course, on grid points. The *maximum metric*, or l_∞-*metric* ,

$$d_\infty(p_1, p_2) = \mathbf{max}\{|x_1 - x_2|, |y_1 - y_2|\} ,$$

corresponds to distance measurements by measuring the length of shortest connecting 8-paths (8-way-stepping).

A set of image points can be described by means of its contour. For this aim, the contour steps are coded using a set of basic steps in the *main directions*. Basic steps can be represented by directional codes 0, 1, 2, ... For step sequences assuming the 8-neighborhood, the following encoding scheme for basic steps is used:

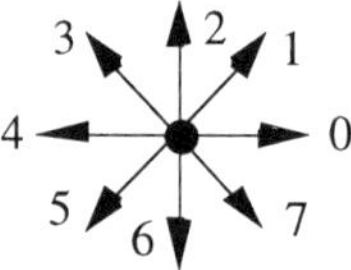

Each code number c corresponds one-to-one to a vector (a, b) describing the basic step in the xy-coordinate system of the image plane:

directional code c	0	1	2	3	4	5	6	7
directional vector (a, b)	(1, 0)	(1, 1)	(0, 1)	(-1, 1)	(-1, 0)	(-1,-1)	(0, -1)	(1, -1)

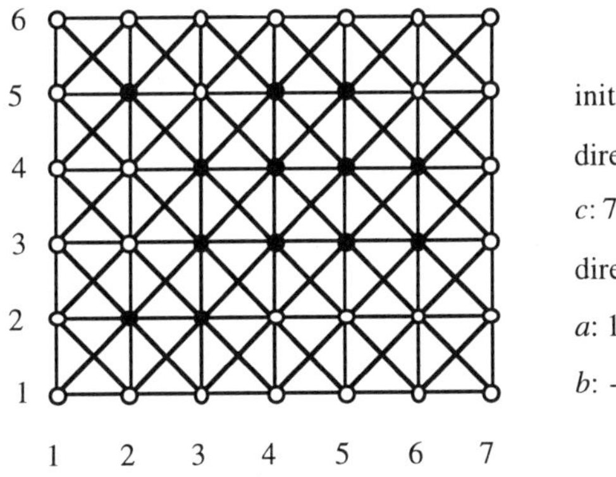

Figure 1.3: An 8-connected set of image points, and a directional encoding of the contour.

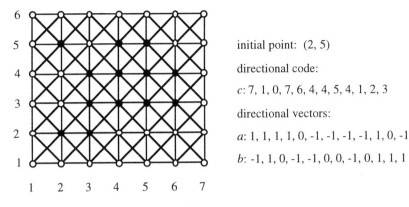

The *directional code of a contour* consists of the start coordinates (x, y) of an initial point, and of a sequence of directional code numbers, cp. Figure 1.3. Additionally, further information on the contour can be included, e.g. by using geometrically impossible sequences of code numbers.

If a contour is closed, then the sums on all a_i 's and on all b_i 's of the directional vectors (a_i, b_i) satisfy

$$\sum a_i = \sum b_i = 0 \quad .$$

This equation can be used as a test for correctness, cp. operator in Section 7.2.1. The *contour length* is often estimated by the number of directional code numbers, i.e. the number of pixels on the contour registered during contour following. In the orthogonal grid, an other estimate of the contour length is given by

$$\mathbf{card}\{\text{even directional codes}\} + \sqrt{2} \cdot \mathbf{card}\{\text{odd directional codes}\}.$$

For calculating the area, the following *area formula for directional codes* can be applied. The area of the polygon defined by the basic steps of the closed contour is equal to the half absolute value of the sum:

$$\sum_{i=1}^{n} a_i \cdot \left(y_{i-1} + \frac{b_i}{2} \right)$$

with $y_0 = 0$ and $y_i = y_{i-1} + b_i$. In general, this area formula for directional codes can be derived from the *area formula for non-crossing closed polygonal chains*. For such a polygon, given by the sequence $p_1, p_2, ..., p_n$ of vertices, with $p_i = (x_i, y_i)$, the area is equal to half of the absolute value of the sum

$$\sum_{i=1}^{n} x_i \cdot \left(y_{i-1} - y_{i+1}\right),$$

where $y_0 := y_n$ and $y_{n+1} := y_1$. The values of *x-width* and *y-width* of a grid point polygon,

$$x\text{-width} = \mathbf{max}\ \{x_i\text{: } 1 \le i \le n\} - \mathbf{min}\ \{x_i\text{: } 1 \le i \le n\} \quad \text{and}$$
$$y\text{-width} = \mathbf{max}\ \{y_i\text{: } 1 \le i \le n\} - \mathbf{min}\ \{y_i\text{: } 1 \le i \le n\},$$

can be calculated by repeated updating of the values

$$x_i = \sum_{j=1}^{i} a_j + x_0 \quad \text{and} \quad y_i = \sum_{j=1}^{i} b_j + y_0, \quad \text{for} \quad i = 1, 2, ..., n$$

during the contour following, where (x_0, y_0) denotes the initial point of the contour. Such shape features as area, width or cotter length have their importance in approaches to contour image analysis.

1.1.2 Discrete Image Values and Functions of Image Values

For image values $f(x, y)$ of a digital image f, it is common to assume a range of G gray values, $G \ge 2$. By *quantization*, continuously distributed gray shades are mapped into a finite set of numerical gray values. For this range of G gray values, a connected interval of non-negative integers is assumed. For gray value u, let

$$0 \le u \le G - 1.$$

In image processing it is practical to use the values $G > 2$ (*gray value image*) and $G = 2$ (*binary image*). $G = 256$ is the default value. Here, for each gray value 8 bits are needed (because $2^8 = 256$), which can be seen as an efficient and simple way for storing image data (one byte represents one gray value). Because the data type *byte* covers integers from – 128 to + 127, a certain adjustment to the domain 0...255 is necessary. With 256 levels of gray also the upper limit of visually separable gray levels is approximately attained. Only two gray levels are accepted as possible gray values in *bilevel images*. Normally, 0 and 255 are selected for better visibility instead of 0 and 1 as in binary images.

Gray value 0 is interpreted by gray shade Black, and gray value $G - 1$ is interpreted by gray shade White:

gray value image f **(scalar image)**
two-dimensional image matrix with image values $f(x, y) = u$
gray value range $\{0, 1, ..., G-1\}$ with $G > 2$

gray value	$u = 0$	corresponds to the gray shade Black
gray value	$u = (G-1)/2$	corresponds to a gray shade "Middle-Gray"
gray value	$u = G-1$	corresponds to the gray shade White

binary image b
two-dimensional image matrix with image values $b(x, y) = u$
gray value range $\{0, 1\}$

gray value	$u = 0$	corresponds to the gray shade Black
gray value	$u = 1$	corresponds to the gray shade White

For image processing, the computer representation of images can be restricted to matrices (image matrices). Following Figure 1.1, x denotes the column index and y denotes the row index.

A binary image can be seen as a special gray value image. For image visualization, a binarization, i.e. a mapping of a variety of gray levels onto two gray shades Black and White, can be of relevance (*halftoning*). This is true for black-and-white screens, or for some printers. Here, in general a higher gray value resolution should be traded off against a higher spatial resolution for obtaining high-quality visualizations. However for a constant spatial resolution, binarization can also be an interesting graphical effect, or an important step within an image analysis procedure. But note that the "complex structure of information" given in a multi-level gray value image gets lost by binarization to a great extent. So, this step should be avoided as far as possible.

The gray value image *BUREAU* of the French test-picture data set of *CNRS*[4] is shown in Figure 1.4. Two windows are marked in this picture. The numerical gray values of these windows are given on the right. As with a magnifying glass, this shows the gray value structure in detail. The lower picture window is characterized by a relatively homogenous distribution of gray values. A strong discontinuity characterizes the upper picture window. The gray value pattern shows an increment of gray values. Orthogonal to this increment there is a *gray value edge* in this image.

Statistical estimations, such as the estimates *AVERAGE* or *VARIANCE* of the mean or of the variance, can be used to evaluate the gray value structures.

---

[4] Test-picture data sets are of value for evaluating results in image processing on identical pictorial data. Here, reference is made to a data set of several 256 x 256 gray value images (256 gray values) distributed by the French *CNRS*.

139	140	136	140	172	221	217	222	219	217
136	137	137	143	169	219	212	218	222	209
139	140	139	138	171	218	217	219	222	213
141	145	145	145	172	219	214	222	217	205
138	139	147	147	172	217	218	219	224	203
143	146	143	143	170	219	220	224	225	201
150	144	144	146	172	217	222	221	222	210
137	132	138	137	166	221	219	224	229	210
138	143	141	146	171	222	221	225	223	196
146	145	141	142	169	221	222	223	218	189
142	141	141	147	172	217	223	225	223	215
150	143	145	147	170	225	225	222	226	220
149	140	144	139	169	223	225	223	227	223
142	139	143	144	172	224	224	225	229	224
147	144	145	147	171	227	229	228	233	225
142	141	140	147	173	226	227	227	233	228

118	122	118	118	121	119	118	118	116	116
119	117	118	118	116	117	115	117	119	112
119	116	119	116	117	118	117	115	119	113
120	121	120	120	116	117	115	117	117	116
117	122	116	118	120	116	114	115	114	120
119	118	117	117	121	118	118	116	116	117
118	118	119	116	118	119	119	115	116	113
118	121	117	119	117	115	118	112	118	116
117	119	115	117	120	117	115	118	116	118
118	119	116	119	116	116	117	118	114	115
118	117	119	119	118	119	118	119	118	118
117	120	117	119	117	117	119	115	119	120
120	117	119	119	115	119	119	119	119	116
119	118	118	121	121	118	116	120	119	119
120	122	118	123	119	121	121	124	118	120
118	118	118	120	120	120	119	120	119	122

Figure 1.4: For the gray value picture *BUREAU*, the numerical gray value structure of two 10×16 picture windows is shown. The lower picture window **F**(*BUREAU*, (141,56)) represents a relatively homogenous region in the picture. The upper picture window **F**(*BUREAU*, (133,175)) is characteristic for a portion of a so-called gray value edge.

These values are defined as

$$AVERAGE = \frac{1}{a} \sum f(p) \text{ and } VARIANCE = \frac{1}{a} \sum [\, f(p) - AVERAGE \,]^2 ,$$

where the sums are taken on all points p of the placed window or the region of interest, and a is the number of elements, i.e. points p. In Figure 1.4, for the lower (homogenous) picture window it is $AVERAGE = 117.9$ and $VARIANCE = 4.1$, and for the upper (inhomogenous) picture window it is $AVERAGE = 184.0$ and $VARIANCE = 1403.8$.

The notion of a *gray value relief* is illustrated by 3-D plots of gray value distributions in Figure 1.5. This intuitive notion assumes that in local windows of digital images, gray values "don't differ too much". Under this smoothness assumption, gray values form *plateaus*, *slopes*, or *valleys*. To be mathematically precise, discrete *functions of image values* have to be characterized which represent different discrete "elevations of a three-dimensional surface".

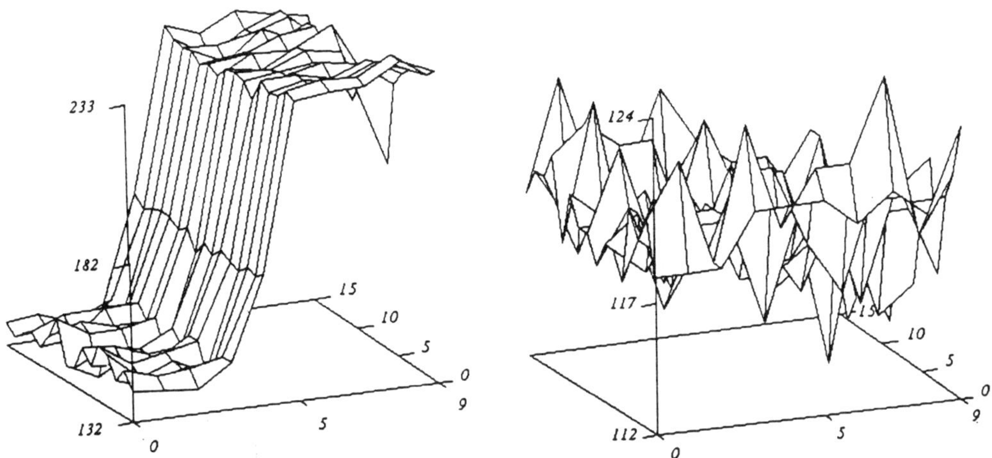

Figure 1.5: 3-D plots of both marked picture windows of Figure 1.4: lower - on the left, upper - on the right.

By continuous interpolation of these discrete image values, a digital gray value image can be transformed into a continuous ("smooth") image value distribution. For a continuous function $z = f(x, y)$ in two variables x, y, the *gradient* **grad** f of f,

$$\mathbf{grad}\, f = \left(\frac{\partial f}{\partial x}, \frac{\partial f}{\partial y} \right),$$

characterizes "changes in the elevation" of this function in the xyz-space. In each point $(x, y, f(x, y))$ of such a continuous image value function, a tangential plane is defined orthogonal to the *normal vector*

$$\mathbf{n} = \left(\frac{\partial f}{\partial x} , \frac{\partial f}{\partial y} , 1 \right) .$$

The direction of this normal vector is characterized by angles α, β, γ between vector and xyz-coordinate axes. The directional angle γ to the z-axis, cp. Figure 1.6, characterizes the inclination of the image value function with respect to the xy-plane in a given point $(x, y, f(x, y))$. This *angle of inclination* can be used to describe transitions between gray value plateaus and gray value slopes (e.g. gray value edges). The absolute value of the gradient,

$$|\,\mathbf{grad}\ f\,| = \sqrt{\left(\frac{\partial f}{\partial x}\right)^2 + \left(\frac{\partial f}{\partial y}\right)^2} ,$$

or the absolute value of the normal vector,

$$|\,\mathbf{n}\,| = \sqrt{\left(\frac{\partial f}{\partial x}\right)^2 + \left(\frac{\partial f}{\partial y}\right)^2 + 1} ,$$

is equal to zero at a pixel belonging to a constant gray value plateau, i.e. if it holds $f(x, y) = const$ in the local window around (x, y). These values can be used for characterizing slopes of the image value function.

For the directional angles α, β, γ of vector $\mathbf{n}$ with the coordinate axes it holds that

$$\cos \alpha = \frac{\frac{\partial f}{\partial x}}{|\,\mathbf{n}\,|} , \quad \cos \beta = \frac{\frac{\partial f}{\partial y}}{|\,\mathbf{n}\,|} \quad \text{and} \quad \cos \gamma = \frac{1}{|\,\mathbf{n}\,|} .$$

For the angle of inclination

$$\gamma = \mathbf{Arc\ cos} \frac{1}{|\,\mathbf{n}\,|}$$

it follows that

$$\gamma = \mathbf{Arc\ tan} \left(|\,\mathbf{grad}\, f\,| \right) .$$[5]

[5] $\mathbf{Arc\ cos} \frac{1}{|\mathbf{n}|} = \mathbf{Arc\ cot} \frac{1}{\sqrt{|\mathbf{n}|^2 - 1}} = \mathbf{Arc\ cot} \frac{1}{|\,\mathbf{grad}\, f\,|} = \mathbf{Arc\ tan} \left(|\,\mathbf{grad}\, f\,| \right)$

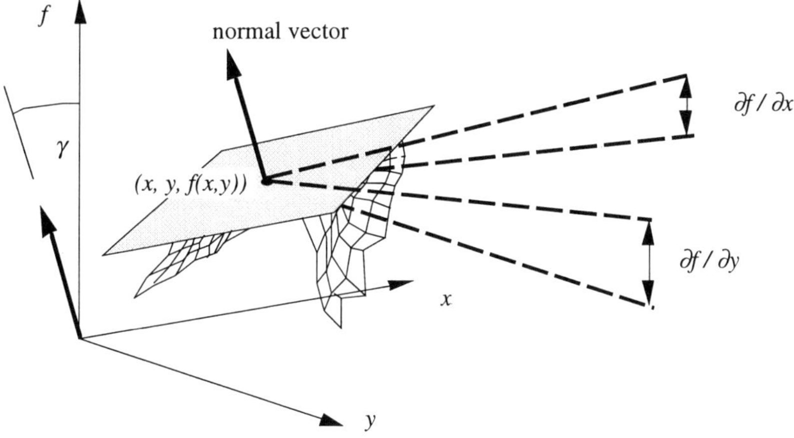

Figure 1.6: A tangential plane to a point $(x, y, f(x, y))$ of an image value function, the corresponding normal vector and the angle of inclination γ.

Gray value edges can be modeled, e.g., by strong local changes of angle γ. For the modeling of gray value edges, more efficiently computable values can also be used as ratios of directional changes of image value functions in x- and y- directions,

$$\frac{\partial f}{\partial x} \Big/ \frac{\partial f}{\partial y} \quad \text{and} \quad \frac{\partial f}{\partial y} \Big/ \frac{\partial f}{\partial x} \quad .$$

The absolute value of the gradient of an image value function is invariant with respect to rotation and translation of the xy-coordinate system. The same is true for the square of this absolute value,

$$\left(\frac{\partial f}{\partial x}\right)^2 + \left(\frac{\partial f}{\partial y}\right)^2 ,$$

for the *Laplacian* of the image f,

$$\frac{\partial^2 f}{\partial x^2} + \frac{\partial^2 f}{\partial y^2} ,$$

and for the *square variation*,

$$\left(\frac{\partial^2 f}{\partial x^2}\right)^2 + 2\left(\frac{\partial^2 f}{\partial x \partial y}\right)\left(\frac{\partial^2 f}{\partial y \partial x}\right) + \left(\frac{\partial^2 f}{\partial y^2}\right)^2 \quad .$$

In digital images, the derivatives $\partial f / \partial x$ and $\partial f / \partial y$ have to be approximated by discrete values, see Section 6.2.

By repeated discrete derivations of a given image value function, and by corresponding arithmetic combination, new image value functions can be computed representing certain local gray value situations.

1.1.3 Color Images, Image Sequences and Multi-channel Images

A *multi-channel image* is a combination of several binary or gray value images. This combination can be a result of the process of image acquisition, or of subsequent comparisons and adjustments of the acquired image data.

By combining images $f_1, f_2, ..., f_n$ into a multi-channel image f, in image f each image point (x, y) is assigned a vectorial image value

$$f(x, y) = (f_1(x, y), f_2(x, y), ..., f_n(x, y)) = (u_1, u_2, ..., u_n).$$

For a multi-channel image f, $f_i(x, y)$ denotes the image value u_i in the channel i, with $i = 1, ..., n$. A multi-channel image can be represented by a three-dimensional image matrix $f(x, y, i)$, with $1 \le i \le n$, where $f(x, y, i) = f_i(x, y)$:

multi-channel image f **(vector-valued image)**
three-dimensional image matrix with image values $f(x, y, i) = u_i$
$u_1, u_2, ..., u_n$ with image value range $\{0, 1, ..., G-1\}$, $G \ge 2$

channel value $u_i = 0$ corresponds to the gray shade Black in channel i
channel value $u_i = (G-1)/2$ corresponds to "Middle Gray" in channel i
channel value $u_i = G-1$ corresponds to the gray shade White in channel i

Such multi-channel images are generated, e.g. by acquiring *color images* or *multi-spectral images*.

Display representations of color images are based on color additions. Nearly all colors can be represented by a weighted sum of three chosen main colors. In general, the colors **R**ed, **G**reen and **B**lue are selected as these main colors. For the resulting *RGB model*, the computer treatment of color images is carried out by dealing with three gray value images corresponding to the Red, Green and Blue channel. Thus, for a (three-channel) color image f, for each image point (x, y) three gray values r, g, b are given,

$$f(x, y) = (r, g, b),$$

also called *tristimulus values*. Colors are specified by concrete value combinations r, g, b, and they are relative, perception-dependent events:

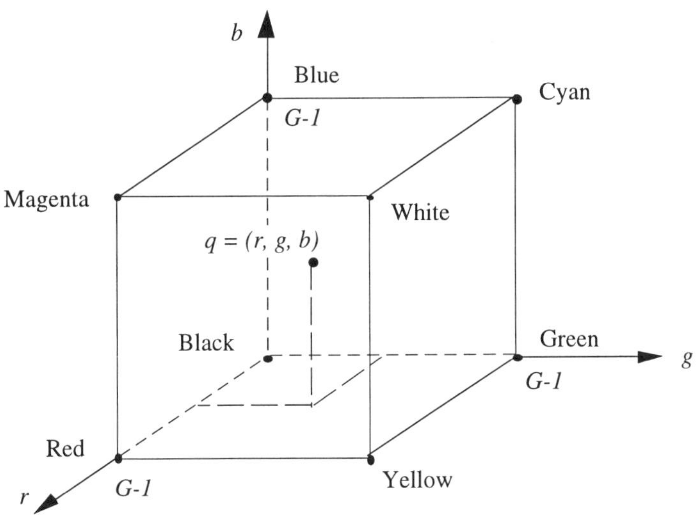

Figure 1.7: In the RGB model, each point $q = (r, g, b)$ inside the color cube represents exactly one color for integer r,g,b with $0 \le r, g, b \le G-1$. Thus, G^3 colors can be represented, i.e. for $G = 256$ there are 16 777 216 different colors.

RGB-color image f

image values $f(x, y, 1) = r$, $f(x, y, 2) = g$ and $f(x, y, 3) = b$

r, g, b in the image value range $\{0, 1, ..., G-1\}$, $G \ge 2$

image value 0	corresponds to the gray shade Black
image value $(G-1)/2$	corresponds to a middle-saturation color shade
image value $G-1$	corresponds to Red, Green, or Blue

The main colors Red $(G-1, 0, 0)$, Green $(0, G-1, 0)$, Blue $(0, 0, G-1)$, and the mixture colors Black $(0, 0, 0)$, Yellow $(G-1, G-1, 0)$, Magenta $(G-1, 0, G-1)$, Cyan $(0, G-1, G-1)$, and White $(G-1, G-1, G-1)$ constitute the corner points of the *color cube,* defined by all possible value combinations of r, g, and b. All integer triples (r, g, b), for $0 \le r, g, b \le G-1$, are characterizing colors in the RGB model. This color cube is illustrated in Figure 1.7. On the main diagonal (u, u, u), $0 \le u \le G-1$, all gray shades are located.

In Figure 1.8, an example of a RGB image is given. The RGB model is the common internal representation of color images for computers. For identical scenes or objects, different cameras or scanners can produce different color values because it is possible that their color characteristics are different.

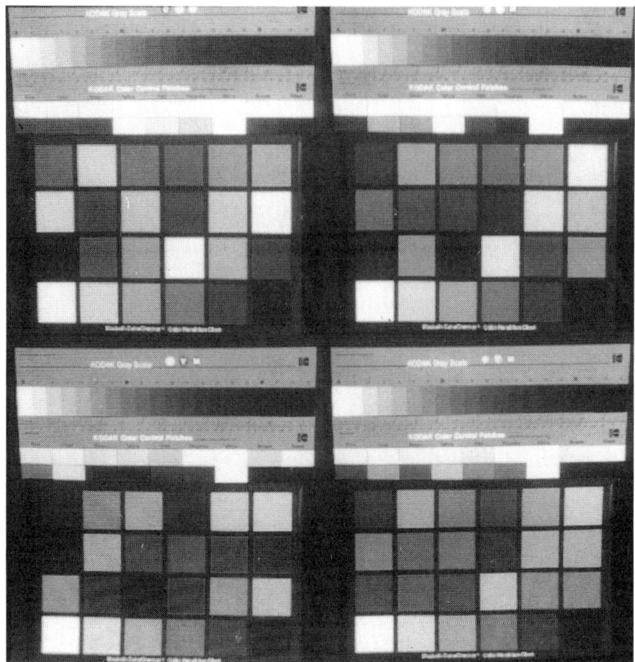

Figure 1.8: Representation of three color channels of an RGB image by gray value images: Red - top left, Green - top right, Blue - below left. Below right, the intensity channel (cp. value i of HSI model) conveys an impression about the combination of the three color channels (cp. *color diagram* in the glossary).

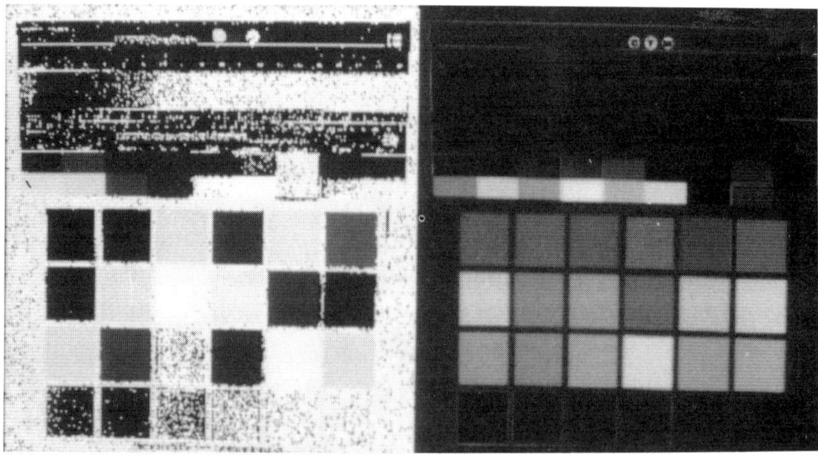

Figure 1.9: Representation of the channels for hue h (left) and saturation s (right) for the RGB image given by Figure 1.8.

For image processing and computer graphics, further image representation models are of interest, which are based on studies about human color perception. In the *HSI model* **h**ue, **s**aturation and **i**ntensity are used as coordinate axes. These coordinate axes are well behaved for defining visually interpretable local features for the processing of color images.

Let $q = (r, g, b)$ be a color in the RGB model. *Hue H* characterizes the dominant color in q. Red, e.g., is chosen as reference color, i.e. $H = 0°$ or $H = 360°$ corresponds to color Red. Formally, H is given by

$$H = \begin{cases} \delta & , \text{if } b \le g \\ 360° - \delta & , \text{if } b > g \end{cases}$$

where

$$\delta = \mathbf{arc\ cos} \frac{\frac{(r-g)+(r-b)}{2}}{\sqrt{(r-g)^2 + (r-b)(g-b)}} .$$

For example, for $g = b = 0$ and $r \neq 0$ it follows that $\delta = \mathbf{arc\ cos}\ 1$, i.e. $H = \delta = 0°$. Let h be the hue value H scaled to range $\{0, 1, ..., G-1\}$.

The *saturation S* of color q can be understood as a measure for the dilution of color q in White. The extreme case $S = 1$ holds for a pure color, and the other extreme case $S = 0$ denotes always a gray shade. Formally, S is defined by

$$S = 1 - 3 \cdot \frac{\mathbf{min}\{r, g, b\}}{r + g + b} .$$

For example, for $r = g = b \neq 0$ it follows $S = 0$. For $r = 0$ or $g = 0$ or $b = 0$, but $r + g + b \neq 0$, it follows that $S = 1$. Let s be the saturation value S scaled to range $\{0, 1, ..., G-1\}$.

The *intensity i* of color q corresponds to its relative brightness in the sense of monochromatic gray values. The extreme case $i = 0$ corresponds to Black. The intensity is defined by

$$i = \frac{r + g + b}{3} ,$$

and by this definition it is scaled to the range $\{0, 1, ..., G-1\}$. Altogether, for color $q = (r, g, b)$ in the RGB model a representation (h, s, i) of the same color is given in the HSI model. This transformation is one-to-one (with a few singularities):

HSI-color image f

image values $f(x, y, 1) = h$, $f(x, y, 2) = s$, and $f(x, y, 3) = i$

h, s, i with range $\{0, 1,..., G-1\}$, $G \geq 2$

hue h	denotes the dominant color
saturation s	degree of non-dilution in White
intensity i	relative brightness

Some boards for image processing on PC's, workstations etc. transform PAL-video images, RGB images etc. in real time into HSI images, and vice-versa.

Besides color images, multispectral images are a more general example of multi-channel images. Here, several gray value images, representing different spectral domains of the light, are acquired from (nearly) the same camera position. For example, with four-channel LANDSAT images these channels take images in spectral domains 500–600 *nm* (Blue-Green), 600–700 *nm* (Yellow-Red), 700–800 *nm* (Red-Infrared), and 800–1100 *nm* (Infrared).

With thematic multi-channel images, different channels of an image f represent different "views" on the subject. For example, channel 1 can be the original gray value picture, channel 2 an edge picture derived from channel 1, channel 3 a segmentation picture (e.g., segments with "similar texture"), channel 4 a range picture (distance camera to object encoded by gray values). For objects shown by channel 1, channel 5 can be an overlay masking specific regions, etc.

A *complex-valued image* f consists of two channels. One channel represents the real part, and the other the imaginary part. For each complex image value $u = u_1 + u_2\sqrt{-1}$, all u_1-values are attributed here to channel 1, and all u_2-values are attributed to channel 2:

complex-valued image f

image values $f(x, y, 1) = u_1$ and $f(x, y, 2) = u_2$

u_1, u_2 in the gray value range $\{0, 1,..., G-1\}$, $G \geq 2$

gray value $u_i = 0$	corresponds to the gray shade Black
gray value $u_i = (G-1)/2$	corresponds to a gray shade "Middle Gray"
gray value $u_i = G-1$	corresponds to the gray shade White

Complex-valued images occur when signal theoretic transforms are applied which are based on Fourier transformation.

Besides multi-channel images, further combinations of images can be given by *image sequences* (e.g. camera in motion, static camera but moving objects). Images in the sequence can have gray values, or even vector values, as e.g. RGB images. For image sequences f_t, with $t = 1, 2, 3,...$, in general a certain time interval can be assumed, i.e. a value T as maximum of the time t. Then, a sequence of gray value images can be represented by a three-dimensional image matrix with values $f(x, y, t)$, for $1 \leq t \leq T$.

To some extent, multi-channel images can be digitally processed by reducing vector information to scalar information, and by applying gray image operators. This approach can be selected according to the field of application. In principle, there are two possibilities,

(A) at first all channels are processed as gray value, or scalar images, then the resultant images are combined into a single scalar resultant picture, or

(B) at first all channels are combined, and then the resultant scalar picture is transformed.

For the combination of several scalar pictures $f_1, f_2, \ldots, f_n$, some simple analytical relations are considered in Sections 5.4 and 5.6.1. The book in general is focusing on transformations of gray value, or scalar images.

The mentioned types of images are listed in the following summary:

Synopsis of Images

- gray value images (scalar)
  - bilevel images
  - binary images
- multi-channel (vector-valued)
  - two-channel
    - complex-valued
  - three-channel
    - RGB-images
    - HSI-images
  - multi-spectral
  - thematic multi-channel images
  - image sequences
    - of gray value images
    - of multi-channel images

For example, statements on gray value images are also true for bilevel or binary images in the sense of this summary.

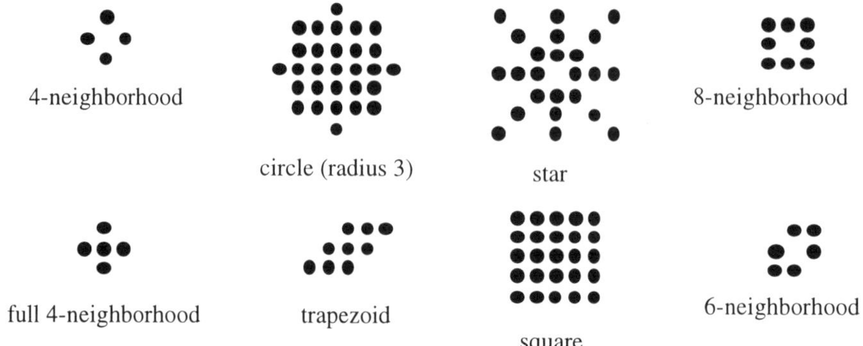

Figure 1.10: Examples of window shapes.

1.2 SUB-IMAGES

For recalculating image values in certain image points, in general not all pixels of the input picture are used as arguments but just certain subsets of the set of all pixels. Reasons for such selective processing are twofold. Firstly the objective nature of the picture elements, e.g. contextual correlation fades away with certain distance between pictorial regions. Secondly because of algorithmic reasons, i.e. the computational complexity should not exceed certain limits.

These subsets of the set of all pixels $(x, y, f(x, y))$ are *sub-images*, or *image cut-off's*. Their shape is defined by a *window*, and the values in the sub-image are defined by the given image function and by the position of the window on the raster.

1.2.1 Windows

In principle, a *window* might by any subset of image points, cp. Figure 1.10. The specification of the window shape depends upon the aims of the image processing. In general, image processing should be orientation-independent. Thus, circular shaped windows should be favored. But, for simplifying the indexing of window elements it is common to use a square or rectangular set of points as a window. Here, the window **F** consists of $a = m \cdot n$ image points,

$$\mathbf{F} = \{(i, j):\ 1 \le i \le m \wedge 1 \le j \le n\},$$

arranged in m columns and in n rows. It is assumed that $m < M$ and $n < N$. The *window size* is given by $a = m \cdot n$.

Practically, common values of m and n are quite small in image processing in general, say 3, 5, or 7. If not specified otherwise, for the operators in this book m is sensibly smaller than M, and n is sensibly smaller than N.

The algorithms presented in this book will necessitate rectangular windows. For symmetry reasons n, m are assumed to be odd in general. But, modifications for different window shapes are quite straightforward. For a square window $n = m$, and n assumed to be odd for symmetry reasons, let

$$k = \textbf{integer}\left(\frac{n}{2} \right).$$

A square window with odd side length contains all image points (i, j) with $1 \leq i, j \leq 2k + 1$. For even side length it follows that $1 \leq i, j \leq 2k$.

Because of their different shapes or of different topological characteristics, the following special window categories can be considered:

SYNOPSIS OF WINDOWS

- rectangular
  - square
    - symmetric (n odd, $I = J = (n + 1)/2 = k + 1$)
  - non-square
    - n, m odd
- non-rectangular
  - connected
  - non connected

For shifting a window on the image raster during a certain process, it can be considered to start with a certain initial position of the window. During image processing, this initial set of points has to be translated, but not rotated. The translation is univocally defined by the motion of a reference point.

A *reference point*(I, J) of window **F** in initial position is a selected image point inside **F**, or inside the image region "specified" by **F**. This reference point is the center of symmetry in an axially symmetric window. Note that this center does not belong to the set **F** itself if **F** is the 4- or 8- neighborhood.

In Figure 1.11 the translation of a window is illustrated for the so-called *non-centered initial position*. The reference point is given by the centroid coordinates:

$$I = \textbf{integer}\left(\frac{m+1}{2} \right) \quad \text{and} \quad J = \textbf{integer}\left(\frac{n+1}{2} \right).$$

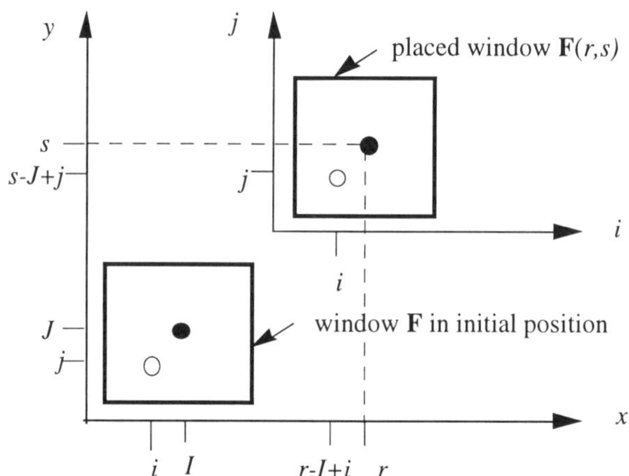

Figure 1.11: Using the non-centered *ij*-coordinate system for placed rectangular windows $\mathbf{F}(p)$, with $p=(r, s)$, a point (i, j) is transformed into a point $(r - I + i, s - J + j)$ by translation with shift vector $(r - I, s - J)$.

For odd n, m, all points (i, j) with $1 \le i \le 2I - 1$ and $1 \le j \le 2J - 1$ segregate the initial position of the rectangular window. For a square window with odd side length it holds $I = J = k + 1$, and it is called a *symmetric window* with *centroid* (I, J).

The *placement of a window* **F** is specified by a translation of the reference point (I, J). A placement of **F** on the *current point*, or the *relative reference point* $p = (r, s)$ is defined by a shift of **F** with the shift vector $(r - I, s - J)$. As result, a *placed window* $\mathbf{F}(p)$ is obtained. For the rectangular default window $\mathbf{F} = \{(i, j): 1 \le i \le m \wedge 1 \le j \le n\}$, the placed window[6]

$$\mathbf{F}(p) = \mathbf{F}(r, s) = \{(r - I + i, s - J + j): 1 \le i \le m \wedge 1 \le j \le n\} \qquad (1.1)$$

is obtained with the relative reference point $p = (r, s)$. Here, i, j are the relative window coordinates of the non-centered *ij-coordinate system.*

For a relative *ij*-coordinate system an alternative approach consists in assuming a rectangular window **F** in *centered initial position* with reference point (0, 0), cp. Figure 1.12. Here, a positioning of the rectangular window

$$\mathbf{F}(0, 0) = \{(i, j): -I + 1 \le i \le -I + m \wedge -J + 1 \le j \le -J + n\}$$

(with centroid coordinates I, J) on the relative reference point $p = (r, s)$ corresponds to a translation of $\mathbf{F}(0, 0)$ by vector(r, s). This leads to the placed window

[6] Simplifying, for $p = (r, s)$ and $\mathbf{F}(p) = \mathbf{F}((r, s))$, double parentheses are omitted.

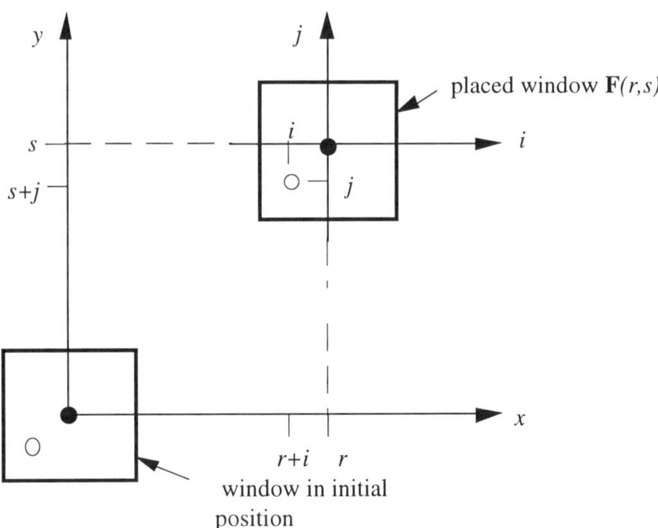

Figure 1.12: In the centered *ij*-coordinate system, relative window coordinates i, j can also take on negative integer values.

$$\mathbf{F}(p) = \mathbf{F}(r, s) = \{(r + i, s + j): -I + 1 \leq i \leq -I + m \wedge -J + 1 \leq j \leq -J + n\} \qquad (1.2)$$

with relative reference point $p = (r, s)$. Relative window coordinates i, j are related to the *centered ij-coordinate system* as illustrated by Figure 1.12.

The representations (1.1) and (1.2) of $\mathbf{F}(r, s)$ specify the same set of points. A coordinate transformation from the non-centered *ij*-coordinate system to the centered *ij*-coordinate system can be realized by a shift of the non-centered *ij*-coordinate system using translation vector (I, J).

In Figure 1.4, a rectangular window $\mathbf{F}_0$ of size 10×16 was used. For those two image windows, this window was placed on reference points (141, 56) and (133, 175), respectively. In Figure 1.4, the actual positions of both placed windows $\mathbf{F}_0(141, 56)$ and $\mathbf{F}_0(133, 175)$ are framed by rectangles in the image.

In this book, the centered *ij*-coordinate system is favored because this allows simple access to that point which will change its value upon the window operation. In the standard situation $n = m =$ odd in (1.2) the index range

$$-k \leq i, j \leq k$$

has to be considered. In general it is suggested to use the same coordinate system for all windows while implementing an image processing library.

1.2.2 Windows on Images

A *picture element* or *pixel* of a scalar image f or of a multi-channel image f, is a triplet (x, y, u) or $(x, y, (u_1, u_2,..., u_n))$ combining an image point (x, y) with its value $u = f(x, y)$ or $(u_1, u_2,..., u_n) = f(x, y)$, respectively. A *sub-image*, or *image cut-off*, is a set of pixels of a given image f.

A *picture window* $\mathbf{F}(f, p)$ is generated by a placed window $\mathbf{F}(p)$ in an image f, cp. Figure 1.13. In Figure 1.4, the image windows $\mathbf{F}_o(BUREAU, (141, 56))$ and $\mathbf{F}_o(BUREAU, (133, 175))$ are shown on the right for a rectangular window $\mathbf{F}_o$.

For a rectangular $m \times n$ -window $\mathbf{F}$ and relative reference point $p = (r, s)$, the image window can be given by

$$\mathbf{F}(f, p) = \{(r - I + i, s - J + j, f(r - I + i, s - J + j)): 1 \leq i \leq m \wedge 1 \leq j \leq n\}$$

using the non-centered ij-coordinate system[7] , or by

$$\mathbf{F}(f, p) = \{(r + i, s + j, f(r + i, s + j)): -I + 1 \leq i \leq I - 1 \wedge -J + 1 \leq j \leq J - 1\}.$$

using the centered ij-coordinate system. An image window is a sub-image with specified reference point.

If an n-channel picture $f = (f_1, f_2,..., f_n)$ is assumed, then an picture window $\mathbf{F}(f, p)$ contains n + 2-tuples

$$(r + i, s + j, f_1(r + i, s + j), f_2(r + i, s + j), ..., f_n(r + i, s + j)).$$

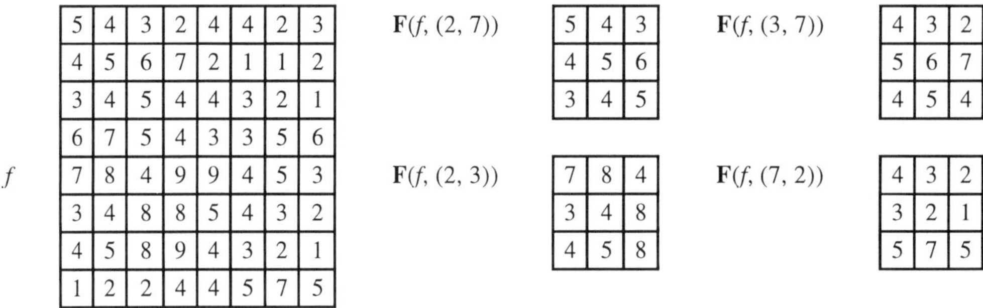

Figure 1.13: For a square window, with parameters $m = n = 3$ and $I = J = 2$, four picture window examples with the assumed 8×8 image f are given.

---

[7] The notation $\mathbf{F}(f, p) = \mathbf{F}(f, (r, s))$ is formally correct for $p = (r, s)$. A brief notation $\mathbf{F}(f, r, s)$ may lead to confusion between image gray values and image point coordinates.

For example, for color pictures, say RGB or HSI, $\mathbf{F}(f, p)$ consists of 5-tuples.

For addressing elements of an picture window either the non-centered *ij*-coordinate system or the centered *ij*-coordinate system can be chosen.

Example 1.1: All gray values of a rectangular picture window $\mathbf{F}(f, p)$ have to be added, for $p = (r, s)$. Here, in

$$SUM(\mathbf{F}, (f, p)) = \sum_{i=1}^{m} \sum_{j=1}^{n} f(r - I + i, s - J + j) \quad = \sum_{i=-I+1}^{I-1} \sum_{j=-J+1}^{J-1} f(r + i, s + j),$$

the first addition corresponds to the non-centered *ij*-coordinate system, and the second addition to the centered *ij*-coordinate system.

For a default picture window, i.e. square with odd side length, using $k = I - 1$ or $I = k + 1$ it follows:

$$SUM(\mathbf{F}, (f, p)) = \sum_{i=1}^{n} \sum_{j=1}^{n} f(r - k - 1 + i, s - k - 1 + j)$$

$$= \sum_{i=0}^{n-1} \sum_{j=0}^{n-1} f(r - k + i, s - k + j) \quad = \sum_{i=-k}^{k} \sum_{j=-k}^{k} f(r + i, s + j).$$

In the first sum of the second row, for simplicity the non-centered *ij*-coordinate system was shifted by vector (1,1). The second sum illustrates a kind of "standard indexing" as used throughout this book in general.

Example 1.2: The leftmost and uppermost point has the coordinates $(r - I + 1, s - J + n)$, for an picture window $\mathbf{F}(f, p)$ and $p = (r, s)$. This follows from $i = 1$ and $j = n$ if the non-centered *ij*-coordinate system is applied, or from $i = -I + 1$ and $j = -J + n$ if the centered *ij*-coordinate system is used.

An image point p of image raster $\mathbf{R}$ is a *border point* with respect to a considered window $\mathbf{F}$, if the placed window $\mathbf{F}(p)$ is not completely contained in the image raster

$$\mathbf{R} = \{1, 2, \ldots, M\} \times \{1, 2, \ldots, N\}.$$

Assume that all pixels of an picture window $\mathbf{F}(f, p)$ contribute to the calculation of a new image value at point p. Then, border points p need a special treatment. Because only some grid points q of a placed window $\mathbf{F}(p)$ are present in the image raster $\mathbf{R}$ in such a situation, and image values $f(q)$ are known only for such points q, for the remaining grid points q of the placed window $\mathbf{F}(p)$ outside the image raster $\mathbf{R}$, it is not simple to define meaningful image values $f(q)$.

Specifications of window-based image operations when dealing with border points can be determined by image contents or the task of image processing. There is no universal "recipe".

In this book, only in some exceptional cases methods for dealing with border points are suggested for image operators. This has to be done, e.g., for large windows, "large" in comparison to image size A.

In general, when dealing with border points one of the following approaches can be selected for specifying image values outside the image raster $\mathbf{R}$:

(1) f is assumed to be constant outside the image raster $\mathbf{R}$, e.g. $f(x, y) = 0$ for all grid points $p = (x, y)$ not in $\mathbf{R}$, or $f(x, y)$ equal to the average of all gray values of image f for all grid points $p = (x, y)$ not in $\mathbf{R}$.

(2) f is assumed to be periodic, and one period is given on $\mathbf{R}$. For example, $f(x + M, y) = f(x, y)$ and $f(x, y + N) = f(x, y)$ following this assumption.

(3) f is mirrored at the "borders" of $\mathbf{R}$, e.g. $f(M + c, y) = f(M - c, y)$ and $f(x, N + b) = f(x, N - b)$, where $0 < c < M/2$ and $0 < b < N/2$ can also be assumed.

In this book, for an picture window $\mathbf{F}(f, p)$ it is assumed that the placed window $\mathbf{F}(p)$ is completely contained in the image raster $\mathbf{R}$, if not explicitly mentioned otherwise.

1.3 FUNCTIONS ON WINDOWS

By means of functions on windows, a value is calculated for an picture window $\mathbf{F}(f, p)$ which should be a gray value in the set $\{0, 1,..., G-1\}$ in the sense of generating "new" gray value images, but, which in general can also be a real, or even a complex number. Functions on windows are also called *operator kernels* because they contribute essentially to the definition of window-based image operators.

A *window function* ϕ is a function defined on picture windows $\mathbf{F}(f, p)$, and mapping into the set of integers, reals, or complex numbers. For windows $\mathbf{F}$, relative reference points p and images f, the value $\phi(\mathbf{F}(f, p))$ can be an integer, a real or a complex number. In general, the values of window functions are restricted to integers, or real numbers. In this book, complex numbers only occur in connection with the Fourier transformation. For a window function ϕ, a univocally defined window $\mathbf{F}$ is assumed which can be translated into different positions inside the image raster so that the window function transforms these placed windows.

1.3.1 Classes of Window Functions

For selecting or defining a window function, it is reasonable to consider first its nature among other window functions. The following classification scheme can be used:

SYNOPSIS OF OPERATOR KERNELS

order independent, or order dependent
position independent, or position dependent

combinatorial (e.g. order statistical)
analytic
- polynomial
  - linear (convolution)
    - separable
- exponential
- trigonometric

logically structured

A value of a window function ϕ can depend not only on the image values $f(x, y)$ but also on their relative positions (x, y) in the placed window $\mathbf{F}(p)$, or on their absolute positions (x, y) in the image raster $\mathbf{R}$. This is true because an picture window $\mathbf{F}(f, p)$ contains triplets and not only image values $f(x, y)$.[8]

For a *position independent window function* it is meaningless when the placed window $\mathbf{F}(f, p)$ is situated in the image raster $\mathbf{R}$. For example, the sum of all four "corner values" of a placed window is position independent, but the order in the placed window, i.e. the relative *ij*-coordinates, also has an influence.

For an *order independent window function*, function values are independent of the relative *ij*-positions of single image values inside the picture window, i.e. they are independent of the relative positions of all $a = n \cdot m$ image values in the picture window $\mathbf{F}(f, p)$. For any permutation of all image values of the placed window, an order independent window function produces a constant function value. For example, the minimum calculation represents an order independent, and also a position inde-

---

[8] A image value may be scalar (gray value image, binary image), or vectorial (multi-channel image, image sequence). For example, for n-channel images $\mathbf{F}(f, p)$ contains $(n + 2)$-tuples. Taking account of the remarks at the end of Section 1.1.3, in the following window functions are defined only for scalar images.

pendent window function. But, in general an order independent window function is not necessarily also a position independent window function. Values of a position dependent window function are also affected by the position of the placed window **F**(p) within the image raster **R**. For example, a position dependent window function can be defined such that only placed windows "close to the image border" lead to new gray values, otherwise gray values remain unchanged.

The values of an *order dependent window function* depend upon the relative positions (i, j) of image values within the picture window. For example, image values "in the left window region" can be subtracted by the window function from values "in the right window region", thus approximating a directional derivative.

An order dependent window function is not necessarily also a position dependent window function. For order dependence, only relative positions of image points within the picture window have to be considered, but absolute coordinates can have no influence at all. Order dependent and position independent window functions are typical in image processing.

Window functions can also be characterized by their functional complexity. *Linear window functions* are defined by a linear combination of image values contained in **F**(f, p).

More complicated window functions can be defined by polynomial, trigonometric or exponential mappings of image values. The combinatorial analysis, as e.g. the determination of the minimum of all image values in **F**(f, p), leads to non-linear (and also non-polynomial, non-trigonometric, or non-exponential) window functions.

Analytic window functions are defined by an analytic function, e.g., arithmetic averaging (order and position independent), variance estimation (order and position independent), or the weighted sum of all image values in the placed window, where weights are determined by the (inverse) distance of a point to the relative reference point (order dependent, but position independent).

A window function can be realized or defined by a decision tree. The leaves of the decision tree are labeled by operations for calculating a functional value, and the remaining (internal) nodes of the tree are labeled by decisions based on the specific data in the picture window. Such a situation is described by a *logically structured window function.*

1.3.2 Auxiliary Functions on Windows or Images

Histograms are important auxiliary functions for defining window functions. In histograms, frequencies of single events are recorded and represented. These single events are considered to appear in a whole image f, or in selected picture windows $\mathbf{F}(f, p)$.

A single event can be: "An image point has gray value u.". These special events are represented in *gray value histograms* which are histograms of the frequencies of gray values. For $0 \leq u \leq G-1$ and $p = (r, s)$,

$$hist(\mathbf{F}(f, p), u) = \mathbf{card}\{\ (i, j) : 1 \leq i \leq m \ \wedge\ 1 \leq j \leq n \ \wedge\ f(r - I + i, s - J + j) = u\ \}$$

denotes the absolute frequency of appearance of gray value u in the picture window $\mathbf{F}(f, p)$, and

$$HIST(\mathbf{F}(f, p), u) = \frac{1}{a} \cdot hist(\mathbf{F}(f, p), u)$$

denotes the relative frequency, for window size $a = m \cdot n$. Here, the set-theoretic function **card** denotes the cardinality (i.e. the number of elements). The relative frequencies $HIST(\mathbf{F}(f, p), u)$ span the interval [0, 1].

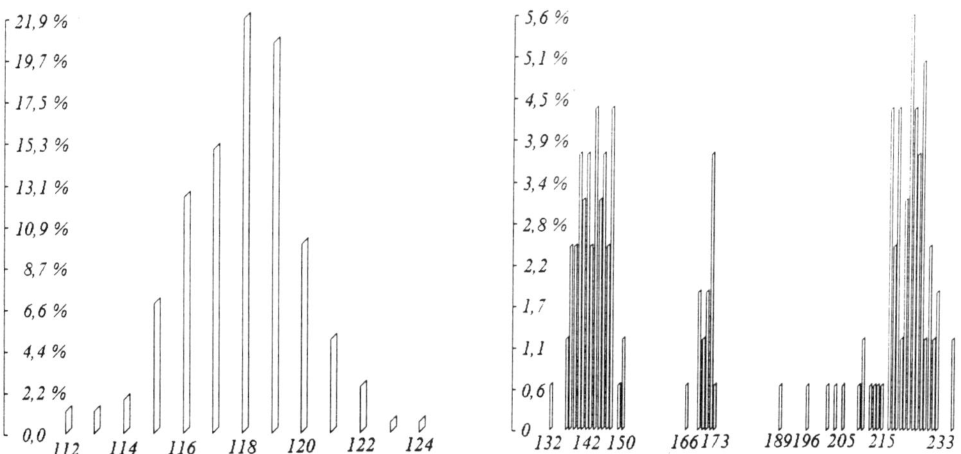

Figure 1.14: Bar diagram representation of gray value histograms for both picture windows of Figure 1.4. On the left: lower (homogenous) picture window. On the right: upper picture window (edge).

Considering all the gray values u, the values of the relative gray value histogram add up to 1:

$$\sum_{u=0}^{G-1} HIST(\mathbf{F}(f, p), u) = 1 \ .$$

For a scalar image f, a gray value histogram can be graphically represented by a bar diagram. In the bar diagram, a percent scale is appropriate. Note that histogram values, i.e. percentages of the single gray values u, are in general quite small.

Formally, histograms of multi-channel images are more complicated. For example, for a two-channel image $f = (f_1, f_2)$, the absolute gray value histogram is defined by

$$hist(\mathbf{F}(f, p), u_1, u_2) = \mathbf{card}\{(i, j): 1 \leq i \leq m \wedge 1 \leq j \leq n \wedge f(r - I + i, s - J + j) = (u_1, u_2)\},$$

with $0 \leq u_1, u_2 \leq G - 1$ and $p = (r, s)$.

Figure 1.15: Examples of two-dimensional histograms. On the left: Original gray value images. On the right: Corresponding histograms for events "gray value" and "result of the Sobel operator" (top) or events "gray value" and "absolute value of average of gray value differences between 4-neighbors in a 3×3-window" (below). The x-axis corresponds to the gray values 0...255, and the y-axis is for the second event. Frequency is represented by brightness.

For two-channel images, histograms can be represented by gray value images of size $G \times G$. For image point (u_1, u_2) a gray value is calculated which is proportional to $HIST(\mathbf{F}(f, p), u_1, u_2)$. If both images f_1 and f_2 are absolutely identical in the placed window $\mathbf{F}(p)$, then non-zero values appear only along the main diagonal. The "intensity of a strong main diagonal" can be used as criterion for evaluating a correlation between two images f_1 and f_2.

For a scalar image f, the result of image processing can be used as a second channel, see Figure 1.15. Instead of a given second channel, for the input image f a certain local feature is calculated and considered as a second image. "Visible clouds" in the histograms (bright regions) correspond to frequently appearing combinations of events.

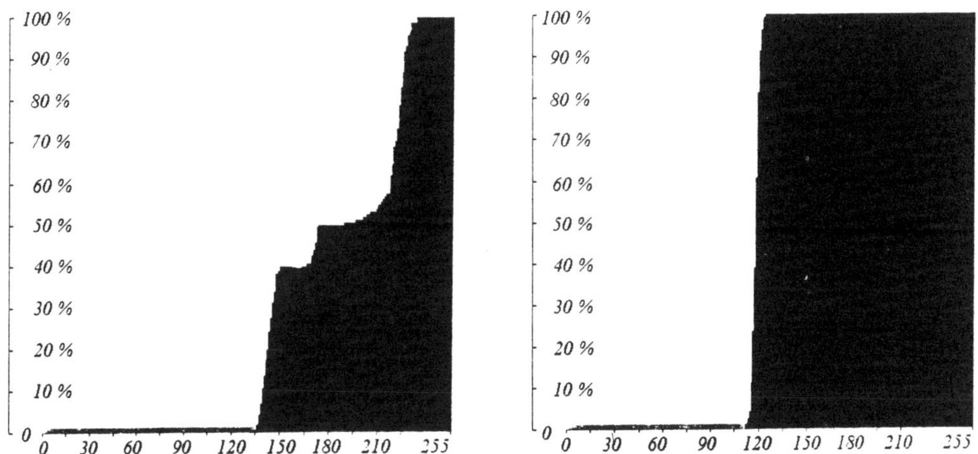

Figure 1.16: Sum histograms for both picture windows shown in Figure 1.4, left: lower (homogenous) picture window, right: upper picture window (edge).

A further single event, for which a histogram can be defined, is: "An image point of an picture window has a gray value smaller or equal to u." These events lead to the relative or absolute *sum histograms*

$$SUMHIST(\mathbf{F}(f, p), u) = \sum_{v=0}^{u} HIST(\mathbf{F}(f, p), v) \quad \text{or}$$

$$sumhist(\mathbf{F}(f, p), u) = \sum_{v=0}^{u} hist(\mathbf{F}(f, p), v) \quad ,$$

with $0 \leq u \leq G-1$. These sum histograms are monotone increasing functions. It holds

$$SUMHIST(\mathbf{F}(f, p), u) \leq SUMHIST(\mathbf{F}(f, p), v) \quad \text{for } u \leq v.$$

Because the relative sum histogram adds up to 1, this histogram is monotonously increasing from 0 to 1 (Figure 1.16).

The histogram *HIST* is an estimation of the density function of gray values, for $a \to \infty$. This histogram *SUMHIST* is an estimation of the distribution function of the gray values.

By means of histograms, the gray value statistics inside an picture window can also be evaluated by functions *AVERAGE* for arithmetic averaging or *VARIANCE* for the averaged square deviation. For a placed window $\mathbf{F}(p)$ with reference point $p = (r, s)$ and window size $a = m \cdot n$, let

$$AVERAGE(\mathbf{F}(f, p)) = \frac{1}{a} \sum_{i=1}^{m} \sum_{j=1}^{n} f(r - I + i, s - J + j)$$

and

$$VARIANCE(\mathbf{F}(f, p)) = \frac{1}{a} \sum_{i=1}^{m} \sum_{j=1}^{n} [f(r - I + i, s - J + j) - AVERAGE(\mathbf{F}(f, p))]^2$$

AVERAGE is the averaged brightness of the given picture window, and *VARIANCE* is the averaged square deviation from this averaged brightness.

Elementary transformations of this definition equation of the *VARIANCE* function lead to[9]

$$VARIANCE(\mathbf{F}(f, p)) = \frac{1}{a} \sum_{i=1}^{m} \sum_{j=1}^{n} f(r - I + i, s - J + j)^2 - [AVERAGE(\mathbf{F}(f, p))]^2.$$

Thus, a single run through the picture window data is sufficient for calculating *AVERAGE* as well as *VARIANCE*. It holds

---

[9] For a function f let $f(x)^2$ be the squared value $f(x)$. But, $f^2 = f \cdot f$ denotes an iterated application of function f, i.e. $f^2(x) = f(f(x))$.

$$AVERAGE(\mathbf{F}(f, p)) = \sum_{u=0}^{G-1} u \cdot HIST(\mathbf{F}(f, p), u)$$

and

$$VARIANCE(\mathbf{F}(f, p)) = \sum_{u=0}^{G-1} [u - AVERAGE(\mathbf{F}(f, p))]^2 \cdot HIST(\mathbf{F}(f, p), u)\ .$$

In image processing systems, basic functions as *AVERAGE, VARIANCE* and *HIST* are often supported by special parallel processing hardware for fast computation.

A *bimodal histogram* curve is characterized by two strong local maximum and one local minimum between them ("valley between two peaks"). The minimum defines a gray value which can be used to separate both modes. For an picture window $\mathbf{F}(f, p)$, a symmetry coefficient *SYMMETRY* can be defined based on a function *ENTROPY*. This coefficient can be used as a bimodality measure for gray value histograms. Defining function *ENTROPY* by

$$ENTROPY(\mathbf{F}(f, p)) = -\sum_{u=0}^{G-1} [HIST(\mathbf{F}(f, p), u) \cdot \mathbf{log}_2\ HIST(\mathbf{F}(f, p), u],$$

this function specifies the average number of bits per image value needed for an exact representation of the first order statistics of the picture window. For example, an picture window with constant gray values leads to the function value zero.
For picture window $\mathbf{F}(f, p)$, define

$$v = \mathbf{min}\{u \colon u = 0, 1, \ldots, G-1 \wedge SUMHIST(\mathbf{F}(f, p), u) \geq 0.5\}$$

as "gray value average" of $\mathbf{F}(f, p)$. Then, the above introduced symmetry coefficient is defined by

$$SYMMETRY(\mathbf{F}(f, p)) = \frac{-\sum_{u=0}^{v} [HIST(\mathbf{F}(f, p), u) \cdot \mathbf{log}_2\ HIST(\mathbf{F}(f, p), u)]}{ENTROPY(\mathbf{F}(f, p))}\ .$$

For example, for a *symmetric histogram* with mirror axis at $G/2$, a *SYMMETRY* value of 0.5 results. For Figure 1.4 it holds *ENTROPY* = 3.039 and *SYMMETRY* = 0.591 in the lower (homogenous) picture window, and *ENTROPY* = 5.188 and *SYMMETRY* = 0.498 in the upper (inhomogenous) picture window.

1.3.3 Some Window Functions

In this subsection, some basic window functions for image processing are provided. In the simplest case of an one-element window, a window function transforms a single pixel into a new image value. For a position independent function, an image value is transformed to a new image value. Such a gray value transformation is defined by a *gradation function* $t(u)$. The diagram of t is called *gray value characteristic* or *gray value curve*.

In Figure 1.17 a gray value characteristic is shown where all image points in the input image with gray value larger than $G/2$ are mapped into gray value $G/2$. On the basis of this gray shade mapping, bright image domains can be "darkened" by a constant value of $G/2$.

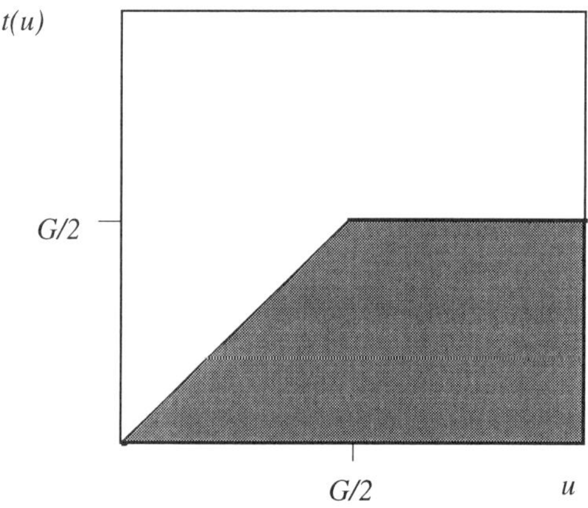

Figure 1.17: Gray value characteristic of gradation function $t(u) = u$, for $0 \leq u \leq G/2$ and $t(u) = G/2$, for $G/2 \leq u \leq G - 1$.

Gradation functions $t(u)$ are special operator kernels for one-element windows. They are dealt with in Chapter 5, e.g. for thresholding, spike noise, or gray-value transforms as in common use in desktop publishing systems. Often they aim at achieving certain *histogram transforms*.

In general, a histogram transformation T is a mapping of digital images f onto digital images, $h = T(f)$. As a result, the histogram $HIST(f,\_)$ of image f is transformed into the histogram $HIST(h, \_)$ of image h. Conditions to be met by the histogram $HIST(h, \_)$ can be helpful in defining the transformation T. A histogram transformation is defined by a gradation function t,

$$h(x, y) = t(f(x, y)) \quad \text{for} \quad T(f) = h \quad \text{and} \quad 0 \le f(x, y) \le G-1.$$

The following three examples illustrate this relation between transformation and gradation function:

(1) *General scaling*: Assume that for image f histogram values are non-zero only within a certain interval. In the resultant image h all gray values should be used as far as possible. A linear gradation function t can be defined as following:
For input image f the minimum and maximum gray value are calculated,

$$u_{\min} = \mathbf{min}\, \{f(x, y) \colon 1 \le x \le M \ \wedge \ 1 \le y \le N\} \quad \text{and}$$

$$u_{\max} = \mathbf{max}\{f(x, y) \colon 1 \le x \le M \ \wedge \ 1 \le y \le N\}.$$

Let $\alpha = -u_{\min}$ and $\beta = \dfrac{G-1}{u_{\max} - u_{\min}}$. Then, t is defined by

$$t(u) = \begin{cases} (u + \alpha) \cdot \beta & , \text{if } 0 \le (u + \alpha) \cdot \beta \le G - 1 \\ 0 & , \text{if } (u + \alpha) \cdot \beta < 0 \\ G-1 & , \text{if } (u + \alpha) \cdot \beta \ge G \end{cases}$$

This scaling allows to exploit the complete gray value scale in image h without loss of image information caused by saturation effects.

(2) *Constrained scaling:* For the input image f the values $a_f = AVERAGE(f)$ and $v_f = VARIANCE(f)$ can be calculated. The values $a_h = AVERAGE(h)$ and $v_h = VARIANCE(h)$ can be imposed to the resultant image h. For example, the objective may be to match several images in contrast and brightness. For this task, the linear gradation function of example (1) can be used with parameters

$$\alpha = a_h \cdot \frac{\sqrt{v_f}}{\sqrt{v_h}} - a_f \quad \text{and} \quad \beta = \frac{\sqrt{v_h}}{\sqrt{v_f}}.$$

(3) *Histogram equalization*: In the resultant image h all gray values should appear "equally often", i.e. an uniform distribution

$$HIST(h, u) = 1/G, \quad \text{for } u = 0, 1, \ldots, G-1,$$

should be obtained "as well as possible".

For the input image f, the relative sum frequency *SUMHIST* can be calculated. The histogram equalization is defined by the gradation function

$$t(u) = (G-1) \cdot SUMHIST(f, u).$$

The uniform distribution in h is approximated by a discrete one. The ideal uniform distribution would be a constant relative gray value histogram $1/G$. But, for discrete image information, this ideal situation cannot be realized because in the original histogram not all the gray values have non-zero entries, or because $M \cdot N$ is not a multiple of G.

For the relative sum histogram, the ideal uniform distribution is given by a (real) straight line segment from point (0, 0) to point $(G-1, 1)$. A measure of uniform distribution can be defined based on the deviation of this ideal straight line segment from all the points $(u, SUMHIST(\mathbf{F}(f, p), u))$.

If a window contains more than a single point, then a value of a window function can reflect gray value configurations of local image domains. The following three examples illustrate this dependency from local domains. In Chapter 6, a wide range of such window functions is presented.

(4) *AVERAGE*: In general, the value $SUM(\mathbf{F}(f, p))$ in Example 1.1 is larger than $G-1$. Instead, the arithmetic average

$$\frac{1}{a} SUM(\mathbf{F}(f, p)) = \frac{1}{a} \sum_{i=-I+1}^{I-1} \sum_{j=-J+1}^{J-1} f(r+i, s+j)$$

can be used for transforming local gray value distributions. Here, $p = (r, s)$ and $a = n \cdot m$ are used. The resultant rational number can be mapped by means of

$$AVERAGE(\mathbf{F}(f, p)) = \mathbf{integer}\left(\frac{1}{a} SUM(\mathbf{F}(f, p)) + 0.5\right)$$

onto the closest integer in the set $\{0, 1, \ldots, G-1\}$. This linear window function is position and order independent. The value of this window function is influenced by all the gray values in the picture window, i.e., also by "outliers" or by spike noise.

(5) *MEDIAN*: For a values $u_1, u_2, \ldots, u_a$, the *median* is the value situated in the middle position of the sorted sequence of these a values. For example, for the given 15 values 3, 8, 5, 2, 6, 7, 8, 1, 2, 2, 3, 8, 2, 5, 1 the median is equal to 3. In the sorted sequence (1, 1, 2, 2, 2, 2, 3, 3, 5, 5, 6, 7, 8, 8, 8) the number 3 is in position 8.

A window function *MEDIAN* is defined for picture windows

$$\mathbf{F}(f, p) = \{(r - I + i,\ s - J + j,\ u_{r-I+i,s-J+j}) \colon\ 1 \leq i \leq m \wedge 1 \leq j \leq n\}.$$

It is used for computing values of *MEDIAN*($\mathbf{F}(f, p)$), i.e. the median of the gray values $u_{r-I+i,\,s-J+j}$, with $p = (r, s)$, $1 \leq i \leq m$ and $1 \leq j \leq n$. This window function is position and order independent. The function value is not influenced by single "outliers".

(6) *SOBEL*: Differences of values of neighboring pixels in row and column direction can be used for defining the discrete approximation of first derivatives of gray value images f. In the simplest case, the incremental ratio in the y-direction" can be specified by

$$h(x, y) = \frac{f(x, y) - f(x, y+1)}{\delta y} = \frac{f(x, y) - f(x, y+1)}{1}$$

$$= f(x, y) - f(x, y+1).$$

In general, arithmetic window functions can be computed with the aid of *masks*. A mask consists of coefficients for additions or subtractions in all window positions. The simple y-derivative has coefficient 1 for reference point (x, y) and coefficient –1 for point $(x, y+1)$. This is represented by the mask

0	–1	0
0	1	0
0	0	0

This simple approximation of the derivative of f in y-direction is very sensitive to noise, and is practically not usable. Some improved "robustness" can be obtained by an arithmetic transformation using the complete 3×3 -window. The coefficients of arithmetic *SOBEL_X* and *SOBEL_Y* are given by the masks

1	0	–1
2	0	–2
1	0	–1

–1	–2	–1
0	0	0
1	2	1

For example, –2 is the coefficient of $f(x+1, y)$) in *SOBEL_X*. These window functions can be applied on placed 3×3 -windows for computing of discrete derivations, i.e. of discrete approximations of the known derivation operations in real analysis.

1.4 OPERATORS

An *image processing operator* is a mapping of one or several images into a resultant image. This mapping can have a restricted domain, e.g. binary images or three-channel images. In the sequel, image processing operators are briefly called *operators*.

The following synopsis can be used for referencing operators. This structure corresponds to the Chapters 4, 6 and 7 of this book. Point operators are dealt in the Chapter 5.

SYNOPSIS OF OPERATORS

- geometrical
  - contracting
  - expanding
  - motion
    - translation
    - rotation
    - inversion
- local (parallel or sequential)
  - homogenous
    - point operator
    - window operator
  - inhomogenous
    - with process control
- global
  - topological
  - geometrical-constructive
  - signal theoretical
    - convolution
      - filtering

Local operators could also be classified with respect to the applied window function. For example, a window operator based on a logically structured window function could briefly be called a *logically structured operator*.

1.4.1 Geometrical Operators

Geometrical operators map picture windows of the input image f onto picture windows of the resultant image h according to a general coordinate transformation K. Examples are rotation, translation or mirroring at specified straight lines.

The coordinate transformation K is the essential part in the definition of a geometrical operator. However in some cases the window function can be defined so as to introduce some minor gray value changes.

A *coordinate transformation* K maps an image element $(K(x, y), f(K(x, y))) = (r, s, f(r, s))$ onto an image element $(x, y, h(x, y))$ of the resultant image h. The image point $K(x, y) = (r, s)$ of the input image f has the value $f(r, s)$, and it moves into position (x, y) of the resultant image h. If K is one-to-one, then in this point (x, y) of h the value is univocally given by $f(K(x, y)) = f(r, s)$. Then it holds $h(x, y) = f(K(x, y))$. Geometrical operators are discussed in Chapter 4.

One-to-one coordinate transforms K are considered in the analytical geometry of real space. For example, let us assume a (square) image f of size $N \times N$. Then, the coordinate transform

$$K(x, y) = (y, x)$$

defines a geometrical operator for diagonal inversion. It holds $h(x, y) = f(y, x)$ for this operator. The coordinate transform

$$K(x, y) = (y, N + 1 - x)$$

defines a 90° rotation of image f. It holds $h(x, y) = f(y, N + 1 - x)$.

Because coordinate transforms K are defined for discrete image point coordinates, K is not one-to-one in general. For a univocal mapping K several image points can be mapped onto the same image point. This holds, e.g., for *expanding operators* where the resultant image h is larger than the input image f. For example, an expanding operator is defined by the coordinate transform

$$K(2x, 2y) = K(2x + 1, 2y) = K(2x, 2y + 1) = K(2x + 1, 2y + 1) = (x, y) ,$$

for $x = 1, 2, \ldots, M$ and $y = 1, 2, \ldots, N$. The resultant image has size $2M$ x $2N$. Image values of h can be equal to $f(x, y)$ in image points $(2x, 2y)$, $(2x + 1, 2y)$, $(2x, 2y + 1)$, $(2x + 1, 2y + 1)$. Then, h can feature a certain "block structure". Improvements are considered in Section 4.2.2.

Assume that the inverse mapping K^{-1} is univocal. Then it can happen that image points are mapped by K onto sets of image points. This holds, e.g., for *contracting operators* where the resultant image h is smaller than the input image f. For example, assume that picture windows $\mathbf{F}(f, (2x, 2y))$, containing 2×2 image points

$$(2x, 2y), (2x + 1, 2y), (2x, 2y + 1), (2x + 1, 2y + 1)$$

and their values, are mapped onto single image points (x, y) in h. The coordinate transformation of this contracting operator is given by

$$K(x, y) = \{(2x, 2y), (2x + 1, 2y), (2x, 2y + 1), (2x + 1, 2y + 1)\}$$

for $x = 1, 2, \ldots, M/2$ and $y = 1, 2, \ldots, N/2$. For ensuring $h(x, y) = f(K(x, y))$, a window function ϕ has to be defined such that $h(x, y) = \phi(\mathbf{F}(f, (2x, 2y)))$, cp. Section 4.2.1.

1.4.2 Point Operators

With a *point operator* T, with $T(f) = h$, a new value $h(x, y) = T(f)(x, y)$ has to be computed for each image point (x, y). This value depends only upon the image element $(x, y, f(x, y))$. In the position independent case, coordinates (x, y) have no influence on the resultant value. Therefore, such an operator is defined by a gradation function t. Function t is defined on the set $\{0, 1, \ldots, G - 1\}$ of all gray values. Finally, t is a mapping into this set of gray values. However for simplicity t can also be considered to be a real-valued function. It holds

$$h(x, y) = T(f)(x, y) = t(f(x, y)).$$

The position independent point operator T is univocally defined by gradation function t. Additionally, the set of image points can be specified where this transformation should take place.

As default, for point operators it is assumed that the gradation function t is applied to each image point (x, y) of the image raster $\mathbf{R}$ for generating a new image $h = T(f)$. Gradation functions are represented by gray value characteristics, cp. Figure 1.17.

Position dependent point operators can realize a certain dependency upon the distance between image points and image border.

Point operators are dealt in detail in Chapter 5. Point operators are the extreme case of local operators for one-element windows.

1.4.3 Local Operators

A *local operator* Z maps an input image f on a resultant image $h = Z(f)$, wherein the image h

- some gray values of f can be directly copied, e.g. for border points p let $h(p) = f(p)$,
- the remaining pixel values $h(p')$ are determined by a window function ϕ on picture windows $\mathbf{F}(f, p)$.

In the second case, for point p' in h window $\mathbf{F}$ is placed into a certain position p in f, and then $h(p')$ is calculated by applying the window function ϕ on the picture window $\mathbf{F}(f, p)$. Thus the complete definition of a local operator includes

(i) selection of window $\mathbf{F}$ and of the reference point inside $\mathbf{F}$,
(ii) specification of the window function ϕ,
(iii) selection of the image points p in f in which the window function ϕ has to be applied to $\mathbf{F}(f, p)$, in which order this application should be performed, and in which points p' of h the results $\phi(\mathbf{F}(f, p))$ should be attributed as new values $h(p')$, and finally
(iv) decision about setting h equal to f, i.e. whether the computed values $\phi(\mathbf{F}(f, p))$ should overwrite the previous values of f, or if the input image f should be saved and a new image h should be generated.

For a *parallel local operator* the results $\phi(\mathbf{F}(f, p))$ are stored in a resultant image h different from f. This notation reflects the action of a parallel computer which computes all the results $\phi(\mathbf{F}(f, p))$ simultaneously and as independent processes, for all the picture windows $\mathbf{F}(f, p)$ which have to be transformed by the local operator. With such a parallel computer, the resultant image h can be determined in a single parallel run. The given image f remains unchanged. Such parallel computers exist with different specifications, and often they are integrated on image processing boards as special processors. When realizing a parallel local operator on a (usual) sequential machine we have the important property: calculating $\phi(\mathbf{F}(f, p))$, the original input image f is still available.

For a *non-parallel* or *sequential local operator* the results $\phi(\mathbf{F}(f, p))$ are entered into the input image f after computing the operator. Thus, the resultant image h is produced via f through a sequence of intermediate images, and the original image f gets lost during this stepwise process. Such sequential operators are used for calculating a resultant image h "in place". During such a sequential process, it happens that some arguments used by the window function ϕ have been already calculated by ϕ itself. A computation "in place" can be useful for saving memory, or because the "basic structure" of the specific local operator is a sequential one.

For the point (iii) mentioned above, it is assumed in general that operator definitions are only given for non-border points to avoid "complex border-point procedures". For a *homogenous local operator* Z the window function ϕ is applied in f on picture windows $\mathbf{F}(f, p))$ for all non-border points p, and the resultant $\phi(\mathbf{F}(f, p))$ is taken as the image value in point p. Otherwise, the operator Z is called inhomogenous. Note that a homogenous operator always identifies the current point p' in h with the relative reference point p of the placed window $\mathbf{F}(p)$.

In general, local operators are assumed to be homogenous. Input image f is transformed into image h of the same size. Homogenous parallel local operators can be implemented on array processors. On such an array, the window $\mathbf{F}$ defines the surrounding of each image point, and all $N \times M$ image points are transformed simultaneously by realizing the window function on their surrounding.

Point operators can be considered as extreme cases of homogenous (parallel or sequential) local operators. For the window $\mathbf{F}$ it holds $n = m = 1$. In (ii), the gradation function t has to be specified. In (iii), all image points of the grid are considered (in any order). In (iv), for point operators the direct entry of result $t(u)$ in f can be accepted because there is no influence on further computations (if f does not need to be saved).

A *process controlled local operator* is an inhomogenous operator where the window function ϕ is performed for such image points p of f in $\mathbf{F}(f, p)$, which are determined stepwise at execution time by a process running on f. For example, the process can be such that only each tenth image row is a *scanning line*, say from left to right. Another process example can be that certain regions have to be detected in f, and then the local operator is only performed at the contour of such regions (*contour tracking*), cp. Section 7.2.1.

Local operators are the main topic of this book. They are dealt with in Chapter 6. In defining a local operator, a certain *control structure* is selected by means of the specification (iii). The window function ϕ is often used as *operator kernel* with this control structure. Control structures of homogenous local operators besides other topics are specified in Chapter 3.

1.4.4 Global Operators

Global operators are a further class of image processing operators, defined by functional dependencies of image values $h(p)$ upon image values in (potentially) any point position of input image f. Here, for an image point p it is not possible to specify a priori a local neighborhood in f such that $h(p)$ depends only upon this neighborhood. In principle, any position in image f can contribute to any position in image h under certain conditions.

Typical representatives of this operator class are Fourier and Walsh transformation, cp. Section 7.3. Geometrical constructions such as image representations of Voronoi diagrams by means of a gray-value image h, based on point patterns in f, are also global operators, cp. Section 7.2.

Global operators are not "local operators with very large windows". Otherwise, image points would be either border points (with all the ad-hoc rules for defining ϕ on such points), or the picture window with reference point p would not contain the complete input image f (with exception of a single point p for which $m = M$ and $n = N$).

Global operators are dealt in Chapter 7. They are often characterized by their requirement of large amounts of computing power (memory and computing time). This is true, e.g., for the signal-theoretic operators in Section 7.3 if they are realized for whole images of size 512×512 or more. But these global transformations can also be considered as window functions ϕ. For example, the Fourier transformation of picture windows $\mathbf{F}(f, p)$, or some evaluations of the high-frequency image content for computing a kind of "edge image" h of f.

Often the implementation of global operators needs more efforts in problem analysis, programming etc. as local operators. This is true for the methods reported in Sections 7.1 and 7.2. However in these cases the resultant programs are quite efficient with respect to computing time.

If it is desired to develop a personal library of image processing routines, global operators have to be understood as more demanding tasks. Here, a systematic study of methodological fundamentals is suggested.

Sometimes global operators can be approximated by iterative realizations of local operators. Here, the problem of "short sightedness" of local operators can be solved by making use of recursive operators, by several iterations through the image, by different scan orders (bottom-up, top-down etc.) in successive runs, or by different resolutions, cp. the *pyramid* in Section 4.2.3.

1.5 BIBLIOGRAPHIC REFERENCES

For the fields of image processing and image analysis there exist several good textbooks. For example, see

Ballard, D.H., Brown, C.M.: *Computer Vision*. Prentice-Hall, Englewood Cliffs, 1982.
Bässmann, H., Besslich, P.W.: *Konturorientierte Verfahren in der digitalen Bildverarbeitung*. Springer, Berlin, 1989.
Gonzalez, R.C., Wintz, P.: *Digital Image Processing*. 2nd ed., Addison-Wesley, Reading, MA, 1987.
Horn, B.K.P.: *Robot Vision*. MIT Press, Cambridge, MA, 1986.
Jain, A.K.: *Fundamentals of Digital Image Processing*. Prentice-Hall, Englewood Cliffs, 1989.
Kanade, T.: *Three Dimensional Vision*. Kluwer, Boston, 1987.
Marr, D.: *Vision*. W. H. Freeman, San Francisco, 1982.
Niblack, W.: *An Introduction to Digital Image Processing*. Prentice-Hall, Hemel Hempstead, 1986.
Rosenfeld, A., Kak, A.C.: *Digital Picture Processing, Vols. I and II*. Academic Press, Orlando, FL, 1982.
Shirai, Y.: *Three-Dimensional Computer Vision*. Springer, Berlin, 1987.
Wahl, F.M.: *Digital Image Processing*. Artech House, Norwood, 1987.

In the sequel, some textbooks, monographs and handbooks published since **1990** are listed.

Topics in Computer Vision (image processing, pattern analysis, image understanding) are dealt with in:

Ernst, H.: *Einführung in die digitale Bildverarbeitung - Grundlagen und industrieller Einsatz mit zahlreichen Beispielen*. Franzis-Verlag, München 1991.
Galbiati, L.: *Machine Vision and Digital Image Processing Fundamentals*. Prentice-Hall, Englewood Cliffs , 1990.
Haralick, R.M., Shapiro, L.G.: *Computer Vision, Volume I*. Addison-Wesley, Reading 1992.
Haralick, R.M., Shapiro, L.G.: *Computer Vision, Volume II*. Addison-Wesley, Reading 1993.
Jähne, B.: *Digital Image Processing. Concepts, Algorithms and Scientific Applications*. Springer, New York, 1991.
Sonka, M., Hlavac, V., Boyle, R.: *Image Processing, Analysis and Machine Vision*. Chapman & Hall, London, 1993.

For image processing see

Baeseler, F., Bovill, B.: *Scanning and Image Processing for the PC*. McGraw-Hill, New York, 1992.
Jaroslavskij, L.P.: *Einführung in die digitale Bildverarbeitung*. Hüthig, Heidelberg, 1990.
Lagendijk, R.L., Biemond, J.: *Iterative Identification and Restoration of Images*. Kluwer, Dordrecht, 1991.

Lindley, C.: *Practical Image Processing in C. Acquisition, Manipulation, Storage.* Wiley, New York, 1991.
Marion, A.: *An Introduction to Image Processing.* Chapman and Hall, London, 1987.
Pitas, I., Venetsanopoulos, A.N.: *Nonlinear Digital Filters.* Kluwer Academic Publishers, Boston, 1990.
Pitas, I.: *Digital Image Processing Algorithms.* Prentice Hall, Hertfordshire 1993.
Pratt, W. K.: *Digital Image Processing.* 2nd ed., Wiley, New York, 1991.

Image processing and pattern analysis are the topics of the following textbooks and handbooks:

Abmayr, W.: *Einführung in die digitale Bildverarbeitung.* B.G. Teubner, Stuttgart, 1994.
Bässmann, H., Besslich, P.W.: *Bildverarbeitung Ad Oculos*, Springer, Berlin, 1991.
Gonzalez, R.C., Woods, R.E.: *Digital Image Processing, 3rd ed.*, Addison-Wesley, Reading 1992.
Haberäcker, P.: *Praxis der digitalen Bildverarbeitung und Musterrekennung.* Carl Hanser Verlag, München, 1995.
Jain, A.K.: *Fundamentals of Digital Image Processing.* Prentice-Hall, Englewood Cliffs, 1990.
Pavlidis, T.: *Algorithmen zur Grafik und Bildverarbeitung*, Verlag Heise, Hannover, 1990.
Russ, J.C.: *The Image Processing Handbook.* CRC Press, Boca Raton, 1992.
Voss, K., Süsse, H.: *Praktische Bildverarbeitung.* Carl Hanser Verlag, München, 1991.
Zamperoni, P.: *Methoden der digitalen Bildsignalverarbeitung*, Vieweg, Wiesbaden, 2. Auflage, 1991.

For these two subfields of computer vision see also the monographs

Gauch, J.M.: *Multiresolution Image Shape Description.* Springer, New York, 1992.
Leavers, V.F.: *Shape Detection in Computer Vision Using the Hough transform.* Springer, Berlin, 1992.
Osten, W.: *Digitale Verarbeitung und Auswertung von Interferenzbildern.* Akademie Verlag, Berlin 1991.
Rao, A.R.: *A Taxonomy for Texture Description and Identification.* Springer, New York, 1990.
Voss, K.: *Discrete Images, Objects, and Functions in* Z^n. Springer, Berlin, 1993.

For signal processing see, e.g.,

Besslich, P.W., Tian, L.: *Diskrete Orthogonaltransformationen.* Springer, Berlin 1990.

and color models are dealt with in

Davidoff, J.: *Cognition Through Color.* The MIT Press, Cambridge 1991.

2 METHODICAL FUNDAMENTALS

In this Chapter some conceptual ideas about the application context in computer vision are sketched. Corresponding operators are discussed later. Some tasks and methodical approaches of computer vision are described such as image segmentation, image enhancement, image restoration, etc. In Chapters 4 to 7, the required operators are discussed with respect to special aims and properties. Therefore these operators and their properties should normally be seen within the context of more complex tasks.

Because of the presence of pictorial information in different areas of the daily life as engineering, medicine, science, art or publishing business, computer vision incorporates various fields of activity. The computer vision's subfield of image processing is often seen as preprocessing for image analysis. Indeed, often the analytical processes are so complicated that they have yet to be realized by human beings. For such situations, image processing also supplies adequately treated picture material to improve efficient interactive image analysis.

To illustrate this situation, some examples of typical applications of image processing procedures are presented in this Chapter (see Sections 2.3 to 2.8). They illustrate that the solution of concrete problems mostly requires several steps, i.e. the application of a sequence of "suitable" basic operations. The operators described in this book are fundamental for carrying out such operations. No general systematic approach can be recommended for the choice of operators and their sequential order. Practical experience is often the only key. Therefore it is very important that the known properties, the impacts and the heuristic knowledge about the specific "tools", i.e., about the image processing operators, should be systematically specified (see Chapters 4 to 7).

Some specific image processing techniques have acquired their own autonomous importance in other fields as well, e.g. in desktop publishing. Transformation of pictures for editing and for obtaining special effects, for pseudo coloring, for contrast enhancement, for solarization or for generating a poster effect are of interest for publishing, for graphic design or for home applications, such as manipulating and combining video images.

2.1 IMAGE SYNTHESIS

Usually, *image input* into the computer is realized by means of special visual sensors such as CCD-matrix cameras with frame grabbers. Image reconstruction techniques such as computer tomography in medicine or material inspection are further examples of image input. Image input can also be realized by scanners. If there is no special image input system available or in cases of interest in special pictorial data computer graphic techniques can be chosen for *image synthesis*. The synthetic pictorial data obtained can be useful, e.g. for evaluating or testing image processing operators.

Normally image synthesis is not considered to be a subfield of image processing but rather of computer graphics. However synthetic textures are often of interest as input data for testing certain image processing operators. Therefore a special fast implementable procedure for the generation of textures is given here. This procedure allows the synthesis of multifarious textures by means of parameter adaptation.

Assume that homogenous textures have to be generated in an $M \times N$ image f. If only a segment of an image has to be textured, then such a segment can be "cut out" of a homogenous textured image f, see Section 5.1.1. The following description of the process follows the structure as specified in the "Instructions to the Reader" section:

(1) The aim consists in generating $M \times N$ gray value images f with G gray levels, containing a *homogenous texture*. At program start texture parameters a_{lb}, a_b, a_{rb}, a_l are given as inputs with values within the interval $[-50, +50]$. Here "*lb*" stands for below left and "*b*", "*rb*" and "*l*" have to be interpreted analogously. Moreover, a parameter *noise* must be specified, with a value within the interval [50, 200] for *additive noise*.

(2) Let $H := (G-1)/2$. In raster scan order[1] a value q is generated for each image point (x, y), with $2 \leq x \leq M-1$ and $2 \leq y \leq N$,

$$q = (f(x-1, y-1) - H) \cdot \frac{a_{lb}}{100} + (f(x, y-1) - H) \cdot \frac{a_b}{100}$$

$$+ (f(x+1, y-1) - H) \cdot \frac{a_{rb}}{100} + (f(x-1, y) - H) \cdot \frac{a_l}{100}$$

$$+ r(x, y).$$

Then this value q is mapped into a gray value according to

---

1 Compare Glossary.

$$f(x, y) = \begin{cases} H + q & , \text{if } 0 \leq H + q \leq G-1 \\ 0 & , \text{if } H + q < 0 \\ G-1 & , \text{if } H + q > G-1 \end{cases}$$

The space-independent additive noise r is defined by the parameter *noise* and by a random variable Z with expected value zero. Let Z be an uniformly distributed variable with values within the interval $[0, G{-}1]$. Then define

$$r(x, y) = (Z - H) \cdot \frac{noise}{200} .$$

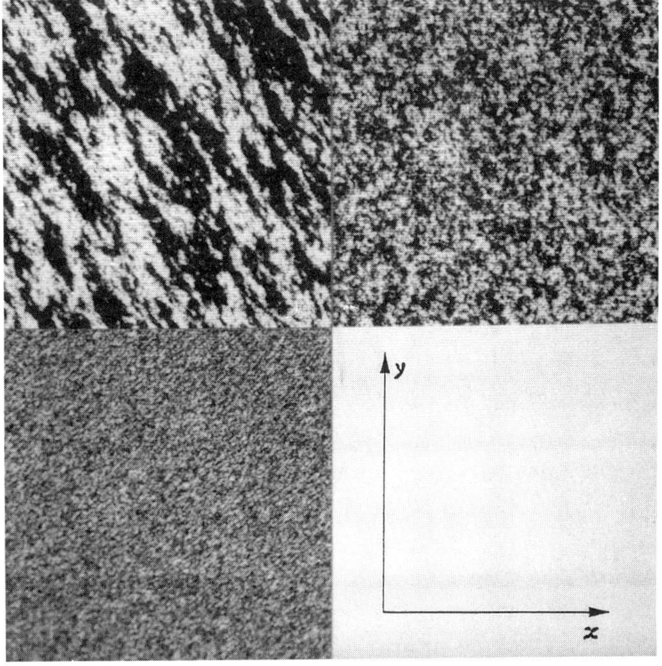

Figure 2.1: Three examples of generated textures, for the parameters a_{lb}= 50, a_b= 40, a_{rb}= 30, a_l = 50 and *noise* = 300 (top left), a_{lb}= 10, a_b= 20, a_{rb}= 30, a_l = 50 and *noise* = 250 (top right), a_{lb}= -10, a_b= 2, a_{rb}= 30, a_l = 1 and *noise* = 200 (below left).

(3) At program start f is initialized to zero at all image points. For generating additive noise the function *RND_EQU* can be selected as a random number generator as described in Section 3.4.1. For simplicity integer quantities are used as parameter values. For the final matching of the value $H + u$ to the discrete and limited gray value range, a general, often applicable procedure

$$ADJUST(q\text{: number, } u\text{: gray value})$$

is used. For an integer or a real number input q this procedure performs first (in case of real q) a rounding to the next integer

$$q := \textbf{integer}(\ q\ +\ 0.5),$$

and then it outputs the value

$$u := \textbf{min}\{\ G-1,\ \boldsymbol{max}\{\ 0,\ q\ \}\ \}\ .$$

The initialization of rows and columns with a constant value can be performed by a procedure for straight line generation, e.g., by the procedure *BRESENHAM* of Section 3.4.10.

(4)
```
procedure TEXTURE(f: gray_value_image);
var  z: real;   a_lb, a_b, a_rb, a_l , noise, q, r, H, Z: integer;
begin {TEXTURE}
        input of parameters a_lb, a_b, a_rb, a_l and noise in the intervals
                                                          as given in (1);
        initialization of row y = 1, column x = 1 and column x = M, with the
                                                constant value H := (G–1)/2;
        for  y := 2  to  N  do
            for  x = 2  to  M – 1  do
            begin
                z := RND_EQU( );  {z uniformly distributed in [0,1]}
                Z := integer(z · (G–1)  +  0.5 );
                r  = (Z  – H) · noise / 200;
                q  :=  (f(x–1, y–1) – H) · a_lb  + (f(x, y–1) – H) · a_b
                    + (f(x+1, y–1) – H) · a_rb  +  (f(x–1, y) – H) ·a_l;
                q := q /100  +  r;
                call ADJUST( q + H,  f(x, y))
            end {for}
end {TEXTURE}
```

(5) The *autoregressive model* which is fundamental for this procedure was originally developed for time sequences. It was considered in the context of computer vision by

Chen, C. H.: *On two-dimensional ARMA models for image analysis*. Proceed. ICPR´80, Miami Beach 1980, pp. 1128 - 1131.

and in several subsequent papers. An extensive description of the above procedure for texture synthesis can be found in

Voss, K., Süsse, H.: *Praktische Bildverarbeitung*. Carl Hanser Verlag, München, 1991.

In Figure 2.1 three examples for generated textures are shown. For selection of the input parameters, several program runs are necessary in order to develop an own "feeling" for correct input settings.

As well as textures, (synthetic) *point patterns* are of interest for program testing. Operators as presented in Section 5.2 can easily be modified such that a point pattern is generated in an originally empty image by a random process. Point patterns can be used as input for the operators of Sections 7.2.2 and 7.2.3 for the analysis of time complexities or for program testing.

2.2 HALFTONE IMAGE REPRESENTATION

On-line display of the operator processes can be of some importance for controlling the effects of image processing operators. Binary displays with high resolution or laser printers often only permit halftone image representations (black/white representation) of gray value images. Nevertheless the visual impression of almost continuously distributed gray values can be obtained through a carefully chosen distribution of black or white image points. The requirement for an ideal halftone image representation is a resolution that is better than the resolution of the eyes. The methodical base is the so-called *halftoning*, developed in a wide variety of different algorithms.

In the sequel, the color values black and white are respectively encoded by the values 0 and 1. More generally, *bilevel images* can be produced as output. For the display of such *half-tone images*, a special class of operators must be considered for mapping gray value images into binary images.

In Section 5.3, some operators are given for this binarization of gray value images. The operators of Sections 5.3.1 to 5.3.3 are oriented towards image analysis. A thresholding of a gray value image f

$$b(x, y) = \begin{cases} 0 & , \text{if } f(x, y) \le T(x, y) \\ 1 & , \text{if } f(x, y) > T(x, y) \end{cases},$$

for a locally or globally defined threshold-function $T(x,y)$ can lead to a visual impression similar to a gray value pattern in case of high spatial resolution. In general it is a classical aim of image processing to label "background" and "object" in an image. To get borders as sharp as possible between black and white areas a large variance of the values 0 and 1 should be avoided at gray value transitions (gray value edges). In the most simple case the threshold function $T(x, y)$ is a constant T, with $0 \leq T < G-1$. It is also possible to generate thresholds $T(x, y)$ point-wise, space-independently by means of a random generator of numbers comprised within the interval $[0, G-2]$ (*dithering*).

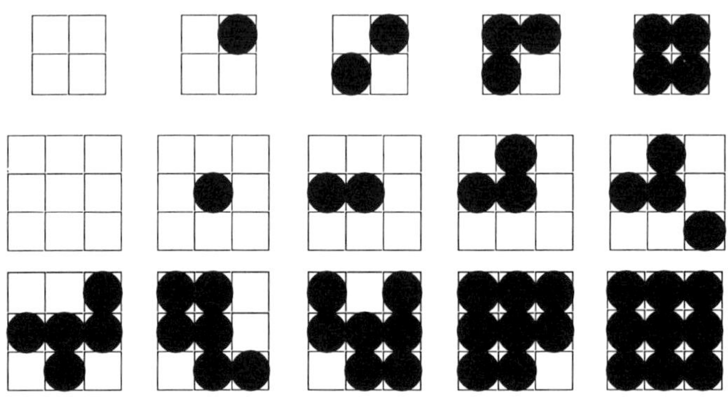

Figure 2.2: A 2×2-set and a 3×3-set of local binary pattern for halftone image representation.

Procedures for halftone image representation traditionally belong to computer graphics rather than to computer vision. Nevertheless the halftone image representation is within the scope of this book because its topic is image processing, i.e. image to image transformations. In Section 5.3.4 an operator is presented for halftone image representation by means of a *dither matrix*, see also the end of this Section. This operator can be recommended for printing images. The operator given in Section 6.1.8, based on the Floyd-Steinberg-algorithm, is especially recommended for the output on displays. This operator offers various possibilities by means of parameter control and it often leads to satisfactory results.

A simple realization of halftone image representation can also be based on a set of *binary patterns*. The resultant image b has a higher spatial resolution than the input image f. With this technique, there is a trade-off between missing gray value resolution and higher spatial resolution. In Figure 2.2, two sets of binary patterns are given. With the above 2×2 set an $M \times N$ gray value image is transformed into a $2M \times 2N$ binary image. The gray values are mapped onto the five different binary patterns. Note that no set of binary patterns is suitable for such a mapping. The generation of periodical stripes inside image regions should be avoided. For example, in Figure 2.2 above the pattern in the middle can lead to diagonal stripes. With the lower 3×3 set of patterns gray values are mapped onto ten patterns. Using this set, an $M \times N$ gray value image is transformed into a $3M \times 3N$ binary image.

Such sets of binary patterns can also be applied for *color images*, where color values are mapped onto binary patterns.

Such *pattern based halftoning* can also be performed with constant spatial resolution, i.e. the gray value patterns are replaced by binary patterns of the same size. In general, this will lead to a certain loss in visual image quality.

For situations where identical image size has to be ensured for the gray-value input image and the bilevel output image, the following threshold binarization is an example of a possible technique, compare also Section 5.3.4 for a more detailed description. This binarization method is based on *ordered dithering* (a different choice could be *random dithering*). An ordered threshold mask contains n^2 different thresholds for the halftoning of $n \times n$ subimages. The n^2 thresholds allow to discriminate between $G_n = n^2 + 1$ different gray value intervals. The threshold mask is defined as an $n \times n$ *dither matrix* $\mathbf{D}_n$.

For halftoning an input image, the thresholds of a chosen dither matrix are applied to correspondingly sized blocks of input gray values. Disjoint positions of the dither matrix are assumed which cover the full pictorial raster **R**. For example, for an initially chosen 2×2 dither matrix

$$\mathbf{D}_2 = \begin{bmatrix} 0 & 2 \\ 3 & 1 \end{bmatrix}$$

with $G_2 = 5$ different gray value intervals, the value $f(1, 1)$ is compared with the threshold 3, the value $f(2, 1)$ with the threshold 1, the value $f(3, 1)$ again with the threshold 3, the value $f(4, 1)$ with the threshold 1, and so on. The following 4×4 dither matrix

$$\mathbf{D}_4 = \begin{bmatrix} 0 & 8 & 2 & 10 \\ 12 & 4 & 14 & 6 \\ 3 & 11 & 1 & 9 \\ 15 & 7 & 13 & 5 \end{bmatrix}$$

allows to discriminate $G_4 = 17$ gray value intervals. In general, for the computation of $n \times n$ dither matrices the recursive generation rule

$$\mathbf{D}_n = \begin{bmatrix} 4\mathbf{D}_{n/2} & 4\mathbf{D}_{n/2} + 2\mathbf{U}_{n/2} \\ 4\mathbf{D}_{n/2} + 3\mathbf{U}_{n/2} & 4\mathbf{D}_{n/2} + \mathbf{U}_{n/2} \end{bmatrix}$$

can be recommended, where the $n \times n$ matrix $\mathbf{U}_n$ has value 1 at all positions. The gray level number G of the input image f can be scaled to G_n, or the values of the dither matrix can be scaled with respect to G.

For example, let $G = 256$. Then, the factor $G/(G_n - 1) = 256/(17 - 1) = 16$ can be used for scaling a $n \times n$ dither matrix. In case of a 4×4 dither matrix, this scaling leads to the threshold matrix

$$\mathbf{T}_4 = \begin{bmatrix} 0 & 128 & 32 & 160 \\ 192 & 64 & 224 & 96 \\ 48 & 176 & 16 & 144 \\ 240 & 122 & 208 & 80 \end{bmatrix}.$$

In general, let $\mathbf{T}_n$ be the $n \times n$ dither matrix which is scaled with respect to the given gray level number G. Then, for halftoning at image point (x, y), at first the matrix coordinates i, j are calculated[2] by

$$i = x \ \mathbf{mod}\ n \quad \text{and} \quad j = y \ \mathbf{mod}\ n\,.$$

Then, the value of a resultant binary image is determined according to

$$b(x, y) = \begin{cases} 0 & , \text{if } f(x, y) \leq \mathbf{T}_4(i, j) \\ 1 & , \text{if } f(x, y) > \mathbf{T}_4(i, j) \end{cases}.$$

The operation of halftoning is an example of a *point operator*: The resultant value at the image point (x, y) is determined by the value $f(x, y)$ and (!) by the image coordinates (x, y). In Section 1.4.2, such a mapping has been called a position-dependent point operator. Such a point operator can be implemented on a parallel processing system (e.g., an array processor).

2.3 GEOMETRICAL FITTING

Images often have to be modified in size. Such a geometrical transformation can be a composition of some "stretching or shrinking" in different directions, of rotations, of

---

2 The coordinates x,y start at 1 and the xy-coordinate system is rotated by 90° with respect to the matrix coordinate system. Therefore, in fact a somehow different transformation follows if this rotation has to be considered. But, since the dither matrix has only the function of providing a systematic juxtaposition scheme of the ordered thresholds, this rotation is not essential.

some mirroring operations etc. For example, assume that different *satellite images* have to be combined to generate very large images, e.g. *map sheets*. For a single map sheet, several satellite images have to be scaled and geometrically transformed. These satellite images have different spatial resolutions, they can differ with respect to angular position (see Figure 2.3) or with respect to specific gray value distributions. The gray value scale can be influenced, e.g., by different lighting conditions during image acquisition. The computer-based design of map sheets is one of the most extensive applications of image processing in the field of *photogrammetry*. Recently, analog techniques for the production of map sheets have been replaced by digital procedures for higher throughput.

In principle, the procedures for *geometrical fitting* can be classified as follows: There are certain procedures that are oriented on just a single image, e.g., geometrical single-image procedures for layout design or for image preprocessing for subsequent image analysis. On the other hand, procedures exist to combine several images into a single image, i.e. into a (large) map. Such procedures are, e.g., geometrical multi-image procedures for the already mentioned map sheet production, or for the combination of several medical images for *whole body representations*.

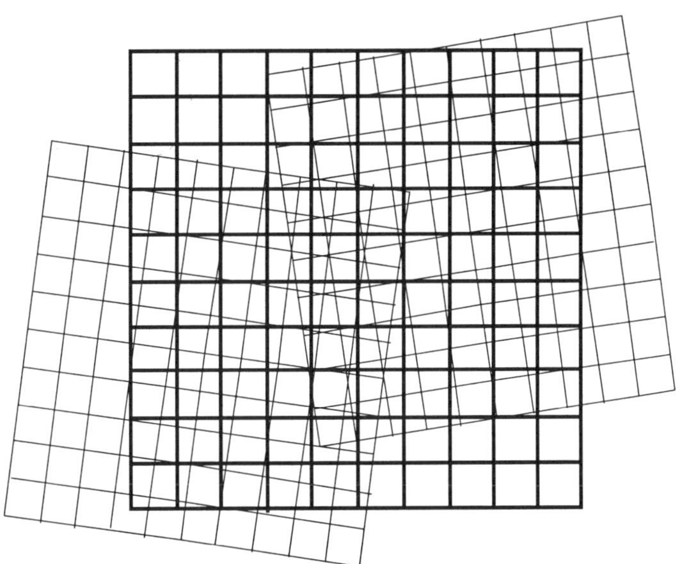

Figure 2.3: Assume that several images (thin grid lines) have to be copied into one image screen. In that case several image values from one or several input images can contribute to a resultant image value.

The processing of single images is illustrated in the following Sections of this Chapter by means of examples, e.g., for contrast enhancement, for the mutual adaptation of the gray value distribution in several images, or for the suppression of noise patterns. For multi-image procedures, the geometrical combination of images without gray value corrections would lead to obviously visible non-homogenous regions, i.e. to certain "gray value transitions or edges". Thus, before selecting a geometrical transformation in the sense of Chapter 4, the given input images should be radiometrically prepared (e.g. matching of the gray value ranges).

A simple example of a geometrical single-image procedure is given by the *magnification* of an $M \times N$ image f into a $2M \times 2N$ image h. A straightforward approach could be to copy each image value $f(x, y)$ into a 2×2 square of the resultant image producing identical image values $h(2x, 2y)$, $h(2x+1, 2y)$, $h(2x, 2y+1)$, $h(2x+1, 2y+1)$ over 2×2 squares. Then a grid pattern will be visible in the magnified image. To refine this simple procedure, values at neighboring point positions of the input image f can be used for calculating the values of the resultant image h,

$$\begin{aligned}
h(2x, 2y) &= f(x, y), \\
h(2x+1, 2y) &= \frac{1}{2}\,(f(x, y) + f(x+1, y)), \\
h(2x, 2y+1) &= \frac{1}{2}\,(f(x, y) + f(x, y+1)), \\
h(2x+1, 2y+1) &= \frac{1}{4}\,(f(x, y) + f(x+1, y) + f(x, y+1) + f(x+1, y+1)).
\end{aligned}$$

By means of this smoothing, block patterns are avoided in the magnified image.

Note that a certain modification of image values can also be included into geometrical transformations, normally in a relatively simple way. In general, a geometrical transformation is characterized by a certain coordinate transformation. In case of the discussed magnification, the coordinate transformation is given by the mapping of a single point (x, y) onto a set $\{(2x, 2y), (2x+1, 2y), (2x, 2y+1), (2x+1, 2y+1)\}$ of four points, see Section 4.2.2.

For geometrical multi-image procedures, first a rough geometrical positioning of the different input images should be performed (*geometrical mosaic formation*). For the resultant mosaic with a fixed coordinate system, the different input images have to be transformed into this unique coordinate system of the final image. These transformations normally cause certain defects characterized by inexact matches of the order of magnitude of about 1-2 image points. The geometrical mosaic formation can be used as reference for the radiometrical preprocessing of the single images, e.g., for the adjustment of contrasts etc. For example, an image placed in the middle of the mosaic and of "average contrast" can be chosen as a *reference image*, and the neighboring images can be gradually matched to the gray value distribution of this reference image. Such a gray value adjustment can be realized with the aid of a certain histogram transformation (cp. Section 5.1).

In the field of photogrammetry several procedures are known for geometrical mosaic formation and subsequent adjustments. The multi-image approach will not be a topic in this handbook because this approach is relevant for a few specific applications, and detailed descriptions would require more space than is possible in this general handbook. The operators in Chapter 4 deal with geometrical single-image procedures.

2.4 IMAGE RESTORATION AND IMAGE ENHANCEMENT

In this Section some examples of tasks taken from the domains of image restoration and of image enhancement are illustrated.

Image restoration aims at the reconstruction of an unknown "original" image based on one or several impaired input images. The "loss of quality" can be due to structured interferences, or to randomly distributed additive, multiplicative or impulse noise. These impairments can result from the input sensor, from the image transmission channel, or from image decoding. *Image enhancement* can also aim at the improved visibility of fine patterns or of fine image structures (*details*). Normally this means more than just a strict reproduction of an unknown original image. Enhancement tasks are typical for the analysis of microscope images in medicine or biology, for material inspection, or for remote sensing. In such application areas, the visibility of significant image structures has to be accentuated for further tasks of image analysis.

Figure 2.4 shows two examples of image restoration by smoothing. The image top left shows a portrait with synthesized additive normally distributed noise. This portrait image has been impaired with the program of Section 5.2.2 with a standard deviation of 20 gray values. The task set is to reduce this noise as much as possible without introducing any essential unsharpness into the resultant image. This task can be solved by an operator for *smoothing in a selected neighborhood*, see Section 6.1.4. The result of this operator is the gray value average in a selected subset of all pixels in the operator window. This subset contains only those image elements that are estimated to belong to the same region as the reference point of the placed window. In the simplest case this estimation can be directly based on a comparison of the gray values inside the window with the gray value at the reference point. This approach aims at preserving the steepness of edge slopes, i.e. at avoiding to smooth edges. If the unweighted mean gray value over the full operator window would be used as the result, then the edge steepness preservation could not be expected. The result of this smoothing in a selected neighborhood is shown in Figure 2.4, top right.

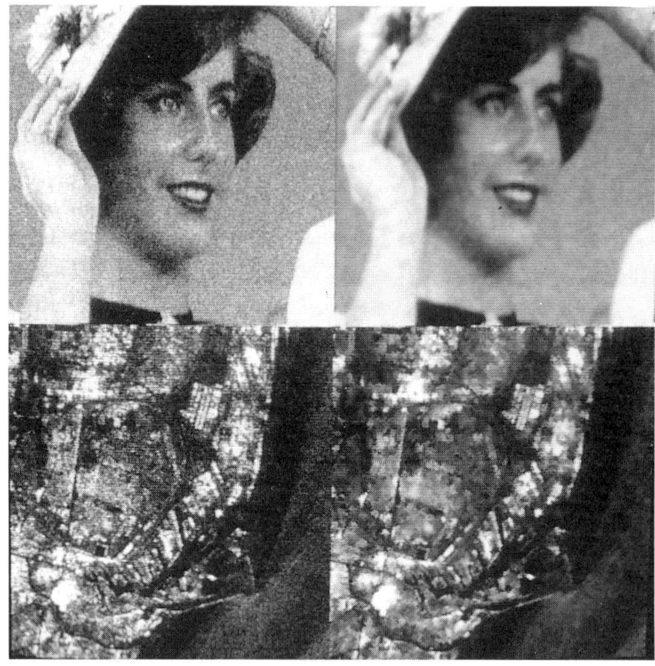

Figure 2.4: Image restoration by smoothing. Above: Suppression of additive noise by gray value averaging in a selected neighborhood. Below: Agglomeration of homogenous regions of a radar image of remote sensing by means of an adaptive rank-order filter.

Below left in Figure 2.4, the original *radar image* of a remote sensing view is impaired by a special kind of a multiplicative noise (*speckle noise*). This is due to the radar imaging process. The restoration of homogenous regions as compact domains of the (unknown) undistorted image can be obtained by smoothing of gray value variations which are due to fine textures. The deletion of meaningful structures, belonging to the image content, has to be avoided during this smoothing process. For solving this task, the smoothing process of an *adaptive rank filter* (see Section 6.1.6) is controlled by local gray value variance. The edge preserving effect of a smoothing filter increases in strength with the intensity of the local gray value variance. In the resultant image, see Figure 2.4 below right, the texture-noise is nearly completely suppressed inside the homogenous regions. The edges and the thin structures are well preserved, which is essential for the reproduction of the image content.

A special task of image enhancement is represented by the emphasizing of image details. In this case the visibility of special image patterns is of main interest, and not an accurate restoration. If the image *details* e.g. texture or patterns with high spatial frequencies are of low contrast then the increase in visibility can be an important task in view of an improved visual image assessment.

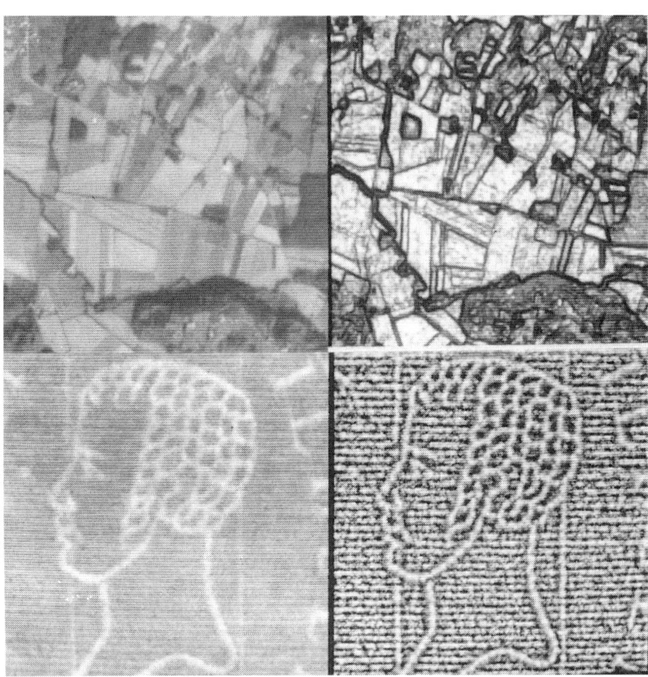

Figure 2.5: Enhancement of details and of the local contrast. Above: left: original image, right: result of applying *inverse contrast ratio mapping*. Below: left: original image, right: result of contrast stretching by means of a *rank order transformation*.

Figure 2.5, top left, shows as example an *aerial view* of low contrast. The task of image enhancement consists in increasing the contrast. The original image in Figure 2.5, below left, has a stripe texture in its background. Here the task of image enhancement consists of first recognizing this texture, and then enhancing it separately from the other image content.

With the upper image, improved definition can be obtained by an enhancement of gray values depending upon the ratio between the local mean gray value and the local standard deviation of the gray values (*inverse contrast ratio mapping*, see Section 6.3.3).

For the lower image, a spatial space-dependent contrast stretching can be performed by means of a rank order transformation, see Section 6.5.8. This operator performs an equalization of the gray value distribution inside the operator window. Within this window, the gray values are spread to the full gray value scale. It can be shown that details are stronger enhanced for smaller window sizes.

2.5 IMAGE SEGMENTATION

The aim of the *image segmentation* is subdivision of an image into homogenous (with respect of a certain *criterion of homogeneity*) and disjoint regions (image segments). The criterion of homogeneity is often of subjective nature and it can be very specific in the different areas of applications.

For image segmentation, two opposite points-of-view can be distinguished. The first approach is the *contour-oriented segmentation*. Here the methods are directed on computing differences between neighboring image points, i.e. of certain differences which have to be specified for the given areas of applications. Between image segments contours are drawn which are characterized by the detected differences.

The second approach is the *region-oriented segmentation*. Unlike the first mentioned approach, this method (e.g., assuming only merging operations) is concerned with detecting similarities between neighboring image points, then similar image points are merged as long as the fundamental criterion of homogeneity is met. Otherwise, the growth of an image segment stops.

In Chapter 6 both points-of-views are discussed, compare Sections 6.2 and 6.4. In the ideal case of image segmentation, both approaches should lead to the same segmentation result or at least it is plausible that they support each other. However in practice there are many reasons why the performances of these two approaches are often strongly different, e.g., noise, difficulties in formulating an analytical criterion of homogeneity, difficulties in the interpretation of texture, etc.

Figure 2.6 at top left shows an *aerial view* of an agricultural area. This image has to be divided into homogenous terrain regions e.g. into crops, meadow, wood, or areas with soil damages etc. For instance, the aim can be the measurement of size or shape of these regions. It seems difficult to formulate an uniform homogeneity criterion for the different kinds of terrain, because some of the ground regions are uniform and other ones contain textures. However, in this case it is possible to characterize the different terrain regions in a first approximation by *modes* (i.e., those gray values where the local density function is a maximum). The density function and the mode can be estimated on the base of the local gray value histogram.

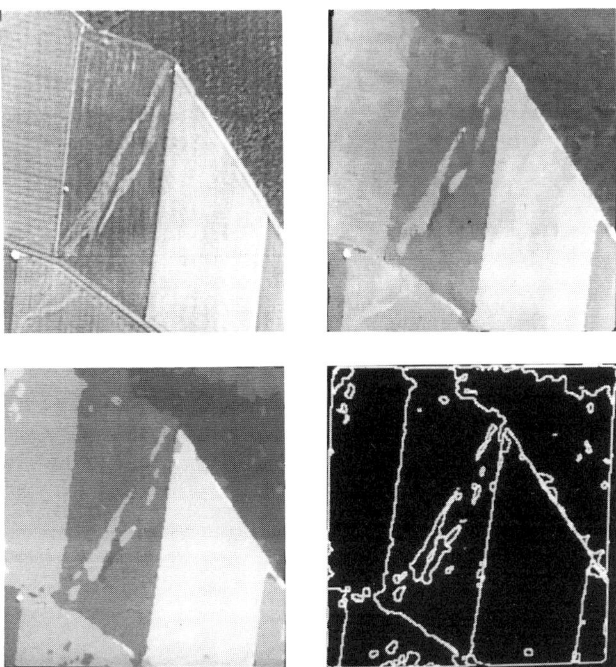

Figure 2.6: Three successive results of an image segmentation process. Top left: original image; top right: intermediate result after performing a gray value agglomeration, below left: next intermediate result after multi-level thresholding, below right: final result after the computation of an edge image.

An image as shown in Figure 2.6 can be obtained by means of *gray value agglomeration* (see Sections 6.4.1 and 6.4.3) at top right. This operation leads to a gray value concentration in proximity of the local histogram modes, with different typical modes for different regions.

This effect can be enhanced by successive applications of *multi-level thresholding* (cp. Sections 5.5.2 and 5.5.3). Such an operator assigns to the image points only one of a few previously determined gray values. The global gray value histogram is analyzed with respect to local minima ("valleys of the histogram"). Each gray value of the input image is between two local minima, and is transformed into that gray value which is defined by the local maximum between these two minima. By assumption the image elements with values between two local minima belong to the same ground region. This process it directed on labeling with a constant gray value all image points belonging to the same type of terrain. This gray value acts as a *region label*. The aim is to obtain a piece-wise constant gray value label image after this step (see Figure 2.6, below left).

After this multi-level thresholding, the different constant gray values should correspond to the different types of ground regions. Then for a resultant image a simple *edge operator* (see Section 6.2, especially 6.2.1) can be used to compute an edge image. This edge image represents the final segmentation result (see Figure 2.6, below right). Note that to solve this segmentation problem, a contour-oriented method (edge operator) was combined with a region-oriented approach (agglomeration and multi-level thresholding).

2.6 ICONIC REPRESENTATION OF LOCAL FEATURES

Image restoration or image enhancement leads to a certain matrix of modified gray values as the resultant image. If the input image represents a certain scene as perceived by a human observer, then in general this is also true (normally, "even better") for the resultant image.

It is also possible to determine at each image point the value of a certain scalar feature by means of a local operator. This computed value can be represented as a symbolic gray value at the same image point. In this way values of the considered feature and its spatial distribution are mapped as in a common physical map. For a human observer such an *iconic local feature map* represents a highly abstract image compared to the original input image. Often the imaged scene is not recognizable anymore.

A typical example of iconic feature representations is the generation of an edge image. This is one of the most frequent tasks in image processing. The computed edge values represent the intensity of the feature "edge element" at the corresponding image points. Since the input images can have different content, noise etc., no universal optimal edge operator exists. There is an ongoing discussion in the field of image processing on new, more effective (with respect to certain properties) edge operators. In general, edge operators should satisfy the following two contrasting requirements:

(a) High spatial frequencies at gray value edges have to be enhanced.

(b) Typically, the input image is affected by noise which is uniformly distributed in the spatial frequency spectrum. The enhancement of such noise should be minimized.

For this reason many edge operators consist of a combination of a high-pass filter and of a low-pass filter (mostly both filters are nonlinear).

An edge operator of this kind is the *morphologic edge detector* (see Section 6.2.3). Its effect upon the original image of Figure 2.7 at the top left, is shown in the same figure, top right. This resultant image illustrates that both conditions **(a)** and **(b)** are quite satisfactorily met, at least for this input image.

Figure 2.7: Edge extraction (above) and an iconic map of a local measure of anisotropy (below). Left: Original image of a natural scene (above) and of an aerial view (below). Top right: edge extraction by means of the morphologic edge detector, below right: iconic representation of a local measure of anisotropy by means of gray values.

The obtained edge image is not yet the final result. In the ideal case, the resultant image should contain only closed contour lines of uniform width (*line image*), and it should not contain *artifacts*, e.g. in the form of isolated points or "dead branches". Taking account of the complexity of the scene displayed in this test image, the quality of the computed edge image (top right) should be sufficient for converting the edge image into a good-quality line image by means of a simple binarization operator.

The example shown in Figure 2.7 below at right, illustrates the iconic representation of a local measure of anisotropy (cp. Section 6.5.9). The input image (below left) is an aerial view of an urban area with straight streets. The enhancement of image points with high values of anisotropy may also be of support for the extraction of roads. In the resultant image, below right, a bright gray value represents a strong local anisotropy. For the chosen definition of anisotropy (cp. Section 6.5.9) this value reaches its maximum at thin lines while it is zero in regions with constant gray values.

2.7 EXTRACTION OF PATTERNS

For some applications of image processing, the task may consist in extracting only specific image segments which are user-defined by means of features of shape, or of the gray value distribution. In the ideal case the resultant image contains all and only the image segments of the specified kind. An example of such *pattern extraction* is the selection of patterns of a certain size, where the size of a pattern is defined by the smaller side of the smallest rectangle (with sides parallel to the coordinate axes) containing the given pattern. This definition is unique for bilevel images. However, the input gray value image often presents such complex scenes as the aerial view in Figure 2.8. Then, it may be not clear which point belongs to a certain pattern, and which does not.

The task can be solved by means of the *top hat transformation*. At first, this transformation carries out a sequence of minimum and maximum operations (see Section 6.5.2). Then, it realizes a point-wise image difference (see Section 5.4.3). Assume that bright patterns on a dark background have to be extracted (as in Figure 2.8).

The minimum operator is first carried out in k iteration steps. Then, k iteration steps of the maximum operator are performed on the resultant image. This sequence of operations deletes all objects that are smaller than $2k + 1$ image points in diameter. Finally, a point-to-point image subtraction between this resultant image (image without "small" patterns) and the original image yields an image containing only segments which had been deleted before. By choosing the parameter k it is possible to extract objects of various sizes from the input image.

The three images in Figure 2.8 have been obtained in this way from the original image (top left) with $k = 1$ (top right), $k = 2$ (below left) and $k = 3$ (below right).

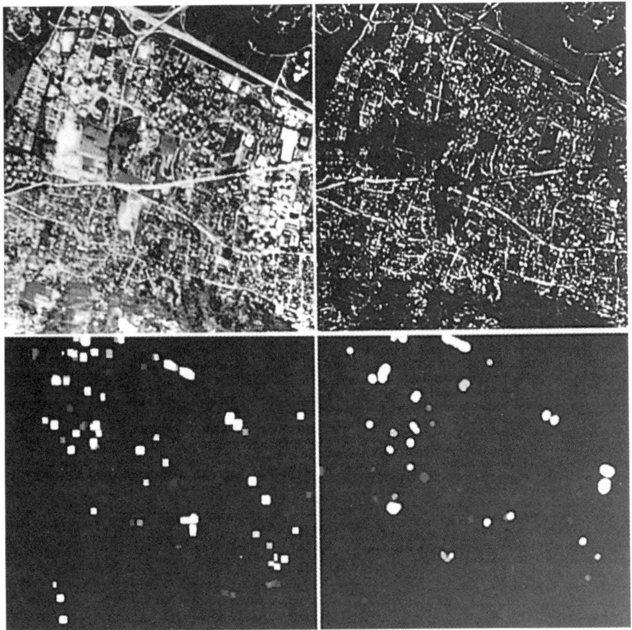

Figure 2.8: Selection of patterns of different size by means of the top hat transformation. Original aerial view (top left) and extracted image segments with a given minimum size t. Top right: $t = 2$, below left $t = 4$, below right $t = 5$.

2.8 ICONIC REPRESENTATION OF GEOMETRICAL RELATIONS

Besides the iconic representation of local features, or the extraction of patterns, for certain tasks of image processing or image analysis it is also desirable to generate iconic representations of geometrical features as points, straight line segments etc., or geometrical patterns as isolated regions of specific geometrical form.

Figure 2.9 shows a microscope image of a liver slice. The chosen technique of slice preparation allows a good visual discrimination between different classes of nuclei, but *cell membranes* are not visible. The task can be that of determining a statistical estimate of the membranes.

Assume that the nuclei are segmented as image regions, e.g. by binarization (Section 5.3) or by contour tracking (Section 7.2.1). For these segments, a central point within the image region can be determined, e.g., the centroid of the region or the center of the smallest circumscribing axes-parallel rectangle. These central points of nuclei represent the geometrical distribution of the nuclei in the image.

For a point set **P**, e.g., all central points of nuclei in an image, the Voronoi-diagram (Section 7.2.2) can be computed. The Voronoi cell of a point p of the set **P** consists of all image points that have no smaller distance to any other point of the set **P** than to point p. The borders of these Voronoi cells lay on the bisection lines of certain point pairs of the set **P**. These borders of the Voronoi cells form the *Voronoi diagram.*

Figure 2.9, right, shows as a graphical overlay the Voronoi diagram for a certain set of selected points. For straight line generation, the procedure *BRESENHAM* (Section 3.4.10) can be used. Such Voronoi cells may be considered as an approximation of the invisible cell membranes. Diagnostic research on cellular tissues may be supported using this method.

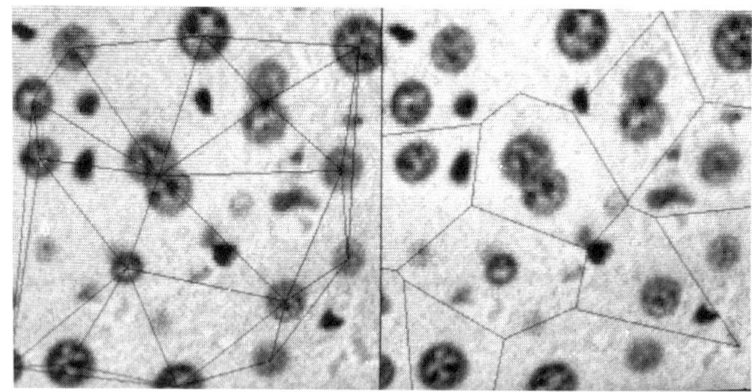

Figure 2.9: Microscope image: left: the Delaunay triangulation is shown as graphical overlay, right: the computed Voronoi diagram for some selected points.

The operators considered in Section 7.2.2 can be used for the generation of such graphical representations of Voronoi diagrams or *Delaunay triangulations.* Such graphs can be used for the interactive, visual evaluation of point distributions, or for the automatic clustering of point sets. These graphs have multifarious applications in the field of image analysis because they allow the definition of different quantitative measures for point positions.

2.9. SPECIAL AND ART EFFECTS

The computer supports the generation of images of esthetic or artistic value. In the field of computer graphics, computer art is already treated as an established application area. Computer animation and computer music are further examples of applications of *computers in the arts*. In image processing, real images (images of scenes etc.) can be used as input, and operators can be chosen for obtaining specific artistic aims. Image processing is becoming a further area of these young art disciplines which are based on applications of digital computers.

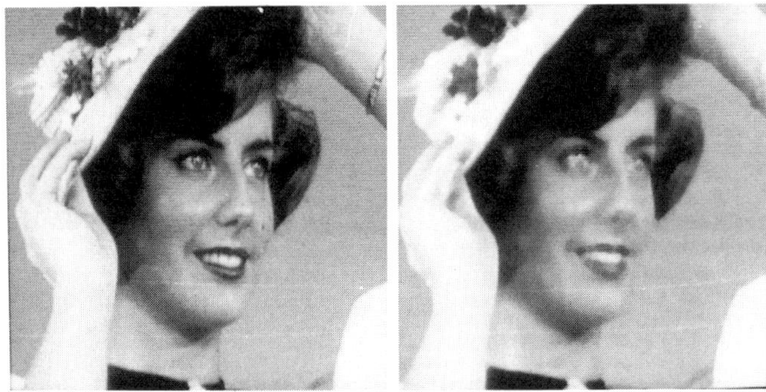

Figure 2.10: The left original image was processed by twice dilation and, subsequently, twice erosion.

In Figure 2.10 an original image is shown as well as the result after the application of the operator explained in Section 6.5.2 (to be exact: at first twice dilation, and then twice erosion with the choice of the octagon-neighborhood). The processed image resembles an impressionistic painting that consist of "dots."

Figure 2.11 shows two examples of possible image manipulations. These examples are apt to demonstrate special effects which can be realized by means of image processing procedures, and which might, e.g., be of interest for DTP-systems. In the upper half, right, an image of *isolines* (see Section 5.5.1) is shown. The input image is top left. Isolines characterize a constant height of the *gray value relief*. For image functions and gray value relief, compare Section 1.1.2.

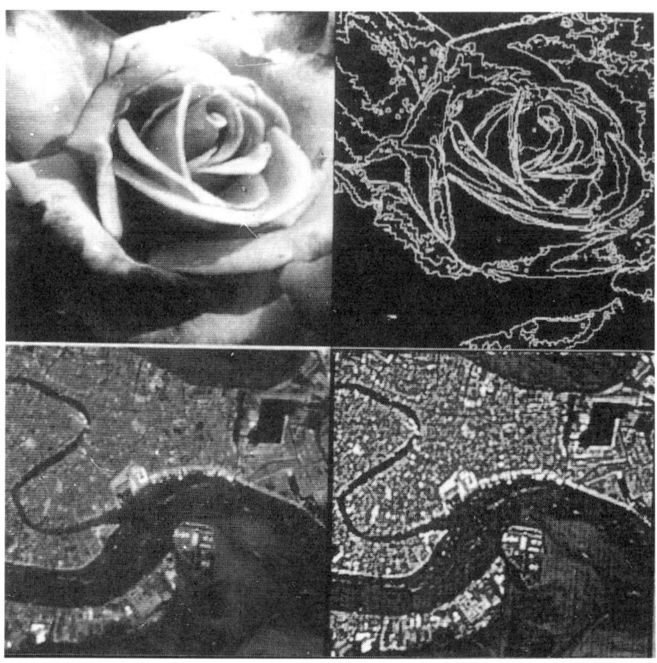

Figure 2.11: Examples of special effects. Left: original images, top right: representation of some constant height lines of the gray value relief by aid of an isoline map; below right: a plastic effect, obtained by adaptive contrast enhancement, can be visible at sufficient image resolution.

In the lower half of Figure 2.11, right, the result of a contrast enhancement operation is shown for the original aerial view on the left. At sufficient resolution, a plastic effect is attained. The procedure of Section 6.3.4 has been used here for contrast enhancement. Experimenting with image processing operators, sometimes such interesting graphical effects can be obtained (often with good luck). For artistic processing of images with speific aims, these operators should be studied systematically (which operator can be applied for which effects etc.). There is a wide field for future studies. A special interesting case of artistic image processing is already well known as *morphing* (i.e., for two given images, intermediate images are computed with a "smooth transition" of one given image into the other) . For some simple special effects (solarization, poster effect, etc.) relevant operators can be found in this book, e.g. by consulting the glossary.

2.10 BIBLIOGRAPHIC REFERENCES

Methodical principles of image processing are described in most of the textbooks listed in Section 1.5. For image synthesis, as referred to in Section 2.1, see the literature for computer graphics, e.g.

Foley, J. D., van Dam, A., Feiner, S. K., Hughes, J. F.: *Computer Graphics - Principles and Practice (Second Edition)*. Addison-Wesley, Reading, 1990.

In Section 2.2 several standard procedures for halftone image representation have been sketched. For more details see

Ulichney, R. A.: *Digital Halftoning*. MIT Press, Cambridge, MA, 1987.

For geometrical multi-image procedures (see Section 2.3) reference is made to the literature of photogrammetry, see e.g.

Jensen, J.R.: *Introductory Digital Image Processing - A Remote Sensing Perspective*. Prentice Hall, Englewood Cliffs, 1986.

For image restoration and image enhancement (see Section 2.4) many suggestions of procedures exist which are based on signal theoretical transformations. For this field, some operations are given in Section 7.3. Extensive material on signal theoretical image transformations can be found, e.g., in

Besslich, P.W., Tian, L.: *Diskrete Orthogonaltransformationen*. Springer, Berlin, 1991.

Pratt, W.K.: *Digital Image Processing*. 2nd ed. Wiley, New York 1991.

Various alternative suggestions of different approaches exist for image segmentation (Section 2.5). The segmentation approaches explained in this book are characterized by algorithmic simplicity. Regarding algorithmic complexity and used models, much more complex proposals exist in literature (see the references given in 1.5 for image processing and image analysis). However in practical experiments the improvement of segmentation quality obtained with more complex methods was often inessential if compared with the methods described in this book.

For a discussion of the definition and applications of anisotropy measures (Section 2.6) see

Zamperoni, P.: *Adaptive rank order filters for image processing based on local anisotropy measures*. Digital Signal Processing, July 1992, pp. 174-182.

The top hat transformation (Section 2.7) is dealt with in

Haralick, R.M., Sternberg, S.R., Zhuang, X.: *Image analysis using mathematical morphology*, IEEE Trans., Vol. PAMI-**9**, July 1987, pp. 532-550.

Pitas, I., Venetsanopoulos, A.N.: *Nonlinear Digital Filters*, Kluwer Academic Publishers, Boston, 1990.

Zamperoni, P.: *Methoden der digitalen Bildsignalverarbeitung*. Vieweg Verlag, Wiesbaden, 2nd ed., 1991.

Section 2.8 gave some comments to a quite new field of computer science. This field has been dynamically growing during the last years. The computational geometry is directed on the efficient realization of geometrical objects or constructions, compare for example

Preparata, F.P., Shamos, M.I.: *Computational Geometry*. Springer, New York, 1985.

Various possibilities for following artistic ambitions (Section 2.9) are shown in

Nake, F.: *Ästhetik als Informationsverarbeitung*. Springer, Wien, 1974.
Völz, H.: *Computer und Kunst*. 2nd ed., Urania-Verlag, Leipzig, 1990.

Stimulations are also given in the interesting photo-book

Krug, W., Weide, G.: *Wissenschaftliche Photographie in der Anwendung*. Akademische Verlagsgesellschaft Geest und Portig K.-G., Leipzig, 1972.

Databases with pictorial information are needed for coping with the increasing manifold of visual information. Compare for example

Chang, S.K.: *Principles of Pictorial Information Systems Design*. Prentice-Hall, Englewood Cliffs, 1989.
Kunii, T.L. (ed.): *Visual Database Systems*. North-Holland, Amsterdam, 1989.

There are many established application fields of image processing. For the preparation of pictorial material for publications with DTP-systems (desktop publishing), compare for example

Pape, U. (ed.): *Desktop Publishing - Anwendungen, Erfahrungen, Prognosen*. Springer, Heidelberg, 1988.

The combination and transformation of pictorial data are important tools in medicine for visualizing anatomical structures, slices of tissues, state and progress of illnesses, etc. Compare for example

Collins, S.M., Skorton, D.J.: *Cardiac Imaging and Image Processing*. Mc Graw-Hill, New York, 1986.
Lemke, H.U., Inamura, K., Jaffe, C.C., Felix R. (eds.): *Computer Assisted Radiology*. Proc. Intern. Symp. CAR'93, Berlin 1993, Springer, Berlin, 1993.

for medical applications of image processing.

3 ALGORITHMIC FUNDAMENTALS

Efficiency is desired for the rapid implementation of image processing operators[1]. This is also an important topic for the run-time behavior of the operators regarding data access, memory requirements, and computing time.

This Chapter starts with measures of computing time efficiency for image processing operators. As for data structures, the scope of this work allows focusing on matrix representations of image data. Generally usable control structures for important operator classes as point or local operators are illustrated. These control structures can be systematically used as a program framework for the implementation of operators specified in Chapters 5 and 6. Section 3.4 provides some algorithms which are used as subroutines in some operators of this book.

3.1 ALGORITHMIC EFFICIENCY

Image data are characterized by their large size. Therefore time-efficiency is an important topic in image processing. A programmer begins to realize this fact e.g. when an image transformation under test takes several minutes. This Section aims to stimulate the programmer's own initiative about analyzing time complexities of operators and optimizing their run-time behavior. The modern computer technology allows interactive image processing in real-time. The operator complexity can be increased following the progress in computer technology. However several image processing operations can easily overload the computer's processing power in spite of their simple definition.

3.1.1 Estimation of Computing Time

Absolute time measurements (e.g.: a problem of which size can be solved in a given time?) or asymptotic estimates characterize the run-time behavior of a program, e.g.

[1] The enclosed reader-service card offers the C-sources of the image operators of this book.

computing a certain window function, or a certain operator. Asymptotic estimates describe a general tendency, and absolute measurements give some quantitative statistics. The forthcoming Chapters contain examples of both ways of efficiency description. This Section deals with a brief introduction into time complexity characterizations of programs.

Absolute measurements are concrete but hard to compute for statistically relevant sample sizes, and also often not clearly interpretable, since they depend upon the specific software environment implementation (language, compiler, with or without run-time optimization, etc.) and upon the computer (processor, frame grabber, multi-user processing, special image processing support, image output device etc.). Often the asymptotic efficiency characterization can be determined faster than absolute measurements, and it provides a global characterization of the program, i.e. for any input situation.

A *computing problem* is a task which has to be solved on a computer, e.g. generating a textured image of size $M \times N$, or sorting a gray values of an picture window. Some *problem parameters* characterize the input sizes of a given computing problem. Often a single parameter a suffices as problem parameter, e.g. for the number of pixels inside a square picture window, for the contour length of a region, or for the cardinality of a point set. In the sequel we assume that a single problem parameter is sufficient for characterizing the input size. Otherwise, if several parameters (e.g. number of iterations, window size and threshold) are relevant for characterizing the data input, then the following definition of asymptotic classes $\mathbf{O}(F)$ can easily be extended.

Asymptotic computing time estimates base on a general, serial computer model (von-Neuman hardware architecture). The computing power of such a theoretical machine can roughly be identified with time models based on languages as C, FORTRAN or Pascal, without allowing parallelism or concurrency in these languages. For each instruction it is assumed that its performance needs uniformly one time-unit (*uniform cost criterion*).

Example 3.1: Assuming the uniform cost criterion, each instruction

$$x := x + 1,\ w := \mathbf{sin}(\beta),\ z := x \cdot y \quad \text{etc.}$$

is assigned one time-unit as time requirement. A simple loop

$$\mathbf{for}\ i := 1\ \mathbf{to}\ a\ \mathbf{do\ begin}\ x := x + 1\ ;\ y := y \cdot x\ \ \mathbf{end}\ \{for\}$$

needs $3a + 1$ time units (one addition for i and x, and one multiplication for y in each iteration of the loop, and one final addition for i).

A certain function $time(a)$ specifies the *time complexity* of a program with problem parameter a, and for any input of size a the algorithm produces a result after at most $time(a)$ time units. All inputs of size a are considered equivalent by this definition. Some "unfavorable" inputs may produce high values of $time(a)$. The defined function *time* is a *worst-case complexity function.*

Example 3.2: Assume a list $\mathbf{L} = [u_1, u_2,..., u_a]$ of a numbers (e.g. a gray values) sorted in increasing order, i.e.

$$u_i \leq u_{i+1} \quad \text{for} \quad i = 1, ..., a-1.$$

The task consists of computing a *mode* v of **L**, i.e. such an element of **L** which appears with the highest frequency in the list ("the dominant gray value in an picture window which determines the value of the resultant image h at the current point p"), and also the number w (*frequency*) how often the mode v appears in **L** (e.g. the value w could be the resultant image value $h(p)$ as a significant feature of the local gray value distribution).

As a straightforward solution, the mode v is initialized by u_1, and the frequency by $w := 1$. A counter *temp* is used to count the number of repetitions of the current element of list **L**. If *temp* exceeds the current value of w then v becomes the current list element, and $w :=$ *temp*. In a more formalized way, this means:

```
v := u1;  w := 1;  temp := 1;
for i := 2 to a do
        if (ui = ui-1) then begin
                temp := temp + 1;
                if (temp > w) then begin v := ui;  w := temp  end {if};
                end {if};
        else  temp := 1
```

This is Algorithm 1. Its time complexity $time_1(a)$ can be estimated by the inequalities

$$3 + 3(a-1) + 1 \leq time_1(a) \leq 3 + 6(a-1) + 1$$

where all instructions are evaluated by the uniform time criterion. However, this straightforward algorithm repeats operations "*temp* := *temp* + 1" and "*temp* > w?" more often than necessary if a list element u_i appears several times. In fact it suffices that updating is only initiated if the previous value of w is exceeded for the first time:

```
v := u1; w := 1;
for i := 2 to a do
        if (ui = ui-w)  then  begin  v := ui;   w := w + 1   end {if}
```

This is Algorithm 2. The inequalities

$$2 + 2(a-1) + 1 \le time_2(a) \le 2 + 4(a-1) + 1$$

give an estimate of the time complexity $time_2(a)$ of this Algorithm 2. This is, in comparison to Algorithm 1, an improved method for computing mode and frequency.

Taking account of the gray value variance of "natural" gray value images, the function of Example 3.2 (for determining the mode v or the frequency w) can be advantageously implemented as window function of a local operator only if segments with constant gray value can be assumed to occur frequently in the input images, or if the image had been previously smoothed.

The Examples 3.1 and 3.2 specify exactly complexity functions or estimate them by means of inequalities. Typically, such exact estimations for programs are not feasible, and in general not useful. Complex program structures make the counting of performed instructions difficult, and anyway the number of performed instructions is an abstract measure of the real time requirement.

Asymptotic function descriptions of a problem parameter a specify the worst-case complexity functions $time(a)$ "up to a scaling factor". The *asymptotic class* $\mathbf{O}(F)$ is defined for functions F and G mapping integers into reals. Formally, it holds

$$G \in \mathbf{O}(F) \leftrightarrow \exists c \exists q \, (c > 0 \wedge \forall a \, (a \ge q \rightarrow G(a) \le c \cdot F(a))).$$

Thus, a function G is a member of class $\mathbf{O}(F)$ if there exists a certain positive constant c such that for almost all a (i.e. for all a with the only exception of a finite number of values smaller than q) it holds that

$$G(a) \le c \cdot F(a),$$

i.e. the function F is an *upper bound* of the function G. The classes $\mathbf{O}(F)$ (read: "big O of F") define upper bounds of algorithmic complexity. The asymptotic constant $c = c(q)$ specifies the ratio between G and F.

Example 3.3: The function $G(n) = \alpha_t \cdot a^t + \alpha_{t-1} \cdot a^{t-1} + \ldots + \alpha_0$ is a member of the class $\mathbf{O}(a^t)$, for $t \ge 0$ and for any reals $\alpha_t, \alpha_{t-1}, \ldots, \alpha_0$. The function $H(a) = 5000 \cdot a + 1000 \cdot a^2 + 0.0005 \cdot a^3$ is in the class $\mathbf{O}(a^3)$, with asymptotic constants $c(q) \ge 6000.0005$ for $q \ge 1$, or $c(q) \ge 1125.0005$ for $q \ge 2$.

Let be $log\ a$ the short notation for the logarithm $log_2\, a$ with the base 2. Since

$$log_b\, a = \frac{log_2\, a}{log_2\, b},$$

for $b > 1$, the base of the logarithm is without relevance for asymptotic values.

The following denotations are used for the asymptotic run-time behavior of programs:

O(a)	linear run-time (is optimum if all a inputs have to be read),
O($a \cdot log\ a$)	"*a-log-a*"-run-time (practically "nearly linear"),
O(a^2), **O**(a^3)	square and cubic run-time,
O(a^t)	polynomial run-time ($t \geq 0$),
O(b^a)	exponential run-time ($b > 1$).

In general, the run-time behavior up to about **O**(a^3) can be seen to be "acceptable for practical purposes". The practical usefulness of each implementation with exponential run-time is out of the question.

3.1.2 Absolute and Asymptotic Evaluations

We discuss the question of the practical relevance of asymptotic complexity characterizations of image transformation programs using the *sorting problem* of a gray values as example. The sorting of gray values inside an picture window is a usual tool for implementing combinatorial window functions. Assuming a homogenous local operator, then sorting has to be carried out at **O**($M \cdot N$) pixel positions, each process for $a = m \cdot n$ integers u, with $0 \leq u \leq G - 1$. Sections 3.4.5 to 3.4.7 provide three different sorting procedures for such a task which can be suggested for specific values of $a = m \cdot n$:

QUICKSORT is an asymptotically fast **O**($a\ log\ a$) method in the sense of expected complexity (**O**(a^3) in the sense of worst-case complexity as defined above), and its use is suggested for "large" values of a,

BUBBLESORT is an asymptotically slow **O**(a^2) method (in the worst-case sense), but it can lead to fast processing for "small" values of a, and

BUCKETSORT is suggest for the sorting of items with values within a limited range, as 0 ... $G - 1$, and this method is suggested for "very large" values of a.

An extensive literature exists on sorting. Several methods have standard denotations. Two of such standard methods were included into the comparative run-time analysis experiments illustrated in Table 3.1: *MERGESORT* as a worst-case **O**($a\ log\ a$) method, and *SELECTIONSORT* as a further worst-case **O**(a^2) method.

The concrete implementation (language, compiler, hardware etc.) influences the results of a run-time analysis. Figure 3.1 shows graphically the measured computing time requirements of the procedures *QUICKSORT*, *BUBBLESORT* and *BUCKETSORT* given later on in this Chapter. The implementation was in C on a

SUN 4/30. The identical source code was used on different machines, leading to certain variations in the measured time requirements. However, the "typical behavior" in the sense of the asymptotic characterization is always visible.

	3×3	5×5	7×7	9×9	11×11
SELECTIONSORT	0.017391	0.090744	0.297959	0.761311	1.627193
BUBBLESORT	0.016176	0.120742	0.453782	1.398689	2.951754
MERGESORT	0.044898	0.153490	0.332293	0.602623	0.935673
QUICKSORT	0.019788	0.086120	0.170948	0.350820	0.513158
BUCKETSORT	0.136946	0.150616	0.175030	0.207213	0.247076

Table 3.1: Average time behavior (in µs) of several sorting routines for the sorting of $a = n \cdot n$ gray values, with $G = 256$. This time table reports about an implementation in C on a SUN 4/50

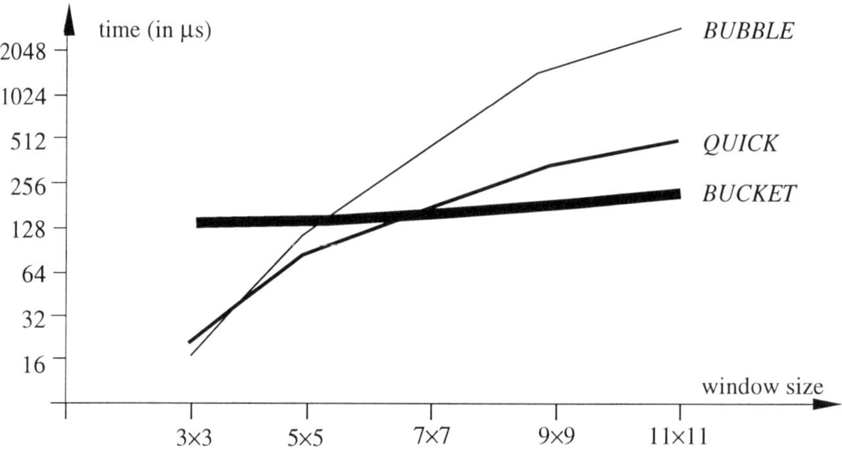

Figure 3.1: Graphical representation of the average time requirements of the sorting routines *QUICKSORT*, *BUBBLESORT* and *BUCKETSORT*: the time axis (in µs) is logarithmically scaled to the base 2, and the axis of the number a of sorted gray values inside an picture window is also scaled in a non-linear way. The implementation on SUN 4/30 was performed with run time optimization.

If an image processing operator subroutine is subject to frequent calls, then a careful study of time-efficient algorithmic solutions is suggested for the involved procedures (e.g. the window function of a local operator). The asymptotic evaluation shows the general behavior, but can not replace a concrete time analysis especially for small values of the problem parameter. The relevant values of the problem parameter and the concrete implementation influence the selection of an algorithm with respect to computing time optimizations.

3.1.3 Decomposition of Homogenous Local Operators

Let us assume time complexity *time*(a) for the computation of a window function of a homogenous local operator. As consequence, the computation of the operator takes about $M \cdot N \cdot time(a)$ computing time for a full $M \times N$ image.

A general speeding-up possibility for homogenous local operators consists in the decomposition of the window functions. This is due to the fact that repeated applications of $m \times m$ window operators, with a "small" value of m, are faster in performance than a $n \times n$ window operator with a "large" value of n. For example, the application of three 3×3 window operators needs less operations than the application of a 7×7 window operator: assume a transformation of $N \times N$ image points, then

$3 \cdot (3 \times 3) \cdot (N \times N)$	$= 27\, N^2$	(three times 3×3)
$(7 \times 7) \cdot (N \times N)$	$= 49\, N^2$	(once 7×7)

estimate the number of operations, if a uniform time complexity is assumed for each image point.

The value of a point p can depend upon $3 \cdot 3$ input image values (a 3×3 *area of influence*) after the transformation by a 3×3 window operator. If an additional 3×3 window operator is performed (at least) for all the 9 points of this area of influence of point p, then the resulting value at point p can depend of $5 \cdot 5$ input image values, cp. Figure 3.2. After three applications of a 3×3 window operator, the area of influence has a size of 7×7, i.e. the same size as for a single application of a 7×7 window operator. However, this does not yet ensure that a specific 7×7 window operator can be carried out exactly by three 3×3 window operators in sequence. Combinatorial estimations already prove that just a few 7×7 operators can be performed via *decomposition*.

Example 3.4: Let us consider the 3×3 window operation *AVERAGE*,

$$AVERAGE(\mathbf{F}(f, p)) = \frac{1}{9} \cdot \sum_{i=1}^{3} \sum_{j=1}^{3} f(r-2+i,\ s-2+j)\ .$$

Each image point $p = (r, s)$ contains the computed value

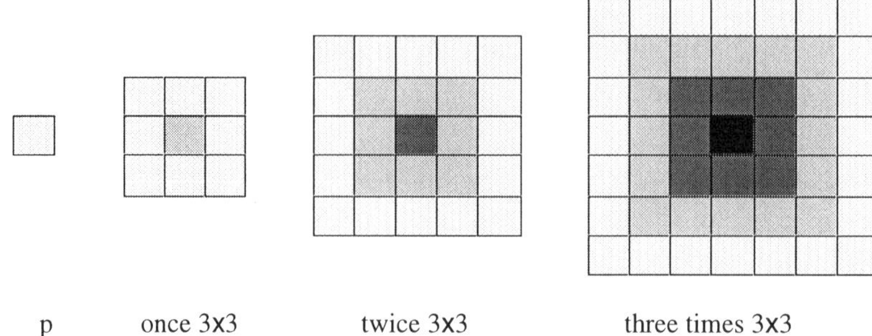

Figure 3.2: Areas of influence of an image point p after a single, a twice repeated or a three times repeated application of a 3×3 window operator.

$$f_1(p) = AVERAGE(\mathbf{F}(f, p))$$

after a single (parallel) application of this window operator, i.e. for a given image point $q = (x, y)$ this is also true for all of its eight 8-neighbors $p_1, ..., p_8$. A second application of this 3×3 window operator $AVERAGE$ produces in position q the average over the previously computed averages in positions $q, p_1, ..., p_8$. Now, the value in position q is equal to

$$f_2(q) = AVERAGE(\mathbf{F}(f_1, q)) = \frac{1}{9} \cdot \sum_{i=1}^{3} \sum_{j=1}^{3} f_1(x - 2 + i, y - 2 + j) \ .$$

A twice repeated application of the average computation leads to a 5×5 window operator

$$AVERAGE^2(\mathbf{F}(f, p)) = \frac{1}{81} \cdot \sum_{i=1}^{5} \sum_{j=1}^{5} a_{ij} \cdot f(r - 3 + i, s - 3 + j)$$

with weighting coefficients a_{ij} as shown in the following kernel, cp. Section 1.3.3:

1	2	3	2	1
2	4	6	4	2
3	6	9	6	3
2	4	6	4	2
1	2	3	2	1

By interpreting this 5×5 operator definition "in the reverse direction", it becomes clear that a decomposition exists into two identical 3×3 operators for this linear 5×5 operator.

A decomposable window operator can be performed by repeated applications of certain window operators with smaller window size. Convolution kernels offer a clear method for representing decompositions of linear window operators. For example,

1	1	1
1	1	1
1	1	1

•

0	1	0
1	0	1
0	1	0

=

0	1	1	1	0
1	2	3	2	1
1	3	4	3	1
1	2	3	2	1
0	1	1	1	0

represents such a decomposition. In general, a linear 5×5 window operator

a_{15}	a_{25}	a_{35}	a_{45}	a_{55}
a_{14}	a_{24}	a_{34}	a_{44}	a_{54}
a_{13}	a_{23}	a_{33}	a_{43}	a_{53}
a_{12}	a_{22}	a_{32}	a_{42}	a_{52}
a_{11}	a_{21}	a_{31}	a_{41}	a_{51}

can be decomposed into two linear 3×3 window operators

a	b	c
d	e	f
g	h	i

and

j	k	l
m	n	o
p	q	r

if the following equations have an unique solution for the kernel coefficients a, ..., i, j, ..., r:

$$a_{15} = a \cdot j$$
$$a_{14} = d \cdot j + a \cdot m$$
$$a_{13} = g \cdot j + d \cdot m + a \cdot p$$
$$a_{12} = g \cdot m + d \cdot p$$
$$a_{11} = g \cdot p$$

$$a_{25} = a \cdot k + b \cdot j$$
$$a_{24} = d \cdot k + e \cdot j + a \cdot n + b \cdot m$$
$$a_{23} = g \cdot k + h \cdot j + d \cdot n + e \cdot m + a \cdot q + b \cdot p$$
$$a_{22} = g \cdot n + h \cdot m + d \cdot q + e \cdot p$$
$$a_{21} = g \cdot q + h \cdot p$$

$$a_{35} = a \cdot l + b \cdot k + c \cdot j$$
$$a_{34} = d \cdot l + e \cdot k + f \cdot j + a \cdot o + b \cdot n + c \cdot m$$

$$a_{33} = g{\cdot}l + h{\cdot}k + i{\cdot}j + d{\cdot}o + e{\cdot}n + f{\cdot}m + a{\cdot}r + b{\cdot}q + c{\cdot}p$$
$$a_{32} = g{\cdot}o + h{\cdot}n + i{\cdot}m + d{\cdot}r + e{\cdot}q + f{\cdot}p$$
$$a_{31} = g{\cdot}r + h{\cdot}q + i{\cdot}p$$

$$a_{45} = b{\cdot}l + c{\cdot}k$$
$$a_{44} = e{\cdot}l + f{\cdot}k + b{\cdot}o + c{\cdot}n$$
$$a_{43} = h{\cdot}l + i{\cdot}k + e{\cdot}o + f{\cdot}n + b{\cdot}r + c{\cdot}q$$
$$a_{42} = h{\cdot}o + i{\cdot}n + e{\cdot}r + f{\cdot}q$$
$$a_{41} = h{\cdot}r + i{\cdot}q$$

$$a_{55} = c{\cdot}l$$
$$a_{54} = f{\cdot}l + c{\cdot}o$$
$$a_{53} = i{\cdot}l + f{\cdot}o + c{\cdot}r$$
$$a_{52} = i{\cdot}o + f{\cdot}r$$
$$a_{51} = i{\cdot}r$$

These equations provide a check on whether a given 5×5 window operator is decomposable or not. From these equations it follows as necessary condition for decomposition that the 5×5 coefficient matrix has to have a rank of at most 3.

The following special case of decomposition is algorithmically of great importance. A $n \times n$ window operator is called *separable* if it can be decomposed into a $n \times 1$ window operator and a $1 \times n$ window operator.

For example, the following two "kernel equations" represent two separable 3×3 window operators:

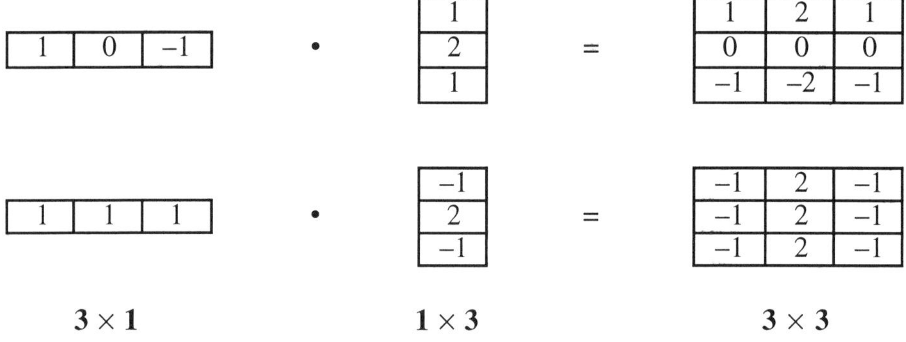

1 0 −1
•
1
2
1
=
1 2 1
0 0 0
−1 −2 −1
1 1 1
•
−1
2
−1
=
−1 2 −1
−1 2 −1
−1 2 −1
3 × 1
1 × 3
3 × 3

In general, the subsequent application of two operators

0	0	0
a	*b*	*c*
0	0	0

and

0	*d*	0
0	*e*	0
0	*f*	0

defines a (separable) 3×3 operator

ad	*ae*	*af*
bd	*be*	*bf*
cd	*ce*	*cf*

It follows that a necessary and sufficient condition for the possibility of separation is that the 3×3 coefficient matrix must have a rank equal to 1. For example, the *Laplace operator*

0	–1	0
–1	4	–1
0	–1	0

is not separable, because its coefficient matrix has rank 2. The window operator

0	1	–2
0	0	0
0	–1	2

is separable, since its coefficient matrix has a rank equal to 1. From $ae = 1$, $af = -2$, $ce = -1$ and $cf = 2$ it follows that $a = 1$, $c = -1$, $e = 1$ and $f = -2$.

Decomposition and separation can be used for speeding-up parallel linear local operators.

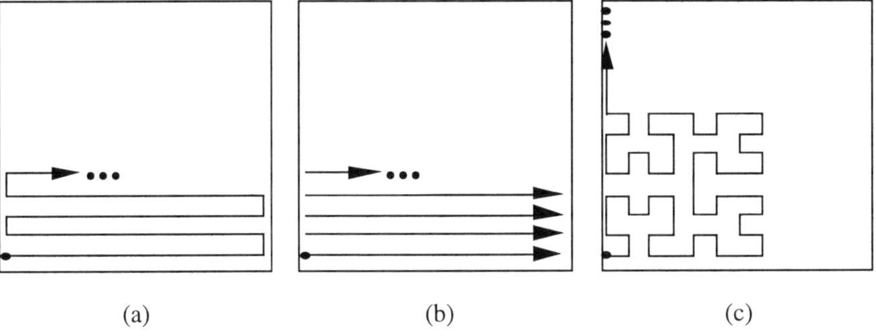

Figure 3.3: Meander-type image scan (a), image scan in raster scan order (b), and Hilbert-curve scanning order (c). The updating method can be based on different scanning orders.

3.1.4 Updating Method for Local Operators

Since the processing windows of subsequently scanned current pixels partially overlap, it is possible to use partial results of window functions computed at the previous window position for a subsequent window position. This technique is called update method. The order of window placements within the image grid, see Figure 3.3, determines the control structure of the updating method.

Let us assume a $n \times n$ window **F**, and assume the meander-type image scan. Let us consider a single step from position p of the current point to the next position q of the placed window $\mathbf{F}(q)$. The previously placed window $\mathbf{F}(p)$ has exactly $n(n-1)$ image points in common with $\mathbf{F}(q)$, n points are newly added, and n image points inside $\mathbf{F}(p)$ have not to be considered any more. Therefore, the computation of a window function on an picture window $\mathbf{F}(f, q)$ may be speeded-up by the use of (partial) results of this function on the previous picture window $\mathbf{F}(f, p)$.

Example 3.5: The window function *MEDIAN* was introduced in Section 1.3.3. A local gray value histogram, see Section 1.3.2,

$$HIST(\mathbf{F}(f, p), u), \quad 0 \leq u \leq G - 1,$$

can be used for the repeated updating of the value $MEDIAN(\mathbf{F}(f, p))$ on an picture window $\mathbf{F}(f, p)$ as follows:

The value $hist(u)$ specifies the absolute frequency of gray value u inside the current window $\mathbf{F}(f, p)$. It holds

$$HIST(\mathbf{F}(f, p), u) = \frac{hist(u)}{a},$$

where a is the number of pixels in the chosen window **F**. A shift of the window into a new position $\mathbf{F}(q)$ means that n gray values must be "deleted" from the function *hist*, and n gray values must be "inserted". Deletion means that the value $hist(f(x, y))$ is decremented by 1 for each image point (x, y) in $\mathbf{F}(p)$ which is not contained in $\mathbf{F}(q)$ any more. Insertion means that the value $hist(f(x, y))$ is incremented by 1 for each image point (x, y) in $\mathbf{F}(q)$ which was not previously contained in $\mathbf{F}(p)$. Let u_m be the median of the previous picture window $\mathbf{F}(f, p)$ gray values. The median of the current picture window $\mathbf{F}(f, q)$ is now determined based on u_m, the n "deleted" gray values, and the n "inserted" gray values.

Let L_p be the number of "deleted" gray values smaller than or equal to u_m, and R_p be the number of such gray values greater than u_m, i.e. $L_p + R_p = n$.

Analogously, let L_q and R_q be these numbers for the "inserted" gray values, i.e. $L_q + R_q = n$.

The median in $\mathbf{F}(f, q)$ is greater than u_m if $L_q < L_p$, and smaller than u_m if $L_q > L_p$. The updated histogram *hist* is used for calculating the shift of the new median (if its value is different u_m) with respect to u_m. The sign of $L_q - L_p$ defines the direction of the shift, and the absolute value $|L_q - L_p|$ allows a count of the number of one-gray-value steps, starting at u_m and leading to the new median.

The meander-type scan allows a simple implementation of the updating method, because the raster scan requires a new computation of the whole histogram at the beginning of each new line.

3.2 IMAGE DATA

The chosen image data representation can influence the efficiency of algorithmic solutions for image processing. For example, a representation can lead to fast row-wise processing but to slow column-wise processing.

3.2.1 Image Data Locations in Image Files

The computer system assigns a certain *image file* to each image. The data of the file are structured according to a chosen format. Text processing systems with options for image inclusion, DTP systems, a manifold of commercial image processing systems, computer graphics software etc. use different *image data formats* as TIFF (Tag Image File Format[2]), DDES (Digital Data Exchange Specifications, from the "Image Technology Committee" or ANSI) etc. Such formats will not be described here. Descriptions can be found in the related commercial software product literature. Instead we describe some standard ways of representing linearly (one-dimensionally) two- or multi-dimensional image data arrays. These basic formulas can be used for transforming arrays into a so-called *raw format*. In some image processing systems, one or several header records are used for general information about the image, after which the image data records start.

(A) A linear representation following the raster scan order gives a first possibility of storing scalar (gray value) images: The image value $f(x, y)$ of array position (x, y) is stored into the file location

$$(y - 1)\cdot M + x$$

[2] TIFF is a trademark of Aldus and Microsoft.

for an image f of size $M \times N$, with the column index x (from left to right), and the row index y (from bottom to top).[3] Usually an image row corresponds to one record of the file. A gray value image needs at least 256 Kbyte memory if $M = N = 512$ and $G = 256$

(B) An ordered set $f_1(x, y) = u_1, ..., f_n(x, y) = u_n$ of scalar images $f_1, ..., f_n$ is a possible representation of a vector-valued image $f(x, y) = (u_1, ..., u_n)$. In general, the data arrays of the images $f_1, ..., f_n$ form a three-dimensional data array of dimension $M \times N \times n$, and an image value $f_i(x, y)$ is stored in position (x, y, i). Different formats are possible for a linear representation of such a three-dimensional image array.

In the *row-alternation format* a record of a vector valued image f consists of M image values of a single image row of a scalar image f_i. A sequence of n records represents a single image row of the vector-valued image. The image value $f_i(x, y)$ is in file location

$$(y - 1)\cdot n\, M \;+\; (i - 1)\cdot M \;+\; x\;.$$

In the *image-alternation format* the channels $f_1, ..., f_n$ are stored in sequence. The image value $f_i(x, y)$ is in file location

$$(i - 1)\cdot M\, N \;+\; (y - 1)\cdot M \;+\; x\;.$$

In the *point-alternation format* a record consists of $n\cdot M$ image values of a single image row of the vector-valued image f The file contains subsequences of n values $f_1(x, y), f_2(x, y),..., f_n(x, y)$. The image value $f_i(x, y)$ is in file location

$$(y - 1)\cdot n\, M \;+\; (x - 1)\cdot n \;+\; i\;.$$

The point-alternation format is suggested for point operators on vector-valued images.

3.2.2 Overwriting or Saving the Original Image

Usually the input image should still be available after image processing. In principle, a distinction is possible between approaches using just a single image file, or approaches using several image files for image processing.

---

[3] This is the orientation of the xy-coordinate system as used throughout the book. It differs, e.g., from the orientation of matrix indices, i.e. x for rows, from top to bottom, and y for columns, from left to right.

(A) The input image and the resultant image share the same allocated memory in the one-image-file approach. At the beginning, the original image data are copied into this image file, and during processing these data are overwritten by the resulting values. For example, in sequential processing with a homogenous local operator computing process the resulting gray values are stored at the reference points of the moving window.

However, if a small loss of image data can be tolerated, also parallel local operators can be carried out using the one-image-file approach. The result of the local operator at a current point $p = (r, s)$ can be stored in the image at a shifted position, outside the picture window $\mathbf{F}(p)$. For example, if $\mathbf{F}$ is an $n \times n$ window and

$$k = \mathbf{integer}((n - 1)/2),$$

then the result can be stored at location $q_1 = (r, s - k - 1)$, cp. Figure 3.4, assuming the window moves in raster scan order. The value at image point $q_2 = (r - k, s - k)$, which is inside $\mathbf{F}(p)$, could also be overwritten because it is not contained in any further picture window after the next shift from $\mathbf{F}(r,s)$ to $\mathbf{F}(r + 1,s)$ in raster scan order. In spite of this displaced storage of resulting gray values, all window elements are original gray values, as it is required for parallel operators. However, this method shifts the resultant image with respect to the original image, $\Delta x = 0$ and $\Delta y = -k - 1$ in case of q_1, and $\Delta x = \Delta y = -k$ in case of q_2. This shift has to be compensated, e.g. by the operator of Section 4.1.2, if a point-to-point operation between resultant and original image has to be subsequently performed, cp. Section 5.4.

(B) Normally several image files are used during image processing. The original image file is subject to a read-only operation. The resultant image is assigned to a special file which will be (often) called *OUT* in the sequel in this book. This multiple-image-file approach needs no shift operation (as discussed under **(A)**), and thus allows faster and correct point-to-point image comparisons. The program control structures in this book are based on this multiple-image-file approach, since it is more clear in presentation and without disadvantages in comparison to the one-image-file approach. This chosen approach allows homogenous local operator implementations with direct access to the original image data.

3.2.3 Reading, Buffering and Writing back into Memory

Normally, recent computer technology allows processing of image data in random access. But still there might be several reasons that only subimages can be processed in the working memory. This can be due to limitations of the working memory (e.g.

after starting several programs), or to the size of the entire images (e.g. images in remote sensing can be of very large size). Therefore the authors decided that such limitations should be considered, and a certain row-wise processing (*row buffering*) of images was chosen as a default solution to cope with such limitations. In general the size and the shape of subimages should suit to the algorithm. Assuming the image file structure as discussed in Section 3.2.1, it is straightforward that records are considered as basic data units for data transfer between the random access memory and the image file. Local operators and their windows are such that all the necessary data of an picture window $\mathbf{F}(f, p)$ are in the random access memory if just a few image rows are transfered to the working memory.

An increasing number of applications allows image processing in random access, without a need for row-wise processing as discussed below under point **(A)**. This direct access to any image point is especially desirable if global operators have to be computed, e.g. for the calculation of Voronoi diagrams, or for global Fourier filtering. In such cases, the image data are, or should be completely available in the random access memory as a two-dimensional array, i.e. a special control structure for dealing with subimages (transfer into working memory etc.) is superfluous in such situations.

We emphasize that a special data structure (i.e. a temporary storing) can be useful for algorithmical purposes inside picture windows. Point **(B)** describes a *spiral linearization* of picture window values which is used several times throughout the book.

(A) Section 3.3 illustrates some control structures for point and local operators. The transfer of image rows between working memory and image file is also considered in these control structures. Assume a record size of 256 times 16 bit words. Then an image row claims exactly one record for the usual gray value image with 512×512 image points, and with 8 bit / pixel value (256 gray levels). Usually data are transferred record by record between the random access memory and further memory locations. Thus, the row-wise transfer between random access memory and image file offers a fast and simple solution to organize the image transformation process.

The windows of local operators cover in general several rows. A move from one row to next one requires a fast and correct supply with the necessary data. Furthermore, at the end of each row all the computed gray values must be transferred into the resultant image. Let us assume a window height of n rows, with odd n, and with the reference point in the center position of the window.

At program start the first n rows $y = 1, ..., n$ are transferred into n buffers $BUF(1...N, j)$ of the random access memory. Such a buffer is also called a *image row store*, with $j = 1, ..., n$.

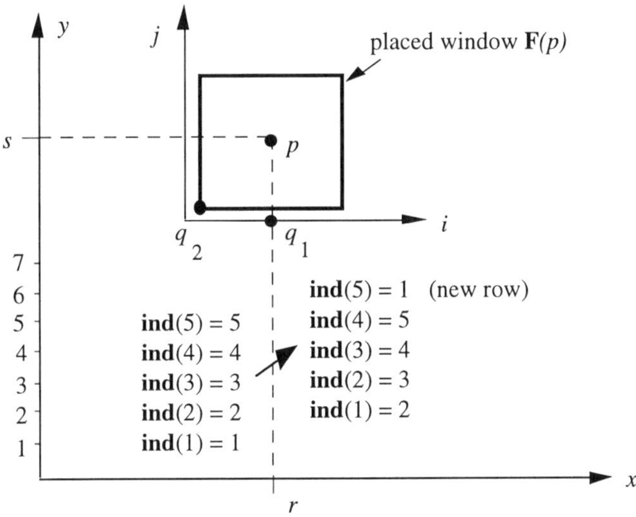

Figure 3.4: A placed $n \times n$ window $\mathbf{F}(p)$ with reference point $p = (r, s)$ within an image array. Below: Explanation of the index permutation at the end of row 3, for $n = 5$.

Figure 3.4 shows an example for $n = 5$. The y- coordinate of the reference point increases from 3 to 4 at the end of the first processing row. At that moment the row $y = 2$ becomes the lowermost row, and the row $y = 6$ becomes the uppermost row of $\mathbf{F}(f, p)$. This new uppermost row has to be transferred from the image file into an image row store. An indirect addressing scheme with indices **ind**(1), ..., **ind**(n) is used for reducing transfer between image row stores. As far as possible a *cyclic permutation of row indices* is used instead of transferring row data between image row stores.

The row indices are initialized by **ind**(j) = j, j = 1, ..., n, for the first processing row (this is row 3 in Figure 3.4). Then, at each row change in the image the indices **ind**(1) to **ind**(n – 1) are assigned the previous values of the indices **ind**(2) to **ind**(n) so that each image row store assumes the position of that image row store which was associated with the row one pixel below. The new row (y-coordinate $n + 1$ after the first row change) is read into that image row store with the previous index **ind**(1). The previous row data are overwritten at that moment After the cyclic permutation, the previous value of **ind**(1) is assigned to the index **ind**(n) which has not yet been assigned a new permutation value. In this way, the image row store with index **ind**(1) denotes always the first (i.e. lowermost) row of $\mathbf{F}(f, p)$, with **ind**(2) the second etc., without transferring any image data besides those into the image row store **ind**(n).

The choice between parallel or sequential processing of a local operator has consequences for the row-wise transfer of image data into the random access memory. The resulting gray values of a processed row are written into a *row output buffer store* $BUFOUT(1...M)$ of the random access memory if the parallel processing mode is selected. At the end of each processed row this row buffer store is transferred into the corresponding row of the resultant image file *OUT*.

If the sequential processing mode is selected then the resulting gray values are directly written into that image row store $BUF(1...N, \mathbf{ind}(k + 1))$ corresponding to the row of the current reference point p in the picture window $\mathbf{F}(f, p)$. At the end of each processed row this image row store is transferred into the corresponding row of the resultant image file *OUT*.

34	33	32	31	30	29	28
35	15	14	13	12	11	27
36	16	4	3	2	10	26
37	17	5	a	1	9	25
38	18	6	7	8	24	48
39	19	20	21	22	23	47
40	41	42	43	44	45	46

$n = 3$

$n = 5$

$n = 7$

Figure 3.5: Spatial arrangement of indices $z = 1, ..., a$, with $a = n^2$, of gray values $F(z)$ inside a placed window $\mathbf{F}(p)$ as used in the case of the centered ij-coordinate system. The continuation of the spiral is possible for windows larger than 7×7. The highest index a is always at the central location (location of the reference point).

(B) Often the algorithmic realization of a window function can be supported by a special data structure representing the current picture window $\mathbf{F}(f, p)$. The temporary storing of all gray values $f(i, j)$ of an picture window $\mathbf{F}(f, p)$, with current point p, into a special one- or two-dimensional data structure simplifies addressing, and improves the clearness of the algorithms. In this book, a two-dimensional array is also used as an alternative to the one-dimensional array.

The gray values inside $\mathbf{F}(f, p)$ are stored into a one-dimensional array $F(z)$, with $z = 1, ..., a$, $a = n^2$ and odd n, as shown in Figure 3.5, if the centered ij-coordinate system is used for addressing window positions.

The gray value $f(p)$ of the current point p is stored in the last (i.e., the ath) position of the linear array independently of the parameter value n. A preloaded look-up table

$$\mathbf{xind}(z) = i,\ \mathbf{yind}(z) = j\ ,\ \text{with } z = 1, ..., a,$$

specifies the correspondence between the centered ij-coordinate system and indices z. The control structure in Section 3.3.1 contains this table as a special data array.

The gray values inside the picture window $\mathbf{F}(f, p)$ are stored into a two-dimensional array $\mathbf{F}(i, j)$ with $i = 1, ..., n$ (columns, from left to the right) and $j = 1, ..., n$ (rows, from bottom to the top), if the non-centered ij-coordinate system is used for addressing window positions, i.e. the array coordinates are identical with the relative column and row coordinates of the window $\mathbf{F}$.

3.3 CONTROL STRUCTURES

General *control structures* as "divide and conquer", backtracking, dynamic programming etc. support the implementation of specific algorithms as known from the field of algorithm design. These control structures consist of sequences or loops connecting abstract program modules. Some classes of image processing operators allow the use of general control structures, and this is true for the classes of point and window operators in a very general sense. Also some global operators (Chapter 7) could be implemented as special cases which make use of general control structure.

3.3.1 Local Operators (centered)

Chapter 6 often presents local operators in such a way that only window functions or operator kernels are given. These operator kernels have to be embedded into a general control structure (*pixel program loop*), organizing the move of placed windows $\mathbf{F}(p)$ into all (non-border) point positions p of a given image according to a selected scan order, ensuring that the entire image domain is processed. Allowing small modifications, such a pixel program loop can be used for different local operators. For example, logically structured operators are characterized by data-dependent and/or location-dependent decisions. These decisions take place within the pixel program loop, and do not influence the general control structure of the program.

The use of predefined control structures allows often to describe the operators of Chapter 6 by giving only the operator kernel, which plays the role of a subroutine called at each pixel program loop. The program control structure is explicitly specified in Chapter 6 only in cases differing from the predefined "standard" control structure for the centered *ij*-coordinate system. For example, this is the case if an operator program uses several runs or iterations (i.e. through the image domain).

In general, the control structure of window operators allows the selection between different square windows of $n \times n$ pixels (odd n), and between parallel or sequential processing, cp. Chapter 1 or Glossary. Special inputs made by the user at the beginning of the control structure, i.e. at program start, specify these selections.

With the input of a special parameter (*PARSEQ* = 0 or = 1) the user chooses parallel or sequential processing. If only parallel processing has to be implemented, then some program lines of the control structure can be eliminated. These rows have the label ⊕ in the following control structures.

The structure requires the following data arrays:

- **ind**(1...*n*): This array contains the indices of indirect row addressing. For example, the initial values are given by **DATA ind**(1, 2, 3, 4, 5), if $n = 5$.
- **xind**(1...*a*), **yind**(1...*a*): These are arrays for the *ij*-coordinates of pixels inside the current placed window **F**(*p*). The gray values of these pixels are stored in the array *F*(1...*a*), and these values are used as arguments of the operator kernel (cp. Figure 3.5). For example, if $n = 5$ then the following arrays are used:

DATA xind(1, 1, 0, –1, –1, –1, 0, 1, 2, 2, 2, 1, 0, –1, –2, –2, –2, –2, –2, –1, 0, 1, 2, 2, 0);

DATA yind(0, 1, 1, 1, 0, –1, –1, –1, 0, 1, 2, 2, 2, 2, 2, 1, 0, –1, –2, –2, –2, –2, –2, –1, 0)

The control structure uses the following parameters and image files:

- The window size n is assumed to be odd. The value $k := (n - 1)/2$ follows after initializing n.
- ⊕ The parameter *PARSEQ* is specified for selecting parallel (*PARSEQ* = 0) or sequential (*PARSEQ* = 1) processing. For label ⊕, see above.
- The original image f (or several input images in certain cases) and the resultant image h (in file *OUT*) have to be allocated as image files.

Processing follows the scheme as given in Figure 3.6. Afterwards all open image files must be closed. Many of the operator kernels of Chapter 6 can be inserted at the location "processing of picture window ..." of this control structure in Figure 3.6. If so then this control structure is mentioned in the operator pseudo-program at the beginning of point (4). In this way repeated listings of identical control structures are avoided.

```
        input of operator parameters and initializations

        for z := 1 to n do
              read row z of f into the image row store BUF(1...M, ind(z));
        for y := k + 1 to N – k do begin
              for x := k + 1 to M – k do begin
                    for z := 1 to a do
                          F(z) := BUF(x + xind(z), ind(k + 1 + yind(z)));

        processing of the picture window F(f, (x, y)) with a specific operator kernel
        arguments:    F(z), with 1 ≤ z ≤ a , or F(i, j), with –k ≤ i, j ≤ k)
        result:       gray value v

⊕                   if (PARSEQ = 0) then
                          BUFOUT(x) := v
⊕                   else BUF(x, ind(k + 1)) := v
                    end {for};
⊕             if (PARSEQ = 0) then
                    write row buffer store BUFOUT(1...M) into row y of the
                          resultant image file OUT
⊕             else
⊕                   write image row store BUF(1...M, ind(k+ 1)) into row y of
⊕                         the resultant image file OUT;
              if (y < N – k) then begin
                    read row y + k + 1 of the input image file into the image row
                          store BUF(1...M, ind(1));
                    LINK := ind(1);
                    for z := 1 to n – 1 do ind(z) := ind(z + 1) ;
                    ind(n) := LINK
                    end {if}
              end {for}
```

Figure 3.6: Control structure for the computation of local operators, using row-wise buffering, the centered *ij*-coordinate system, and the spatial arrangement of Figure 3.5 for the window content if the one-dimensional array $F(z)$ is used. Program lines labeled with ⊕ are superfluous if a parallel operator has to be implemented.

input of operator parameters and initializations

```
for y := k + 1 to N − k do
        for x := k + 1 to M − k do begin
                for z := 1 to a do
                        F(z) := f(x + xind(z), y + yind(z));
```

processing of the picture window $\mathbf{F}(f, (x, y))$ with a specific operator kernel	
arguments:	$F(z)$, with $1 \le z \le a$ (or $F(i, j)$, with $-k \le i, j \le k$)
result:	gray value v

```
⊕               if (PARSEQ = 0) then
                        h(x, y) := v
⊕               else f(x, y) := v
                end {for}
```

Figure 3.7: Control structure for the computation of local operators, using the centered *ij*-coordinate system, and the spatial arrangement of Figure 3.5 for array $F(z)$, and without using row-wise buffering. Program lines labeled with ⊕ are superfluous if a parallel operator has to be implemented.

Operations which are necessary for opening or allocating an image file, acquiring an image via a CCD-camera etc., are not contained in this control structure.

The control structure includes row-wise buffering as discussed in Section 3.2.2. The control structure results simplified without such a buffering, i.e. if random access to image data is possible, cp. Figure 3.7. The data array **ind** is not necessary any more.

3.2.2 Local Operators (non-centered)

A decision for exactly one local operator control structure is possible and also suggested, cp. Chapter 1 (discussion of *ij*-coordinate systems). The centered *ij*-coordinate system is favored in this book. However, for reasons of completeness we also present a general control structure for the non-centered *ij*-coordinate system assuming a temporary storing of window values within a two-dimensional array $F(1...m, 1...n)$. Here, the control structure utilizes only the data array

```
┌──────────────────────────────────────────────────────────────────────┐
│ input of operator parameters and initializations                     │
└──────────────────────────────────────────────────────────────────────┘

for z := 1 to n do
      read row z of f into the image row store BUF(1...M, ind(z));
for y := k + 1 to N – k do begin
      for x := k + 1 to M – k do begin
            for j := 1 to n do
                  for i := 1 to n do
                        F(i, j) := BUF(x – k – 1 + i, ind(j));

┌──────────────────────────────────────────────────────────────────────┐
│ processing of the picture window F(f, (x, y)) with a specific operator kernel │
│ arguments:   F(i, j), with 1 ≤ i ≤ m, 1 ≤ j ≤ n                      │
│ result:      gray value v                                            │
└──────────────────────────────────────────────────────────────────────┘

⊕                 if (PARSEQ = 0) then
                        BUFOUT(x) := v
⊕                 else BUF(x, ind(k + 1)) := v
                  end {for};
⊕           if (PARSEQ = 0) then
                  write row buffer store BUFOUT(1...M) into row y of the
                        resultant image file OUT
⊕           else
⊕                 write image row store BUF(1...M, ind(k+ 1)) into row y of
⊕                       the resultant image file OUT;
            if (y < N – k) then begin
                  read row y + k + 1 of the input image file into the image row
                        store BUF(1...M, ind(1));
                  LINK := ind(1);
                  for z := 1 to n – 1 do ind(z) := ind(z + 1) ;
                  ind(n) := LINK
                  end {if}
            end {for}
```

Figure 3.8: Control structure for the computation of local operators, using row-wise buffering and the non-centered *ij*-coordinate system. Program lines labeled with ⊕ are superfluous if a parallel operator has to be implemented.

input of operator parameters and initializations

for $y := 1$ **to** N **do begin**

read row y of input image f into the row buffer store $BUF(1...M)$;

for $x := 1$ **to** M **do begin**

processing of the gray value $u := BUF(x)$, e.g., with a gradation function result: gray value v

$BUF(x) := v$

end *{for}*;

write row buffer store $BUF(1...M)$ into row y of the resultant image OUT

end *{for}*

Figure 3.9: Special control structure for point operators, using row-wise buffering.

• **ind**$(1...n)$ for indirect row addressing (as in Section 3.3.1) for index computations.

The following parameters and image files are used:

• the window size n (odd) and $k := (n-1)/2$,
• ⊕ the parameter *PARSEQ* for selecting parallel (*PARSEQ* = 1) or sequential (*PARSEQ* = 0) processing, and
• image files of the original image f (or of several input images), and of the resultant image h (file *OUT*).

The control structure is given in Figure 3.8. Processing follows this structure. Afterwards, all open image files have to be closed.

As for the centered *ij*-coordinate system (cp. Figure 3.7) this control structure can be substantially simplified if row-wise buffering can be avoided.

3.3.3 Point Operators

With point operators the placed window $\mathbf{F}(p)$ is identical to the simple point p. A choice between parallel or serial processing is pointless here. For example, the control structure of the centered *ij*-coordinate system covers this situation, with $n = 1$ and $k = 0$. In this way, a unique control structure can be used for local operators and for the special case of point operators.

However, the implementation of a special control structure for point operators is suggested for reasons of time efficiency and clearness. At the beginning, the image files of the input image f (or of several input images) and of the resultant image h (file *OUT*) have to be opened. If row-buffering is selected, then a single row buffer store suffices. Figure 3.9 gives a control structure for point operators. The input of a new row takes place at the beginning of the processing of each new row. Finally, the open image files have to be closed.

3.4 PROCEDURES

This Section combines together several subroutines which can be called for running frequently recurring subprocesses in the implementation of some operators. The structure of each one of these program presentations follows the general scheme described in the "Instructions to the Reader".

3.4.1 Procedure *RND_EQU*

(1) Many programming languages comprise a standard function for generating random numbers (e.g. *RANDOM*). If this is not available, or if an own *random number generator* is desired, then the following procedure *RND_EQU* can be used. The output are uniformly distributed real random numbers between 0 and 1.

(2) Generally, random number generators are based on congruences

$$r_{n+1} := \mathbf{mod}(a \cdot r_n + b, p),$$

with constants a, b and p characterizing the pseudo-random process. Here, $mod(z, n)$ denotes the remainder of the division z/n. The process starts with a seed $r_o = c$. The output

$$z = \frac{r_{n+1}}{p}$$

assumes values in the range between 0 and 1. Suggestions for choosing a, b, c and p are as follows:

- p should be as large as possible, and powers of 10 or of 2 are suggested (e.g. $p = 100{,}000{,}000$),

- a should be one decimal digit smaller than p, it should not contain a special digit pattern (as, e.g., periodicities), and it should end with ...$k21$, with even k (e.g. $a = 31{,}415{,}821$),
- b should be relative prime to p, e.g. $b = 1$ (cp. procedure below),
- c can be chosen in any reasonable way, e.g. using the current computer clock time value for a "random" initialization. Of course, identical values of c will result into identical sequences of pseudo-random numbers.

The input to this process is restricted to a randomly selected seed c. During the process, the previous pseudo-random number is taken for generating the next pseudo-random number.

(3) Even if a, r and b are within the representable number range it may happen that $a \cdot r + b$ leads to an overflow. Therefore, let

$$q = \sqrt{p}.$$

Then, the sum

$$a \cdot r + 1 \text{ (modulo } p)$$

can be carried out as follows: Let $a = q \cdot a_{mult} + a_{rem}$ and $r = q \cdot r_{mult} + r_{rem}$. It holds

$$a \cdot r \;=\; p \cdot a_{mult} \cdot r_{mult} + q \cdot (a_{mult} \cdot r_{rem} + a_{rem} \cdot r_{mult}) + a_{rem} \cdot r_{rem},$$

and thus

$$\begin{aligned} \textbf{mod}(a \cdot r + 1, p) &= \textbf{mod}(q \cdot (a_{mult} \cdot r_{rem} + a_{rem} \cdot r_{mult}) + a_{rem} \cdot r_{rem} + 1, p) \\ &= \textbf{mod}(q \cdot \textbf{mod}(a_{mult} \cdot r_{rem} + a_{rem} \cdot r_{mult}, q) + a_{rem} \cdot r_{rem} + 1, p). \end{aligned}$$

(4)

```
function RND_EQU() : real;
var  rand: integer;                          {a global variable, initialized by c }
begin
    a_mult  := integer(a / q);               a_rem     := mod(a, q);
    r_mult  := integer(rand / q);            r_rem     := mod(rand, q);
    s       := q·mod(a_mult·r_rem + a_rem·r_mult, q) + a_rem·r_rem;
    rand    := mod(s + 1, p);                RND_EQU   := rand / p
end {RND_EQU}
```

(5) Schroeder, M.R.: *Number Theory in Science and Communication. Second enlarged ed.*, Springer, Berlin, 1986.

3.4.2 Procedure *RND_NORM*

(1) This program generates pairs (x, y) of independently, Gaussian or normally distributed real pseudo-random numbers x, y with expected value 0 and variance 1.

(2) The function of a *Gaussian distribution*,

$$\Phi(z) = \frac{1}{\sigma \cdot \sqrt{2\pi}} \cdot \mathbf{exp} \left(- \frac{(z - \mu)^2}{2\sigma^2} \right),$$

is specified by an expected value μ and by the variance σ^2. A real variable z with this distribution can be approximately obtained by summing many uniformly distributed and independent random numbers (Central Limit Theorem of probability theory), say - at least six. However, here a different approach is taken.

The procedure *RND_EQU* (cp. Section 3.4.1) is used for generating uniformly distributed random numbers r and s between -1 and 1. These numbers are used as input for generating the normally distributed pairs.

(3) The generation of pairs x, y of independent, normally distributed numbers x, y starts with two centrally and uniformly distributed numbers r, s:

$$r = 2 \cdot RND\_EQU(\,) - 1, \qquad s = 2 \cdot RND\_EQU(\,) - 1.$$

These values are recomputed if $t = r^2 + s^2 > 1$, or $t = 0$. Then, for $t \leq 1$ it follows that

$$x = \frac{\sqrt{-\frac{2}{t} \ln t}}{2} \cdot r \qquad \text{and} \qquad y = \frac{\sqrt{-\frac{2}{t} \ln t}}{2} \cdot s$$

are approximations of normally distributed numbers with expected value 0 and variance 1. Correlated values can be derived by linear combinations of x and y.

(4)
```
procedure RND_NORM(x, y: real);
begin
    repeat
        r := 2·RND_EQU( ) – 1;     s := 2·RND_EQU( ) – 1;
        t := r² + s²
    until ( t ≤ 1   and   t > 0 );
    temp := (√(– 2 ln(t))/t) / 2;
    x := temp·r;   y := temp·s
end {RND_NORM}
```

(5) Press, W.H., Flannery, B.P., Teukolsky, S.A., Vetterling, W.T.: *Numerical Recipes*. Cambridge University Press, Cambridge, USA, 1988.

3.4.3 Procedure *MAXMIN*

(1) The procedure computes minimum a and maximum b of n numbers $u_1, u_2, \ldots, u_n$ using only $[3n/2] - 2$ comparisons.

(2) $a = \mathbf{min}\{u_1, \ldots, u_n\}$ and $b = \mathbf{max}\{u_1, \ldots, u_n\}$.

(3) The straightforward solution, i.e. to compare successively each number with the previous minimum as well as with the previous maximum, results into $2n - 2$ comparisons. The program given here needs only the least possible number of comparisons, i.e. $[3n/2] - 2$. If this problem has repeatedly to be solved for different window positions, then this fast solution may be of interest.

The input is given in a list **L**, with $\mathbf{L}(i) = u_i$. The program computes the positions i_1 of a, and i_2 of b in **L**, i.e. the desired values are located in $\mathbf{L}(i_1) = a$ and $\mathbf{L}(i_2) = b$.

(4)
```
procedure MAXMIN (L: list_of_numbers; n, i1, i2 : integer);
begin
    i1 := 1;    i2 := 1;    a := L(1);    b := L(1);    j1 := 1;
    if ( mod(n, 2) ≠ 0 ) then j1 := 2;
    for i := j1 to n step 2 do
        begin
            k1 := i;    k2 := i + 1;    x1 := L(k1);    x2 := L(k2);
            if (x1 > x2 ) then
                begin
                    k1 := k2; k2 := i;
                    x1 := x2; x2 := L(k2)
                end {if};
            if (x1 < a) then
                begin
                    a := x1;    i1 := k1
                end {if};
            if (x2 > b) then
                begin
                    b := x2; i2 := k2
                end {if}
        end {for}
    end {MAXMIN}
```

(5) Aho, A.V., Hopcroft, J.E., Ullman, J.D.: *The Design and Analysis of Computer Algorithms*. Addison-Wesley, Reading, 1974.

3.4.4 Procedure *SELECT*

(1) This program computes the element with rank k, $1 \le k \le n$, in a given list $\mathbf{L} = [u_1, u_2, \ldots, u_n]$ of n numbers (in arbitrary order at the time of input). The value $k = 1$ means the computation of the minimum of $\mathbf{L}$, $k = n$ means the maximum, and $k = [n/2]$ means the median. The desired number $ord_k(\mathbf{L})$ is in position k of $\mathbf{L}$ after running this program. The program performs some permutations of the numbers in $\mathbf{L}$ but no changes in value.

For example, assume $\mathbf{L} = [5, 2, 3, 7, 2, 5, 4, 8, 5]$ with $n = 9$. Then it holds $ord_1(\mathbf{L}) = 2$, $ord_2(\mathbf{L}) = 2$, $ord_3(\mathbf{L}) = 3$, ..., $ord_9(\mathbf{L}) = 8$.

(2) Let be $\mathbf{L} = [u_1, u_2, \ldots, u_n]$ and $1 \le k \le n$. Then it holds $ord_k(\mathbf{L}) = u$, with $u \in \mathbf{L}$, if

$$\mathbf{card}\{i : u_i < u \wedge 1 \le i \le n\} < k \quad \text{and} \quad \mathbf{card}\{i : u_i > u \wedge 1 \le i \le n\} \le n - k.$$

The program computes $ord_k(\mathbf{L})$.

(3) A straightforward solution would be a complete sorting of the list $\mathbf{L}$; then $ord_k(\mathbf{L})$ is in position k of the sorted list. However, such an approach includes many unnecessary computing steps. The following method produces no completely sorted list $\mathbf{L}$. After processing, in positions with index smaller than k there are elements of $\mathbf{L}$ smaller than or equal to $ord_k(\mathbf{L})$, and in positions with index greater than k there are elements of $\mathbf{L}$ greater than or equal to $ord_k(\mathbf{L})$. The desired number is in position k of $\mathbf{L}$.

In the following program, the indices *left* and *right* define the borders of the "search space" in list $\mathbf{L}$. The desired element of $\mathbf{L}$ is always in the range between u_{left} and u_{right}. If the situation $right \le left$ is reached, then the search space contains exactly one element of $\mathbf{L}$, and this is the desired number. If $left < right$ is still true, then an element is chosen in the search space for comparisons. In the program, at the beginning the element u_{right} is chosen for this purpose.

(4)
```
procedure SELECT(L: list_of_numbers; n, k: integer);
begin
        if (n ≤ 1) then
                return                                    {L is sorted}
        else    begin
                left := 1;          right := n;
                while (right > left) do
                        begin
                                v := u_right;       i := left – 1;     j := right + 1;
                                repeat
```

```
                    repeat i := i + 1 until u_i ≥ v;
                    repeat j := j − 1 until u_j ≤ v:
                    temp := u_i;    u_i := u_j;    u_j := temp
               until j ≤ i;
               u_j := u_i;    u_i := u_right;    u_right := temp;
               if (i ≥ k) then right := i − 1;
               if (i ≤ k) then left := i + 1
          end {while}
     end {else}
end {SELECT}
```

(5) Originally the procedure *SELECT* has been proposed by

Hoare, C.A.R.: *Partition (Algorithm 63), Quicksort (Algorithm 64), and Find (Algorithm 65).* Comm. ACM **4** (1961), pp. 321 - 322.

The books

Knuth, D.E.: *The Art of Computer Programming. Vol.* 3: *Sorting and Searching.* Addison-Wesley, Reading, USA, 1975.

Press, W.H., Flannery, B.P., Teukolsky, S.A., Vetterling, W.T.: *Numerical Recipes.* Cambridge University Press, Cambridge, USA, 1988.

contain several variants of this procedure.

3.4.5 Procedure *QUICKSORT*

(1) The program sorts n numbers $u_1, u_2, ..., u_n$ into ascending order. For example, input **L** = [5, 2, 3, 7, 2, 5, 4, 8, 5] and $n = 9$ leads to the ascending sorted list **L** = [2, 2, 3, 4, 5, 5, 5, 7, 8]. The list **L** and the number n are given as input. The program is asymptotically fast.

(2) The task consists of computing a certain permutation π of n numbers $u_1, u_2, ..., u_n$ such that

$$u_{\pi(1)} \leq u_{\pi(2)} \leq ... \leq u_{\pi(n)}.$$

(3) With respect to worst-case complexity the procedure *QUICKSORT* has time complexity $\mathbf{O}(n^2)$. Assuming a uniform distribution of n input numbers (numbers within a certain range) then the expected time complexity is $\mathbf{O}(n \log n)$. Essentially the procedure *QUICKSORT* consists of a repeated application of the *SELECT* procedure. By choosing a comparison element v, the list is transformed in such a way that in a left part there are only elements smaller than or equal to v, and in the right part there are only elements greater than v. This process is repeated recursively for the left and for the right part until a sorted list is obtained. A stack **S** is used for controlling the recursion. In the program a stack depth of 50 is assumed, what should be sufficient for image processing tasks.

(4)
```
procedure QUICKSORT(L: list_of_numbers; n: integer);
var   S(1...50): list_of_integers;
begin
   if (n > 1) then begin
         left := 1;     right := n;     p := 2;
         repeat
            if (right > left) then begin
                  v := u_right;     i := left – 1;    j := right + 1;
                  repeat
                        repeat i := i + 1 until u_i ≥ v;
                        repeat j := j – 1 until u_j ≤ v;
                        temp := u_i;     u_i := u_j;        u_j := temp
                  until j ≤ i;
                  u_j := u_i;       u_i := u_right;     u_right := temp;
                  if ((i – left ) > (right – i)) then begin
                        S(p) := left;    S(p+1) := i – 1;     left := i + 1;
                        end {then}
                  else begin
                        S(p) := i + 1;  S(p+1) := right;   right := i – 1;
                        end {else};
                  p := p + 2
                  end {then}
            else begin
                  p := p – 2;
                  if (p > 0) then begin
                        left := S(p);    right := S(p+1)
                        end {if}
                  end {else};
         until p = 0
   end {if}
end {QUICKSORT}
```

(5) The procedure *QUICKSORT* has been proposed by

Hoare, C.A.R.: *Partition (Algorithm 63), Quicksort (Algorithm 64), and Find (Algorithm 65)*. Comm. ACM **4** (1961), pp. 321 - 322.

Several variants, e.g. with respect to the selection of the comparison element, are dealt with in

Knuth, D.E.: *The Art of Computer Programming. Vol. 3: Sorting and Searching*. Addison-Wesley, Reading, USA, 1975.

Press, W.H., Flannery, B.P., Teukolsky, S.A., Vetterling, W.T.: *Numerical Recipes*. Cambridge University Press, Cambridge, USA, 1988.

3.4.6 Procedure *BUBBLESORT*

(1) The program sorts n numbers $u_1, u_2, ..., u_n$ into ascending order. The method is simple, and n is assumed to be small. The list **L** of n numbers and the number n are given as input.

(2) The program computes a permutation π of n numbers $u_1, u_2, ..., u_n$ such that

$$u_{\pi(1)} \leq u_{\pi(2)} \leq ... \leq u_{\pi(n)}.$$

(3) The program *BUBBLESORT* has a worst-case time complexity of $\mathbf{O}(n^2)$. It can be suggested for small values of n, say $n \leq 16$.

(4)
```
procedure BUBBLESORT(L: list_of_numbers; n: integer);
begin
        for j := n – 1 to 1 step – 1 do
                for i := 1 to j do
                        if (u[i+1] < u[i]) then swap u[i] and u[i+1]
end {BUBBLESORT}
```

(5) Aho, A.V., Hopcroft, J.E., Ullman, J.D.: *The Design and Analysis of Computer Algorithms*. Addison-Wesley, Reading, 1974.

3.4.7 Procedure *BUCKETSORT*

(1) This procedure is especially advantageous for sorting of numbers $w_1, w_2, ..., w_a$ into ascending order, with $a = n \cdot n$, e.g. gray values inside a placed window $\mathbf{F}(p)$). It is assumed that the size of the possible value range of numbers w_i is not very large in comparison to a. Let $\{0, 1, ..., G-1\}$ be the value range, with $G \geq 2$. As input it is assumed that the a numbers $w_1, w_2, ..., w_a$ are given in an array $f(1...n, 1...n)$. The number n, which is also known, specifies the array size.

(2) The task consists in computing a certain permutation π of a numbers $w_1, w_2, ..., w_a$ given in a list **W**, such that

$$0 \leq w_{\pi(1)} \leq w_{\pi(2)} \leq ... \leq w_{\pi(n)} \leq G - 1.$$

(3) This sorting method is advantageous if the value range of the input numbers is not very large. This is true for the number G of gray levels. A single program run needs $2a + G$ operations, i.e. the computing time complexity is linear in a. For example, assume $G = 256$ and $a = 9$, 25 or 49. The procedure *BUCKETSORT* performs 274, 306 or 354 operations for these inputs, respectively. The n^2-method *BUBBLESORT* performs 81, 625 or 2401 operations, respectively, for the same inputs. The method has advantages over *BUBBLESORT* if the window size is 5×5 or larger.

The program uses an array *hist*(1...G) for the (gray value) histogram during the sorting process. This array has to be initialized with 0 in all positions just once, at the beginning of the main program loop. The array **W**(1...a) contains the numbers which have to be sorted in place, i.e. the aim is to obtain **W**(1) ≤ **W**(2) ≤ ... ≤ **W**(a) at the end of the program.

For example, assume that an operator window **F**(f, p) is scanned. The gray values $u = f(i, j)$, $(i, j) \in$ **F**(p), are used as addresses for the histogram array to increment the value in position *hist*(u). Then, the array *hist* is scanned from *hist*(1) to *hist*(G), and all positions *hist*(n) with value greater than 0 are searched. For each found position the address u is deposited in **W**(r), starting at **W**(1) up to **W**(a). At the same moment, the value *hist*(u) is decremented, and the process repeats until *hist*(u) is equal to 0. Then, *hist*(u + 1) is addressed etc. In this way, in the array **W** are stored the a original values in ascending order, and the array *hist*(u) is reset to 0 in all positions, i.e. it is initialized for the next run of sorting, e.g. for the next image point.

(4)
```
procedure BUCKETSORT(f(1...n, 1...n): array_of_integers;
                            W(1...a): list_of_integers; n: integer);
var hist(1...G): list_of_integers;
begin
                                {part 1: computation of histogram hist }
      for j := 1 to n do                                   { a = n·n }
            for i := 1 to n do
                  hist(1 + f(i, j)) := hist(1 + f(i, j)) + 1 ;
                                  { hist(u) corresponds to value u – 1}
```

```
                        {part 2: ordered storage into the buckets W(z)}
      z := 1;
      for u := 1 to G do
           while (hist(u) > 0 ) do begin
                W(z) := u – 1;    {value u – 1 corresponds to hist(u)}
                z := z + 1;       hist(u) := hist(u) – 1
           end {while}
   end {BUCKETSORT}
```

(5) Aho, A.V., Hopcroft, J.E., Ullman, J.D.: *The Design and Analysis of Computer Algorithms*. Addison-Wesley, Reading, 1974.

3.4.8 Procedure *FFT*

(1) This program transforms a vector **f** of n complex or real numbers $f(0), f(1),..., f(n-1)$ into its Fourier transformed vector **F**, consisting of n complex numbers $F(0), F(1),..., F(n-1)$. The vector size n is assumed to be a power of 2. The program input are the n numbers $f(0), f(1),..., f(n-1)$ as well as the size parameter n.

(2) The *Fourier transformation* is defined by the equations

$$F(u) = \frac{1}{n} \cdot \sum_{x=0}^{n-1} f(x) \,\mathbf{exp}\left(\frac{-i2\pi xu}{n}\right), \qquad u = 0, 1, ..., n-1$$

for transforming n complex numbers $f(0), f(1),..., f(n-1)$, with

$$i = \sqrt{-1}.$$

The *inverse Fourier transformation* of n complex numbers $F(0), F(1),..., F(n-1)$ is defined by the equations

$$f(x) = \sum_{u=0}^{n-1} F(u) \,\mathbf{exp}\left(\frac{i2\pi xu}{n}\right), \qquad x = 0, 1, ..., n-1.$$

(3) The Fourier transformation FT and the inverse Fourier transformation FT^{-1} transform vectors of reals into vector of complex numbers. Therefore it is necessary to perform the computations in the field of complex numbers $z = z_1 + i \cdot z_2$, with

$$i = \sqrt{-1}.$$

For arbitrary input vectors $\mathbf{f} = (f(0), f(1), ..., f(n-1))$ and corresponding resultant vectors $\mathbf{F} = (F(0), F(1), ..., F(n-1))$, with $\mathbf{f} = FT^{-1}(\mathbf{F})$ and $\mathbf{F} = FT(\mathbf{f})$, it holds

$$\mathbf{f} = FT^{-1}(FT(\mathbf{f})) \quad \text{and} \quad \mathbf{F} = FT(FT^{-1}(\mathbf{F})).$$

This (theoretically) ideal inversion allows testing of the implemented Fourier transformation procedure up to a certain limit. However, some numerical inaccuracies occur in calculating unity root powers

$$W^k = \mathbf{exp}\left(\frac{-i2\pi k}{n}\right) \quad \text{or} \quad W^{-k} = \mathbf{exp}\left(\frac{i2\pi k}{n}\right).$$

If the vector length is no power of two, then vectors can be filled up with zero elements until reaching the length of the next power of two.

The program given here transforms "in place" n complex numbers $f(x) = f_1(x) + i \cdot f_2(x)$, with $x = 0, 1, ..., n-1$, into the Fourier transform $\mathbf{F}$ with time complexity $\mathbf{O}(n \log n)$. In this program, first the input values are permutated (equivalent to a bit inversion of the indices) so that, in the second step, the transformation itself can be carried out in a simplified way. In the program it is renounced to divide by n in comparison to the above transformation equations.

```
(4) procedure FFT(f: vector_of_complex_numbers; n: power_of_2);
    var temp, unit, root: complex_number;
    begin
     j := 0;
     for i := 0 to n − 2 do   begin            {permutation of input values}
        if (i < j) then begin
            temp := f(j);   f(j) := f(i);   f(i) := temp
        end {if};
        k := n/2;
        while (k ≤ j) do begin
            j := j − k;          k := k/2
        end {while};
        j := j + k
     end {for}
     m := log₂ n;
     for k := 1 to m do begin                        {transformation}
        p := 2^k;
        root := 1;  unit := cos (2π/p) − √−1·sin(2π/p);    {complex numbers}
```

```
        for j := 1  to p/2  do begin
            for i := j – 1  to n – 1  step p  do begin
                ip := i + p/2;                temp := f(ip)·root;
                f(ip) := f(i) – temp;         f(i) := f(i) + temp
            end {for};
            root := unit·root;
        end {for};
    end {for};
end {FFT}
```

Attention: This procedure uses vector position indices 0 to $n-1$, while indices in image rows, or in image columns go from 1 to n. If indices 1 to n are used then the permutation starts with $j := 1$, then index i runs between 1 and $n-1$, and "$k < j$?" must be used as test in the **while** loop. In the transformation part index i has to run between j and n.

(5) The design of the fast Fourier transformation has an interesting history, see

Cooley, J.M., Lewis, P.A., Welch, P.D.: *History of the fast Fourier transform.* Proc. IEEE **55** (1967), pp. 1675 - 1677.

The algorithm became popular by the paper

Cooley, J.M., Tukey, J.W.: *An algorithm for the machine calculation of complex Fourier series.* Math. Comp. **19** (1965), pp. 297 - 301.

3.4.9 Procedure *FWT*

(1) This program transforms a vector **f** of n reals $f(0), f(1),..., f(n-1)$ into the Walsh transformed vector **W** consisting of n reals $W(0), W(1),..., W(n-1)$. The vector length n is assumed to be a power of two, $n = 2^m$. As input the n reals $f(0), f(1),..., f(n-1)$ are given, as well as the length n.

(2) The *Walsh transformation* is defined by the equations

$$W(u) = \frac{1}{n} \cdot \sum_{x=0}^{n-1} f(x) \prod_{i=0}^{m-1} (-1)^{B(m,\, i,\, x,\, u)}, \qquad u = 0, 1, ..., n-1,$$

with $n = 2^m$ reals $f(0), f(1),..., f(n-1)$ as input, and with the exponent given by the function

$$B(m, i, x, u) = b_i(x) b_{m-1-i}(u),$$

where $b_k(v)$ denotes the k-th bit in the binary representation of a non-negative integer v. For example, we consider $n = 8$, i.e. $m = 3$, and $v = 3$, i.e. binary 011. Thus, in this example we have

$$b_0(v) = 1,\ b_1(v) = 1 \text{ and } b_2(v) = 0.$$

In the given transformation equations the values of vector **f** are multiplied only by +1 or −1.

The *inverse Walsh transformation* is defined by the equations

$$f(x) = \sum_{u=0}^{n-1} W(u) \prod_{i=0}^{m-1} (-1)^{B(m,\,i,\,x,\,u)}, \qquad u = 0, 1, ..., n-1$$

where n reals $W(0), W(1),..., W(n-1)$ constitute the input.

(3) The Walsh transformation WT and the inverse Walsh transformation WT^{-1} map vectors of reals into vectors of reals. The Walsh transformation requires only half of the memory in comparison to the Fourier transformation, because the latter needs arrays for the real and the imaginary part. The Walsh transformation can also be computed with a fast algorithm (*FWT* - fast Walsh transformation) of time complexity $\mathbf{O}(n \log n)$, and practically it needs only about half of the time as a run of the Fourier transformation, or even less, since integer arithmetic is sufficient for the Walsh transformation of integer vectors. In practical experiments it is also interesting to note that transformation results are comparable with results of the Fourier transformation for small values of n, say $n \leq 64$.

The coordinates in Fourier space are called frequencies, and the coordinates in Walsh space are called *sequencies*.

For an arbitrary input vector $\mathbf{f} = (f(0), f(1), ..., f(n-1))$ and a corresponding resultant vector $\mathbf{W} = (W(0), W(1), ..., W(n-1))$, with $\mathbf{f} = WT^{-1}(\mathbf{W})$ and $\mathbf{W} = WT(\mathbf{f})$, it holds that

$$\mathbf{f} = WT^{-1}(WT(\mathbf{f})) \qquad \text{and} \qquad \mathbf{W} = WT(WT^{-1}(\mathbf{W})).$$

Rounding errors are possible in the division by n. For obtaining the original gray value scale after one direct and one inverse transformatiuon altogether, a division by n is necessary. (It is also possible to divide by $\sqrt{n}$ in each transformation direction.)

The given procedure transforms in place n reals or integers $f(x)$, with $x = 0, 1, ..., n-1$, into the Walsh transformed vector **W** with time complexity $\mathbf{O}(n \log n)$. As in the fast Fourier algorithm, the program starts with a permutation of input values according to the bit-inversion of indices. The actual gray value transformation follows then in a second program part.

(4)
```
procedure FWT(f: vector_of_reals; n: power_of_2);
 begin
  j := 0;
  for i := 0 to n – 2 do   begin          {permutation of inputs, cp. FFT}
     if (i < j) then begin
         temp := f(j);  f(j) := f(i);    f(i) := temp
     end {if};
     k := n/2;
     while (k ≤ j) do begin
         j := j – k;             k := k/2
     end {while};
     j := j + k
  end {for}
  m := log2 n;
  for k := 1 to m do begin                                {transformation}
         p := 2^k;
         for j := 1 to p/2 do begin
            for i := j – 1 to n – 1 step p do begin
               ip := i + p/2;           temp := f(ip);
               f(ip) := f(i) – temp;   f(i) := f(i) + temp
            end {for};
         end {for};
     end {for};
  end {FWT}
```

Attention: The indices of the input and output vector run from 0 to $n - 1$, as in the fast Fourier algorithm. For indexing between 1 and n, the same modifications as mentioned there apply.

(5) There are different variants of the Walsh transformation definition in literature. The presentation given here follows

Gonzalez, R.C., Wintz, P.: *Digital Image Processing*. Addison-Wesley, Reading, 1977.

3.4.10 Procedure *BRESENHAM*

(1) In many image processing programs, a *digital straight line segment* connecting two points p_1 and p_2 of the image domain must be traced on a screen. The points p_1 and p_2 have integer coordinates.

(2) Two points $p_1 = (x_1, y_1)$ and $p_2 = (x_2, y_2)$ define (in the Euclidean plane) a straight line

$$y = \frac{y_2 - y_1}{x_2 - x_1} \cdot x + \frac{y_1 \cdot x_2 - y_2 \cdot x_1}{x_2 - x_1}$$

$$= s \cdot x + c.$$

We consider the straight line segment p_1p_2. A *digital straight line segment* is a finite set of grid points of the square raster grid, individuated by a straight line segment of the Euclidean plane by making use of the *grid intersection digitization criterion*: the grid point closest to each intersection between the grid and the Euclidean segment is an element of the digital straight line segment. Normally this point set has two unique end points, it can be ordered as a 8-path from one end point to the other, and the digital straight line segment can be identified with this 8-path.

(3) The computation of a digital straight line segment uses consecutive "small" steps, from pixel to pixel, starting at $p_1 = (x_1, y_1)$ and ending at $p_2 = (x_2, y_2)$. These elementary steps can be restricted to be parallel to the coordinate axes (4-path), or can also include diagonal steps (8-path). The latter case is mostly used for representing a digital curve, and it is also used in the straight line algorithm given below. Besides the given points p_1 and p_2 the algorithm generates additional

$$\mathbf{max}\ \{\ |x_1 - x_2|, |y_1 - y_2|\ \} - 1$$

"new" points on the screen for a digital straight line from p_1 to p_2. Drawing of an image point $p = (x, y)$ on a screen is accomplished by means of a procedure **marker** (x, y), assumed to be known.

As for the time complexity, all operations are of integer type. During the initialization, there are four subtractions and a single multiplication by 2. In the loop, each new point requires the drawing of the point (with **marker**), a simple test, one addition, and, eventually, one increment **inc**( ) and another addition.

Altogether, this is an optimized procedure for generating digital straight line segments.

(4) **procedure** *BRESENHAM* (p_1, p_2: *integer_point*);

{$p_1 = (x_1, y_1), p_2 = (x_2, y_2)$ }

var $x, y, dx, dy, error, c_1, c_2$: *integer*;
begin

{problem reduction to octants 1, 2, 7, and 8}

```
if (x1 > x2) then begin
        x := x1;        x1 := x2;       x2 := x;
        y := y1;        y1 := y2;       y2 := y
end {if};
dx := x2 - x1;    dy := y2 - y1;
x := x1;    y := y1;
if ( dy > 0 ) then
        if ( dx ≥ dy ) then begin                          {octant 1}
                c1 := 2 · dy;  error := c1 - dx;  c2 := error - dx;
                repeat
                        marker(x,y);
                        inc (x);
                        if error < 0  then  error := error + c1
                        else   begin
                                inc(y);  error := error + c2
                        end {if}
                until  x > x2
                end {then}
        else  begin                                        {octant 2}
                c1 := 2 · dx;  error := c1 - dy;  c2 := error - dy;
                repeat
                        marker(x,y);
                        inc (y);
                        if error < 0  then  error := error + c1
                        else   begin
                                inc(x);  error := error + c2
                        end {if}
                until  y > y2
                end {else}
else  begin
        dy := -dy;
        if ( dy > dx ) then begin                          {octant 7}
                c1 := 2 · dx;  error := c1 - dy;  c2 := error - dy;
                repeat
                        marker(x,y);
                        dec (y);
                        if error < 0  then  error := error + c1
                        else   begin
                                inc(x);  error := error + c2
```

```
                        end {if}
                  until  y < y2
                  end {then}
            else  begin                                    {octant 8}
                  c1 := 2 · dy;   error := c1 – dx;   c2 := error – dx;
                  repeat
                        marker(x,y);
                        inc (x);
                        if error < 0  then  error := error + c1
                        else   begin
                              dec(y);  error := error + c2
                        end {if}
                  until  x > x2
                  end {else}
      end {BRESENHAM}
```

(5) The algorithm has been proposed by

Bresenham, J.E.: *Algorithm for computer control of a digital plotter*. IBM System J. **4** (1965), pp. 25 - 30.

3.5 BIBLIOGRAPHIC REFERENCES

Specific literature is already listed under point **(5)** at the end of each procedure section. The design of time-efficient algorithms for (mostly mathematical) computational problems is the topic of the textbooks

Aho, A.V., Hopcroft, J.E., Ullman, J.D.: *The Design and Analysis of Computer Algorithms*. Addison-Wesley, Reading, 1974.

Aho, A.V., Ullman, J.D.: *Foundations of Computer Science*. Computer Science Press, New York, 1992.

Horowitz, E., Sahni, S.: *Fundamentals of Computer Algorithms*. Computer Science Press, Rockville, 1978.

Knuth, D.E.: *The Art of Computer Programming. Vol.* 3: *Sorting and Searching*. Addison-Wesley, Reading, USA, 1975.

Preparata, F.P., Shamos, M.I.: *Computational Geometry*. Springer, New York, 1985.

Townsend, M.: *Discrete Mathematics*: *Applied Combinatorics and Graph Theory*. Benjamin/Cummings, Menlo Park, 1987.

The time-efficiency of image processing algorithms has been considered in

Miklosko, J., Vajtersic, M., Vrto, I., Klette, R.: *Fast Algorithms and their Implementation on Specialized Parallel Computers*. North-Holland, New York, 1989.

Voss, K., Klette, R.: *Zeiteffektive Algorithmen zur Objektisolierung mittels lokaler Operatoren*. EIK **17** (1981), pp. 539 - 553.

A very useful and vast collection of procedures is given in

Press, W.H., Flannery, B.P., Teukolsky, S.A., Vetterling, W.T.: *Numerical Recipes*. Cambridge University Press, Cambridge, USA, 1988.

There are language versions in C, Pascal and in FORTRAN of this book. The basic idea to the decomposition of homogeneous linear local operators was published in

Kruse, B.: *Design and implementation of an image processor*. Linköping Studies in Science and Technology Dissertations No. **13**, Linköping University, 1977.

i.e., linear operator kernels can be represented one-to-one by polynomials, and decomposition corresponds to the factorization of these polynomials. The updating method for local operators has been used by

Klette, R., Sommer, G.: *Ein Filtersystem für die automatisierte Szintigrammverarbeitung*. Proceed. "Digitale Bildverarbeitung", Weißig bei Rathen, Wiss. Beiträge der TU Dresden, 1984, pp. 36-37.

Tyan, S.G.: *Median filtering, deterministic properties*. in: Huang, T.S. (ed.), Two-Dimensional Digital Signal Processing II. Transforms and Median Filters, Springer, Berlin, 1981.

Graphical formats have been described by

Born, G.: *Referenzhandbuch Dateiformate*. 2nd ed. Addison-Wesley, Bonn, 1992.

Pohl, M., Eriksdotter, H.: *Die wunderbare Welt der Grafikformate*. Wolframs Fachverlag, München, 1991.

The storage of pictorial data, especially in the context of image processing, is the topic of

Schlicht, H.J.: *Digitale Bildverarbeitung mit dem PC*. Addison-Wesley, Bonn, 1993.

Völz, H.: *Komprimierung von Bilddateien*. Nachrichtentechnik - Elektronik **43** (1993), S. 72 - 74 und S. 126 - 128.

4 COORDINATE TRANSFORMATIONS AND GEOMETRICAL OPERATORS

Several image processing operators aim at geometrical mappings of a given image or subimage, and image value modifications are only a consequence of geometrical mappings. These can be size reductions, translations, rotations, mirrorings, geometrical matchings of two images (e.g., for adjusting pictorial structures into identical geometric locations), or fittings of neighboring images into a mosaic. A certain amount of image values correction or fitting is often a partial task within the scope of such mappings.

A *coordinate transformation* defines a *geometrical operator* with respect to the image point mappings, cp. Section 1.4.1. It holds $h(x,y) = f(K(x,y))$, if K is a one-to-one coordinate transformation, f denotes the input image and h is the resultant image. Usually the coordinate transformation is given directly for the resultant image. For example, this means for a vertical mirroring (i.e. the mirroring takes place in vertical direction by exchanging rows), defined by the coordinate transformation $ver(x,y) = (x, N - y + 1)$, the resultant image h is described as

$$h(x,y) = f(ver(x,y)) = f(x, N - y + 1)$$

with $1 \leq x, y \leq N$. This approach specifies a definition scheme

$$"h(x,y) = f(K(x,y))"$$

for geometrical image transformations: For all image points (x,y) in the resultant image h, at first the coordinate transformation is performed for (x,y), and then the value $h(x,y)$ is computed based on the value of f at the transformed location. This approach is the so-called *backward coordinate transformation*. The method ensures that all image points (x,y) in h are assigned an image value.

A different approach is represented by the scheme "$h(x^*,y^*) = h(K^{-1}(x,y)) = f(x,y)$": For all image points (x,y) in the input image f, at first the new h-coordinates $K^{-1}(x,y) = (x^*,y^*)$ are determined, and then it has to be decided how the f-values at (x,y) have to be "distributed" to these new locations. This approach is the so-called *forward coordinate transformation*. The method does not ensure that a new value is assigned to all image points of h.

The operator descriptions given in this Chapter follow the backward transformation approach.

A transformation in place, i.e. input and output image share the same image data array, needs special attention. It has to be ensured that the original gray values are not overwritten if they are still needed for coordinate transformations later on. The operator descriptions make a clear statement whether two image data arrays are used for f and h, cp. Section 3.2.2, or just one for a transformation in place.

A random access to the input image data array f, and (if a second data array is used) to the resultant image data array h allows a direct coordinate transformation as specified by the mapping K. Such a direct geometrical transformation is also possible in the random access memory if "small" subimages (rectangular picture windows, circular shaped picture windows etc.) must undergo geometrical transformations. If the image data are not fully available in the random access memory then Section 3.2.3 applies, i.e., as discussed there under point **(A)**, only some image rows (or image columns) of f or h are stored in the random access memory at each time. This situation complicates the implementation of a transformation, but it is assumed for the geometrical transformations of Section 4.1 and 4.2. If random access is possible to the full data arrays, then these algorithms can be simplified in a straightforward way.

4.1 ONE-TO-ONE COORDINATE TRANSFORMATIONS

Let us assume that the resultant image h is obtained from the input image f only by means of a coordinate transformation K. The image values of f are mapped unchanged into the image h, i.e. only certain permutations take place.

These strong assumptions restrict the coordinate transformations on the image domain **R**, which is a rectangular subset of the orthogonal grid, to only a few possible mappings. Let us assume $M = N$ so that rotations and diagonal mirrorings are also possible.

The following is a complete list of all *one-to-one coordinate transformations* from the $N \times N$ image grid **R** into this grid **R**:

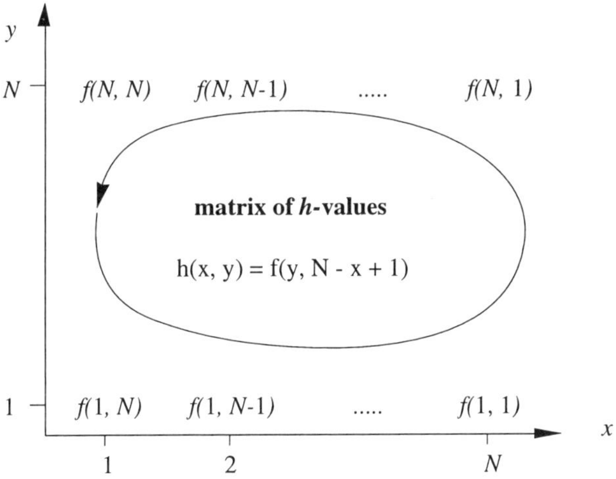

Figure 4.1: Let us consider a 90° rotation $h(x, y) = f(rot(x, y))$. After this transformation, the f-values are in positions in image h as shown in the figure. This illustrates that the rotation is performed in a counter-clockwise sense.

$ver(x, y)$	$= (x, N - y + 1)$	*vertical mirroring*	(cp. 4.1.1)
$hor(x, y)$	$= (N - x + 1, y)$	*horizontal mirroring*	(cp. 4.1.1)
$dia(x, y)$	$= (y, x)$	*diagonal mirroring*	
$rot(x, y)$	$= (y, N - x + 1)$	*rotation* by 90°	(cp. 4.1.3)
$rot^2(x, y)$	$= (N - x + 1, N - y + 1)$	rotation by 180°	(cp. 4.1.1)
$rot^3(x, y)$	$= (N - y + 1, x)$	rotation by 270°	
$tor(x, y)$	$= (N - y + 1, N - x + 1)$	rotation and mirroring.	

The rotation center is the central point of the image grid (it can have non-integer coordinates), and rotation is performed in a counter-clockwise sense. To complete the discussion an identical mapping $id(x, y) = (x, y)$ is also mentioned.

These eight coordinate transformations of the $N \times N$ image grid form a *group* in the sense of the mathematical algebra. Any consecutive application of some of these transformations is identical to one of these transformations[1]. For example, it holds $rot = hor \cdot dia$ (i.e. at first *hor*, then *dia* as a second transformation) because of

---

[1] This group has been considered by [Klette, R.: *A parallel computer for digital image processing*. EIK **15** (1979), pp. 237 - 263]. It holds that

$$id = ver \cdot dia \cdot rot = ver^2 = hor^2 = dia^2,$$

$$ver = dia \cdot rot = dia \cdot hor \cdot dia = hor \cdot rot^2,$$

$$hor = rot \cdot dia = dia \cdot ver \cdot dia = ver \cdot rot^2,$$

$$hor \cdot dia\ (x, y) = dia(hor(x, y)) = dia(N - x + 1, y) = (y, N - x + 1) = rot(x, y)\ .$$

These coordinate transformations are given as backward transformations. Figure 4.1 puts into evidence that the coordinate transformation *rot* gives rise to a counter-clockwise rotation. If the forward transformation scheme $h(rot(x, y)) = f(x, y)$ is used, then a clockwise rotation results.

4.1.1 Image Mirroring

(1) The image *f* is mirrored in place about its horizontal and/or its vertical axis.

Attributes:

Image: arbitrary with even *M* and *N*
Operator: geometrical (mirroring) without image value modifications

Inputs:

Variant $VAR = 1$ if a horizontal mirroring is computed, $VAR = 2$ for a vertical mirroring, and $VAR = 3$ if mirrorings in both directions are carried out.

(2) The resultant image *h* is generated in place, i.e. in the original image data array of the input image *f*, and it holds that

$$h(x, y) = \begin{cases} f(M - x + 1, y) & \text{, if } VAR = 1 \quad (\text{function } hor) \\ f(x, N - y + 1) & \text{, if } VAR = 2 \quad (\text{function } ver) \\ f(M - x + 1, N - y + 1) & \text{, if } VAR = 3 \quad (\text{function } rot^2). \end{cases}$$

(3) The mirroring in horizontal direction is carried out by a data exchange within an image row store $BUF1(1...M)$. A second image row store $BUF2(1...M)$ acts as temporary memory if a vertical mirroring is performed.

(4) *Control structure:* here explicitly given, a transformation in place

```
if (VAR = 1 or VAR = 3) then
    for y := 1 to N do begin
```

---

dia	$= rot \cdot ver$	$= hor \cdot rot,$	
rot	$= dia \cdot ver$	$= hor \cdot dia,$	
rot^2	$= ver \cdot hor$	$= hor \cdot ver,$	
rot^3	$= ver \cdot dia$	$= dia \cdot hor$	and
tor	$= rot \cdot hor$	$= ver \cdot rot$	$= dia \cdot ver \cdot hor.$

```
                    read row y  into the image row store BUF1(1...M);
                    for  x = 1  to M/2  do begin
                              u := BUF1(x);          BUF1(x) := BUF1(M − x + 1);
                              BUF1(M − x + 1) := u
                    end {for};
                    write BUF1(1...M) into the row y
          end {for}
end {if};
if  (VAR = 2 or VAR = 3)  then
          for  y := 1  to  N/2  do begin
                    read row y  into the image row store BUF1(1...M);
                    read row N − y + 1 into the image row store BUF2(1...M);
                    write BUF2(1...M) into the row y ;
                    write BUF1(1...M) into the row N + 1 − y
          end {for}
end {if}
```

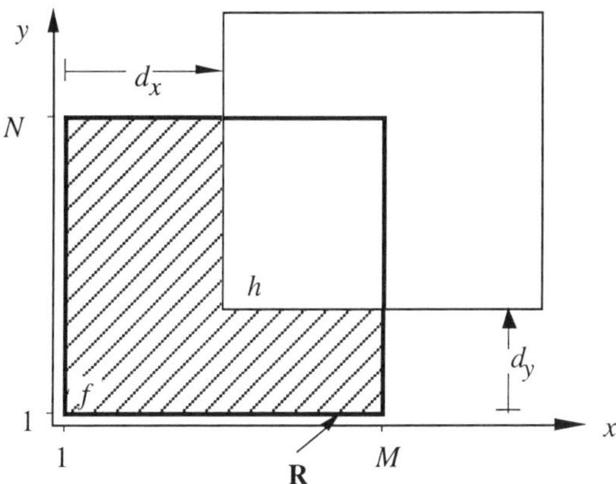

Figure 4.2: Illustration of the shifts d_x and d_y, where f denotes the input image and h is the shifted image.

4.1.2 Image Shifting

(1) The whole image is shifted by d_x pixels into horizontal direction, and by d_y pixels into vertical direction. The input image is read row-by-row, and the shifted image is stored into a (different) resultant image data array.

Attributes:

Images: arbitrary
Operator: geometrical (shifting) without image value modification

Inputs:

shift amounts d_x and d_y , with $-M \le d_x \le M$ and $-N \le d_y \le N$

(2) The coordinates $(x, y) = (1, 1), ..., (M, N)$ of the shifted image h satisfy the equation

$$h(x, y) = \begin{cases} f(x - d_x, y - d_y) & \text{, if } 1 \le x + d_x \le M \text{ and} \\ & \quad 1 \le y + d_y \le N \\ 0 & \text{otherwise.} \end{cases}$$

(3) An axis-parallel image shift (cp. Figure 4.2) is often useful as for instance when generating a mosaic of different images, making location-dependent comparisons of results of different image processing operators, or correcting parallel image transformations in-place (cp. point **(A)** in Section 3.2.2). The $M \times N$ image matrix is assumed to be a submatrix of a larger matrix having value 0 outside of the $M \times N$ array, and it is shifted within this larger matrix. All values shifted out of the image domain **R** are lost after this shift since only the resulting $M \times N$ values are stored. Thus the use of two image data arrays, for f and h, is suggested. Gray values of points of the fictitious larger matrix can move into the image domain **R**, and they arc assumed to be identical to 0 in the algorithm. A row buffer store *BUFNUL*(1...*M*), having value 0 in all of its positions, is initialized for this aim. Also those positions of the image row store *BUF*(1...*M*), which are not in use for storing of image data, are used for storing this default gray value 0.

(4) *Control structure:* here explicitly given
Required data arrays:

BUF(1...*M*) as image row store for the resultant gray values,
BUFNUL(1...*M*) stores an empty row, i.e. it contains 0 in all of its positions.

```
for z := 1 to M do begin
        BUF(z) := 0;   BUFNUL(z) := 0
        end {for};
if (dx ≥ 0 and dy ≥ 0) then begin
        for y := N – dy to 1 step –1 do begin
                read the first M – dx gray values of row y into the image row store
                        BUF(dx + 1...M);
```

```
            write the content of the image row store BUF(1...M) into the row
                    y + d_y of the resultant image h
            end {for};
      for y := 1 to d_y do
            write the content of the row buffer store BUFNUL(1...M) into the
                    row y of the resultant image h
      end {then}
else if (d_x < 0 and d_y ≥ 0) then begin
      for y := N – d_y to 1 step –1 do begin
            read M + d_x gray values of row y , starting at position x = 1 – d_x, into
                    the image row store positions BUF(1...M + d_x);
            write BUF(1...M) into the row y + d_y of the resultant image h
            end {for};
      for y := 1 to d_y do
            write BUFNUL(1...M) into the row y of the resultant image h
      end {then}
      else if (d_x ≥ 0 and d_y < 0) then begin
            for y:= 1 to N + d_y do
                  read the first M – d_x gray values of the row y – d_y into the
                        image row store positions BUF(d_x + 1...M);
                  write BUF(1...M) into the row y of the resultant image h
                  end {for};
            for y := N + d_y + 1 to N do
                  write BUFNUL(1...M) into the row y of image h
            end {then}
            else if (d_x < 0 and d_y < 0) then begin
                  for y := 1 to N + d_y do
                        read M + d_x gray values of row y – d_y , starting at
                              position x = 1 – d_x, into the image row store
                              positions BUF(1...M + d_x);
                        write BUF(1...M) into row y of the resultant image h
                        end {for};
                  for y := N + d_y + 1 to N do
                        write BUFNUL(1...M) into row y of image h
                  end {then}
```

4.1.3 90° Image Rotation

(1) A square image is assumed. The entire $N \times N$ image f is rotated by 90° in counter-clockwise sense around the center point of the image. The image point (x, y) of the resultant image h contains the image value $f(y, N - x + 1)$ after this rotation. Two different image data arrays, one for f and one for h, are used.

Attributes::

Images:	arbitrary with $N = M$
Operator:	geometrical (rotation) without image value modifications

(2) It holds $h(x, y) = f(N - y + 1, x)$, for $(x, y) = (1, 1), ..., (N, N)$ after the rotation.

(3) The program uses an image row store $BUF(1...N)$ for one column of the input image or for one row of the resultant image. It may be of interest that the coordinate transformation *rot* of the image rotation can also be carried out by consecutive applications of the functions *hor*, *ver*, or *dia*. It holds

$$hor \cdot dia = dia \cdot ver = rot \ .$$

Analogously, the transformation *dia* can be carried out by functions *hor*, *ver*, or *rot* according to

$$rot \cdot ver = hor \cdot rot = dia.$$

All the eight one-to-one coordinate transformations defined at the beginning of Section 4.1 are identical to certain opportune operation sequences of *hor* and *rot*, i.e. these two transformations constitute an example of a generating system for this transformation group.

(4) *Control structure*: here explicitly given

for x := 1 **to** N **do begin**
read image column x of the input image f, top to bottom, into the store $BUF(1...N)$;
write store $BUF(1...N)$, left to right, into the row x of the resultant image h
end {*for*}

4.2 SIZE REDUCTION AND MAGNIFICATION

A *image size reduction* maps several image points onto a single point. Therefore, a single image value has to be computed from a set of several image values. Vice-versa, a *image magnification* maps a single point onto several points, e.g., following the simplest approach, onto a square picture window. If the magnification is obtained by juxtaposing equal image values over all the magnified window's locations, the magnified image features a blocking effect, whose intensity grows with the magnification factor. This blocking effect can be suppressed by interpolating between neighboring magnification windows.

The functions describing size reductions, or magnifications can assume the form of coordinate transformations. For example, the transformation

$$K^{-1}(x, y) = (\textbf{integer}(x/2), \textbf{integer}(y/2)) \quad \text{or}$$

$$h(x, y) = \phi(f(2x, 2y), f(2x+1, 2y), f(2x, 2y+1), f(2x+1, 2y+1)),$$

with even values of M and N, $1 \le x \le M/2$ and $1 \le y \le N/2$, defines a size reduction shrinking an input image f onto the lower left quadrant. The new image value $h(x,y)$ is specified by a certain window function ϕ having four original f-values as arguments (cp. Section 1.4.1).

The tasks of magnification and size reduction can also be solved if row-wise buffering constrains the process. However, the following algorithms can be simplified if direct data access is used.

4.2.1 Image Size Reduction onto a Quadrant

(1) The program shrinks a square image f by the factor $2 \cdot 2 = 4$. The resultant image's position is identified by its lower left corner. Averaging or undersampling are the techniques used for avoiding the blocking effect in the reduced image h.

Attributes:

Images:	gray value images, $M = N$, even
Operator:	geometrical (contraction) with image value modification, window operator

Inputs:

lower left corner (XA, YA) where XA and YA have to be smaller than, or equal to $N/2$, and

variant $VAR = 1$ if undersampling is used, or $VAR = 2$ if averaging is chosen.

(2) The values XC, YC, XE, and YE define the central position and the upper right corner of the resultant image. For $M = N$ it holds

$$XC = 2 \cdot XA, \quad YC = 2 \cdot YA, \quad XE = XA + \frac{N-2}{2}, \quad \text{and} \quad YE = YA + \frac{N-2}{2}.$$

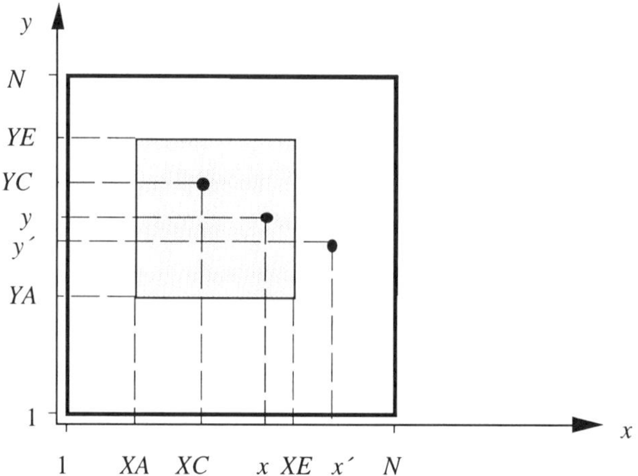

Figure 4.3: Image size reduction onto the size of a quadrant. The program maps an image point $p' = (x', y')$, with $x' = 2x - XC + 2$ and $y' = 2y - YC + 2$, into the image point $p = (x, y)$.

The new image value is given by

$$h(x, y) = \begin{cases} f(2x - XC + 2,\ 2y - YC + 2) & \text{, if } XA \leq x \leq XE \text{ and } YA \leq y \leq YE \\ 0 & \text{otherwise} \end{cases}$$

if *undersampling* ($VAR = 1$) is chosen, or by

$$h(x, y) = \begin{cases} \frac{1}{4}\,[f(2x - XC + 1,\ 2y - YC + 1) \\ \qquad + f(2x - XC + 2,\ 2y - YC + 1) \\ \qquad + f(2x - XC + 1,\ 2y - YC + 2) \\ \qquad + f(2x - XC + 2,\ 2y - YC + 2)] & \text{, if } XA \leq x \leq XE \text{ and } YA \leq y \leq YE \\ 0 & \text{otherwise,} \end{cases}$$

if *averaging* ($VAR = 2$) is selected.

(3) Figure 4.3 illustrates how the entire image domain is mapped onto the square picture window

$$\mathbf{F} = \left\{ (x,y):\ XA \le x \le XE = XA + \frac{N-2}{2}\ \&\ YA \le y \le YE = YA + \frac{N-2}{2} \right\}.$$

A necessary constraint is that XA and YA have to be smaller than, or equal to $N/2$.

The described operation of size reduction can be carried out as follows, using only one image data array (i.e. the same for the input as well as for the output image): The image rows are processed in two successive phases, starting at the uppermost and, respectively, at the lowermost row, and stopping at the row $YC = 2 \cdot YA$. In a similar way the image points in a single row are processed in two successive phases, one starting at $x = 1$ and one at $x = N$, and stopping at the location $XC = 2 \cdot XA$. If undersampling is chosen, then the gray value in each second position of each second row is used as the gray value in the resultant image. If averaging is selected, then the average value is computed over disjoint 2×2 groups of pixel values. All gray values of the resultant image outside of the domain of the shrunk image are set to 0.

(4) *Control structure:* here explicitly given

Required data arrays:

image row stores $BUF1(1...N)$ and $BUF2(1...N)$ of the currently processed rows,

$BUFNUL(1...N)$ as an empty row (i.e., value 0 in all positions), and

$BUFOUT(1...N)$ for the resulting gray values.

```
begin
   input of coordinates XA, YA of the lower left corner of the reduced image;
   input of VAR := 1 (undersampling), or of VAR := 2 (averaging);
   initialize BUFNUL(1...N) as equal to 0 in all positions;
   XE := XA + (N – 2)/2;          YE := YA +(N – 2)/2;
   XC := 2·XA;                    YC := 2·YA;
   for y := 1 to N do
         if (YA ≤ y ≤ YE) then begin
               read row 2y – YC + 1 into the image row store BUF1(1...N);
               read row 2y – YC + 2 into the image row store BUF2(1...N);
               for x := 1 to N do
                     if XA ≤ x ≤ XE then
                           if (VAR = 1) then
                                 BUFOUT(x) := BUF1(2x – XC + 2)
                           else
   BUFOUT(x) := 1/4 · [BUF1(2x – XC + 1) + BUF1(2x – XC + 2)
                           + BUF2(2x – XC + 1) + BUF2(2x – XC + 2)]
                     else
```

```
                    BUFOUT(x) := 0;
              write BUFOUT(1...N) into the row y of the resultant image h
              end {then}
        else
              write BUFNUL(1...N) into the row y of the resultant image h
end
```

4.2.2 Image Magnification by Scale Factor 2

(1) The lower left quadrant of the image domain **R** is magnified onto the entire image domain, i.e. the area scale factor is 2·2 = 4. A nearest neighbor interpolation generates the image values in new locations of the magnified image.

Attributes:

Images:	gray value images, M can differ from N
Operator:	geometrical (expanding) with image value modification, window operator

(2) The following equations specify the interpolation carried out,

$$h(x, y) = f(\frac{x+1}{2}, \frac{y+1}{2}), \qquad \text{if } x \text{ and } y \text{ are odd,}$$

$$h(x, y) = \frac{1}{2}\,[f(\frac{x}{2}, \frac{y+1}{2}) + f(\frac{x+2}{2}, \frac{y+1}{2})], \qquad \text{if } x \text{ is even and is } y \text{ odd,}$$

$$h(x, y) = \frac{1}{2}\,[f(\frac{x+1}{2}, \frac{y}{2}) + f(\frac{x+1}{2}, \frac{y+2}{2})], \qquad \text{if } x \text{ is odd and } y \text{ is even,}$$

$$h(x, y) = \frac{1}{4}\,[f(\frac{x}{2}, \frac{y}{2}) + f(\frac{x+2}{2}, \frac{y}{2}) + f(\frac{x}{2}, \frac{y+2}{2}) + f(\frac{x+2}{2}, \frac{y+2}{2})], \qquad \text{if } x \text{ and } y \text{ are even.}$$

Only image points (x, y) in the lower left quadrant of the input image f are transformed, with $1 \le x \le M/2$ and $1 \le y \le N/2$. The linear interpolation in new image positions uses as input one, two or four nearest neighbor gray values of the original subimage. The coordinates x and y in the given four transformation equations cover all positions in the entire $M \times N$ image.

(3) This program magnifies $M/2 \times N/2$ pixels of the lower left image quadrant. If necessary, the $M/2 \times N/2$ subimage to be magnified can be positioned into this standard initial quadrant position by means of an image shift (cp. Section 4.1.2).

The process uses two image row stores $BUF1(1...M/2)$ and $BUF2(1...M/2)$ for the input of gray values, and two row buffer stores $BUFOUT(1...M, i)$, with $i = 1, 2$, for building up the resultant gray values.

(4) *Control structure*: here explicitly given

```
for y := 1 to M/2 do begin
    read the left half of row y of the input image f into BUF1(1...M/2);
    read the left half of row y + 1 of the image f into BUF2(1...M/2);
    for x := 1 to M/2 do begin
    BUFOUT(2(x – 1) + 1, 1)    := BUF1(x);
    BUFOUT(2x, 1)              := [BUF1(x) + BUF1(x + 1)] / 2;
    BUFOUT(2(x – 1) + 1, 2)    := [BUF1(x) + BUF2(x)] / 2;
    BUFOUT(2x, 2)              := [BUF1(x) + BUF1(x + 1)
                                      + BUF2(x) + BUF2(x + 1)] / 4
        end {for}
    write BUFOUT(1...M, 1) into row 2y – 1 of the resultant image h;
    write BUFOUT(1...M, 2) into row 2y of the resultant image h
end {for}
```

4.2.3 Image Pyramids

An *image pyramid* consists of a set of images with different resolutions: the original image at the ground level, the same image with lower resolution at the next level above etc., up to the top of the pyramid representing a (theoretical) one-pixel image, cp. Figure 4.4.

Such an image pyramid allows the choice of an optimum resolution, e.g., if the resolution has to be chosen as low as possible with regard to the computation time, and yet as high as necessary for ensuring the proper resolution of image details. It can also be used for computer animations if all levels are magnified to $N \times N$ (cp. Section 4.2.2), and these levels are shown, e.g., on a screen in sequence by stepping top-down or bottom-up in the image pyramid. An useful approach in implementing an image processing system can be to transform first each input image into an image pyramid, or at least into a collection of several levels of such a pyramid. Then, subsequent processes of the system can offer optional access to a specified level of resolution. Image pyramids are also a standard image representation scheme for several approaches in image analysis (edge detection, segmentation etc.).

(1) We assume an $N \times N$ input image f, with $N = 2^n$. The resultant image h has dimensions $M \times N$, with $M = N + N/2$. It combines the original image f and its subsequent size reductions. It is also possible that single size reductions are stored in separate data arrays. However, here a one-image representation containing all levels of resolution is chosen. The resultant image *h* contains the original image *f* in its original position, i.e. with $1 \leq x, y \leq N$. Furthermore it contains in its upper right corner the shrunk image *f* with the size of a quadrant, i.e. with $N + 1 \leq x \leq M$ and $N/2 + 1 \leq y \leq N$, then the twice size-reduced image *f* in a position below the latter one etc., cp. Figure 4.4, and finally the image *f* reduced to a single pixel (the top of the pyramid) in its lower right corner.

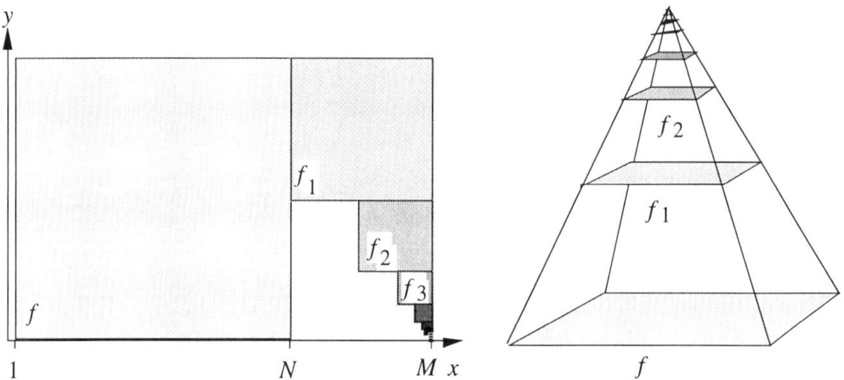

Figure 4.4: Image storage locations of the different levels of the image pyramid as generated by the program (at left), where a $N \times N$ image *f* is mapped into a $M \times N$ image *h*, with $M = N + N/2$. Spatial 3-D model of an image pyramid (at right).

The image pyramid contains image *f* at its ground level, and the subsequent levels of size reduction at the levels above. The uppermost level of a 1×1 image corresponds to the top of the pyramid. Undersampling or averaging (cp. Section 4.2.1) can be used for the calculation of image values within the process of stepwise image shrinking.

Attributes:

Images:	$N \times N$ gray value images and N is a power of two
Operator:	geometrical (expanding) with image value modification

Inputs:

variant VAR = 1 if undersampling is chosen, or VAR = 2 if averaging is selected.

(2) Let be $f_0 := f$. The symbols $f_1, f_2, ..., f_n$ denote the images at the consecutive pyramid levels, with $n = \log_2 N$. If undersampling (VAR = 1) is chosen then it holds

$$f_{i+1}(x, y) = f_i(2x, 2y),$$

with $i = 0, 1, ..., n-1$ and $x, y = 1, 2, ..., 2^{n-i}$ at each level $i+1$. These recursive equations define the images f_{i+1} in the order $i = 0, 1, ..., n-1$. If averaging ($VAR = 2$) is selected then these images are recursively defined as follows:

$$f_{i+1}(x, y) = \frac{1}{4}\,[f_i(2x, 2y) + f_i(2x-1, 2y) + f_i(2x, 2y-1) + f_i(2x-1, 2y-1)],$$

again with $i = 0, 1, ..., n-1$, and $x, y = 1, 2, ..., 2^{n-i}$ for image f_{i+1}. These images are placed by the algorithm into their positions in image h as specified in Figure 4.4.

(3) If the numbers of rows and columns of the input image are not identical, and/or if these numbers are not powers of two, then it is suggested to embed the input image into the next higher resolution $N \times N$, with $N = 2^n$. It should also be mentioned that different methods can be used for computing f_{i+1} based on f_i.

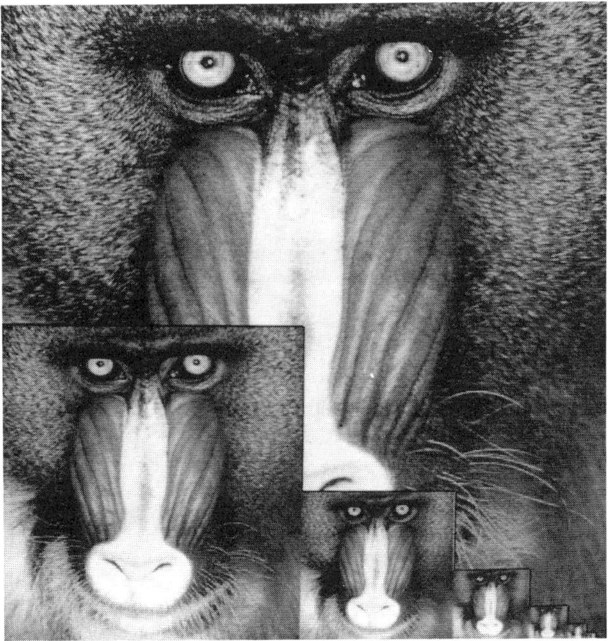

Figure 4.5: Pictorial representation of all levels of an image pyramid of the gray level image *MANDRILL*. The arrangement of levels in this image differs from that produced by the algorithm.

Two image row stores $BUF1(1...N)$ and $BUF2(1...N)$ are used for buffering the input image data, and a row buffer store $BUF(1...N/2)$ for the computed image values. We assume that $N \geq 2$.

In Figure 4.5 the different size-reduced versions of the input image are shown at locations within the original square image domain differing from those produced by the program (this is possible by a simple modification of the given algorithm).

All levels of an image pyramid comprise a total of

$$N^2 \cdot (1 + 1/4 + 1/16 + \dots) = 4/3 \cdot N^2$$

pixels, i.e. the additional memory size required for the pyramid exceeds only one third of the requirements for the $N \times N$ input image.

(4) *Control structure*: here explicitly given

```
begin
    z := N/2 + 1;                                   {computation of image f1}
    for y := 1 to N step 2 do begin
        read row y of the input image f into BUF1(1...N) as well as into the
                positions 1,..., N of row y of h;
        read row y + 1 of the image f into BUF2(1...N) as well as into the
                positions 1,..., N of row y + 1 of h;
        for x := 1 to N/2 do
                if (VAR = 1) then  BUF(x) := BUF1(2x);
                else
   BUF(x) := 0.25 · [BUF1(2x – 1) + BUF1(2x) + BUF2(2x) + BUF2(2x – 1)];
        write BUF(1...N/2) into the positions N + 1, N + 2,..., M of row z of
                image h;
        z := z + 1
        end {for}
    ZB := N/2;    s := 0;                           {computation of f2, f3,..., fn }
    ZA := integer(N/4);          {ZA = number of rows within the current level}
    z := ZA + 1;
    while ZA ≥ 1 do begin
        for x := s + 1 to s + ZA do
                BUF(x) := 0;                        {background value 0}
        for y := ZB + 1 to ZB + 2·ZA step 2 do begin
                read the values in positions N + s + 1,..., M of row y of h into
                        BUF1(1...2·ZA);
```

```
                read the values in positions N + s + 1,..., M of row y + 1 of h
                    into BUF2(1...2·ZA);
                i := 1;
                for x := s + ZA + 1 to N/2 do begin
                    if (VAR = 1) then
                        BUF(i) := BUF1(2i);
                    else
        BUF(i) := 0.25 · [BUF1(2i – 1) + BUF1(2i) + BUF2(2i) + BUF2(2i – 1)];
                    i := i + 1
                    end {for}
                write row buffer store BUF(1...N/2) into the positions N + 1,
                    N + 2,..., M of row z of image h ;
                z := z + 1
                end {for};
            ZB := ZB – ZA;              s := s + ZA;
            ZA := integer(ZA/2);        z := ZA + 1
        end {while}
                                        {filling of the last (lowermost) row}
        BUF(N/2) := 0;                                  {background value 0}
        write BUF(1...N) into the positions N + 1, N + 2,..., M of row 1 of the
            resultant image h
    end
```

(5) Ballard, D.H., Brown, Ch.M.: *Computer Vision*. Prentice-Hall, Englewood Cliffs, 1982.

4.3 AFFINE TRANSFORMATIONS

Here we consider for a given coordinate transformation only a backward transformation, i.e. from the resultant image h to the input image f. An *affine coordinate transformation* $K(x, y) = (x^*, y^*)$ is a linear mapping, i.e. here of image coordinates (x, y) in image h into image coordinates (x^*, y^*) in image f.

In general, the computed coordinates (x^*, y^*) are not necessarily integers, and also not necessarily within the $M \times N$ image domain **R**. Therefore, after transforming the h-coordinates, a one-to-one mapping (*resampling*) of real point coordinates onto integer image coordinates of the image f must be carried out.

This Section 4.3 is organized as follows: Different affine transformations and their combinations are the topics of Section 4.3.1. For specifying an affine transformation a 3×3 transformation matrix must be determined. This matrix can be given in an a-priori way, or can be calculated based on selected fitting points, cp. Section 4.3.2. Finally, the affine transformation, as specified by a transformation matrix, is actually performed with an algorithm given in Section 4.3.3. Different ways of resampling are implemented by this algorithm. This transformation procedure in Section 4.3.3 does not contain row-wise buffering of image data, i.e. random access is assumed for the image data arrays.

4.3.1 Products of Transformation Matrices

(1) Let us assume that two affine transformations, e.g. basic transformations as rotation, translation or mirroring, are given by their homogenous transformation matrices. These transformations have to performed in a certain order in the image plane. The matrices are given with respect to a backward transformation, i.e. from *h* to *f*. The task consists in computing the resultant transformation matrix **C**. A non-singularity test of **C** is included in this task.

Inputs:

the elements $a_{11}, a_{12}, a_{13}, a_{21}, a_{22}, a_{23}$ and $b_{11}, b_{12}, b_{13}, b_{21}, b_{22}, b_{23}$ of two 3×3 transformation matrices **A** and **B**, and

the order: first **A**, then **B**.

(2) Each affine coordinate transformation is defined by an uniquely solvable linear equation system

$$x^* = a_{11} \cdot x + a_{12} \cdot y + a_{13}$$
$$y^* = a_{21} \cdot x + a_{22} \cdot y + a_{23}.$$

Here, x and y denote integer coordinates. If backward transformation is assumed then these are coordinates of the resultant image *h*.

Formally, the treatment of sequences of affine transformations can be simplified if each transformation is uniquely represented by a 3×3 transformation matrix

$$\mathbf{A} = ((aij))_{1 \le i, j \le 3}.$$

Then, the product of 3×3 matrices corresponds to the successive application of the represented transformations. An additional equation

$$1 = a_{31} \cdot x + a_{32} \cdot y + a_{33}$$

is used for obtaining this 3×3 matrix representation. Since transformation equations must be true for any values of x and y it follows that $a_{31} = a_{32} = 0$ and $a_{33} = 1$.

Altogether, an affine transformation is uniquely defined by the *homogenous matrix equation*

$$\begin{pmatrix} x^* \\ y^* \\ 1 \end{pmatrix} = \begin{pmatrix} a_{11} & a_{12} & a_{13} \\ a_{21} & a_{22} & a_{23} \\ 0 & 0 & 1 \end{pmatrix} \times \begin{pmatrix} x \\ y \\ 1 \end{pmatrix} .$$

Translation, scaling, and rotation are three basic types of affine transformations. A translation of an image f is specified by a translation vector $(d_x, d_y)^T$, with the linear equation system

$$\begin{aligned} x^* &= 1 \cdot x + 0 \cdot y + d_x \\ y^* &= 0 \cdot x + 1 \cdot y + d_y , \end{aligned}$$

or by the homogenous transformation matrix

$$\begin{pmatrix} 1 & 0 & d_x \\ 0 & 1 & d_y \\ 0 & 0 & 1 \end{pmatrix} .$$

A *scaling* of an image f is uniquely defined by two scaling factors s_x and s_y (both greater than 0), by the linear equation system

$$\begin{aligned} x^* &= s_x \cdot x + 0 \cdot y + 0 \\ y^* &= 0 \cdot x + s_y \cdot y + 0 , \end{aligned}$$

or by the homogenous transformation matrix

$$\begin{pmatrix} s_x & 0 & 0 \\ 0 & s_y & 0 \\ 0 & 0 & 1 \end{pmatrix} .$$

A *rotation* of an image f is specified by a rotation angle δ (counter-clockwise, with the coordinate origin as rotation center), by the linear equation system

$$\begin{aligned} x^* &= \mathbf{cos}\delta \cdot x - \mathbf{sin}\delta \cdot y + 0 \\ y^* &= \mathbf{sin}\delta \cdot x + \mathbf{cos}\delta \cdot y + 0 , \end{aligned}$$

or by the homogenous transformation matrix

$$\begin{pmatrix} \cos\delta & -\sin\delta & 0 \\ \sin\delta & \cos\delta & 0 \\ 0 & 0 & 1 \end{pmatrix}.$$

The unity matrix

$$\begin{pmatrix} 1 & 0 & 0 \\ 0 & 1 & 0 \\ 0 & 0 & 1 \end{pmatrix}$$

characterizes the identical mapping. *Shearings* of an image f are defined by the homogenous matrices

$$\begin{pmatrix} 1 & a & 0 \\ 0 & 1 & 0 \\ 0 & 0 & 1 \end{pmatrix} \quad \text{or} \quad \begin{pmatrix} 1 & 0 & 0 \\ b & 1 & 0 \\ 0 & 0 & 1 \end{pmatrix}.$$

For a *mirroring* of an image f assume a straight line with angle γ to the x-axis as mirroring axis. This mirroring is specified by the homogenous matrix

$$\begin{pmatrix} \cos 2\gamma & \sin 2\gamma & 0 \\ \sin 2\gamma & -\cos 2\gamma & 0 \\ 0 & 0 & 1 \end{pmatrix}.$$

The transformation *ver* of Section 4.1 (a mirroring about the x-axis), and a subsequent translation with translation vector $(0, N)^T$ are defined by the homogenous matrix

$$\begin{pmatrix} 1 & 0 & 0 \\ 0 & -1 & 0 \\ 0 & 0 & 1 \end{pmatrix}.$$

A mirroring with $\gamma = \pi/4$, identical to the transformation *dia* in Section 4.1, leads to the homogenous matrix

$$\begin{pmatrix} 0 & 1 & 0 \\ 1 & 0 & 0 \\ 0 & 0 & 1 \end{pmatrix}.$$

In general, specific transformation matrices for image transformations can be defined by a-priori given parameters, or can be (approximately) calculated by means of the fitting point method described in Section 4.3.2.

Now we return to the original task given under point **(1)**. Two 3×3 homogenous transformation matrices $\mathbf{A} = ((a_{ij}))_{1 \le i,j \le 3}$ and $\mathbf{B} = ((b_{ij}))_{1 \le i,j \le 3}$ are given, and the transformation characterized by **A** has to be performed first, followed by the transformation characterized by **B**. This results into a certain image transformation $T(f) = h$, i.e. an image f is mapped into an image h.

Affine transformations are not commutative in general, i.e. their order can make a difference. The resulting transformation is defined by the homogenous transformation matrix $\mathbf{C} = ((c_{ij}))_{1 \le i,j \le 3}$, with

$$\mathbf{C} = \mathbf{A} \times \mathbf{B}.$$

The matrix multiplication is also not commutative. The given procedure computes matrix **C**.

All the examples of transformation matrices given above are non-singular, i.e. there exists an inverse matrix, and the specified coordinate transformation is one-to-one (in the Euclidean plane!). A consecutive application of one-to-one transformations is again a one-to-one transformation. Singular 3×3 matrices, i.e. matrices with determinants equal to zero, are not usable as transformation matrices.

A backward transformation matrix **A** uniquely corresponds to a forward transformation matrix $\mathbf{A}^{-1}$, and vice-versa. The resulting matrix product **C** can be mapped into the corresponding forward transformation matrix by means of a matrix inversion.

(3) The program is a multiplication procedure for special 3×3 matrices containing values 0, 0, 1 in their lowermost row. It can also be used if more than two affine transformations have to be performed in sequence. Then, the correct order of repeated calls of this procedure is essential. The resulting transformation of image f into image h uses just one transformation matrix as a result of multiplying all the transformation matrices. This simplification has also the essential effect that one approximation process (cp. Section 4.3.3) suffices for calculating the values of the resultant image h. Otherwise, to perform several transformations would also mean to introduce a sequence of several approximations. If it can be taken for sure that **A** and **B** are non-singular, then the singularity test is not necessary in the program.

(4) **begin**

{computation of the product matrix}

$c_{11} := a_{11}b_{11} + a_{12}b_{21};$ $c_{12} := a_{11}b_{12} + a_{12}b_{22};$
$c_{13} := a_{11}b_{13} + a_{12}b_{23} + a_{13};$ $c_{21} := a_{21}b_{11} + a_{22}b_{21};$
$c_{22} := a_{21}b_{12} + a_{22}b_{22};$ $c_{23} := a_{21}b_{13} + a_{22}b_{23} + a_{23};$
$c_{31} := 0;$ $c_{32} := 0;$ $c_{33} := 1;$

{singularity test}

$det := c_{11}c_{22} - c_{12}c_{21};$
if ($det = 0$) **then**
the matrix **C** is singular (i.e., not usable for a transformation)

end

(5) Foley, J.D., van Dam, A., Feiner, S.K., Hughes, J.F.: *Computer Graphics*. 2nd ed., Addison-Wesley, Reading, 1990.

4.3.2 Computation of Transformation Matrices

(1) The task consists of performing a geometrical fitting for two $M \times N$ images f_1 and f_2. More exactly, an input image f_1 has to be affinely transformed into a resultant image h such that image h and the (template) image f_2 are geometrically fitted (as well as possible) after that transformation. The task consists in computing the *homogenous transformation matrix* of this affine transformation.

To solve this task, three fitted *corresponding point pairs* (p_{1k}, p_{2k}), $k = 1, 2, 3$, are given as input. The points $p_{1k} = (x_{1k}, y_{1k})$ are located in the image f_1, and the points $p_{2k} = (x_{2k}, y_{2k})$ are located in the image f_2. The geometrical fitting operation must map a point p_{1k} onto the corresponding point p_{2k}, for each pair $k = 1, 2, 3$. These three point pairs can be determined, e.g., by means of an interactive corresponding point selection process or by an analytic coordinate transformation, e.g. for selecting *size reductions* or *magnifications* of arbitrary scaling factor. Note that the operators in Sections 4.1 and 4.2 are special affine transformations, i.e. they can be defined by three point pairs in each case.

Inputs:

coordinates $x_{11}, y_{11}, x_{21}, y_{21}, x_{12}, y_{12}, x_{22}, y_{22}, x_{13}, y_{13}, x_{23}, y_{23}$ of the three fitted point pairs.

(2) An affine transformation is uniquely characterized by six transformation parameters $a_{11}, a_{12}, a_{13}, a_{21}, a_{22}, a_{23}$ defining its homogenous transformation matrix. These parameters and the three pairs of fitted points satisfy the following two linear equation systems,

$$x_{21} = a_{11} \cdot x_{11} + a_{12} \cdot y_{11} + a_{13}$$
$$x_{22} = a_{11} \cdot x_{12} + a_{12} \cdot y_{12} + a_{13}$$
$$x_{23} = a_{11} \cdot x_{13} + a_{12} \cdot y_{13} + a_{13}$$

and

$$y_{21} = a_{21} \cdot x_{11} + a_{22} \cdot y_{11} + a_{23}$$
$$y_{22} = a_{21} \cdot x_{12} + a_{22} \cdot y_{12} + a_{23}$$
$$y_{23} = a_{21} \cdot x_{13} + a_{22} \cdot y_{13} + a_{23} .$$

These equations suffice for the unique calculation of the six transformation parameters if both equation systems have rank 3 (without discussing here eventual numerical instabilities). Therefore, the program starts with a test whether the given pairs of fitted points satisfy this necessary condition of an unique solution. These pairs are not usable if the three points in f_1 are geometrically collinear, i.e. on a straight line. Then, a usual solution procedure for linear equation systems (often available in a math library) leads to the desired transformation parameters.

(3) Let us assume that two $M \times N$ images f_1 and f_2 represent the same scene, e.g. two aerial views of the same landscape but taken at different height, or at different days. Image f_2 is used as *reference image* (template). The task consists of transforming f_1 in such a way that the fitting point pairs selected in f_1 and f_2 are in identical positions after this transformation. An affine transformation is sufficient if three pairs of fitted points are given. A *least-square error optimization* of the sum

$$\sum_{k=1,2,3\ldots} [(a_{11} \cdot x_{1k} + a_{12} \cdot y_{1k} + a_{13} - x_{2k})^2 + (a_{21} \cdot x_{1k} + a_{22} \cdot y_{1k} + a_{23} - y_{2k})^2]$$

is a possible approach in situations with more than three pairs of fitted points.

(4) The solution consists in two calls of the following *LES* procedure (linear equation system), with input

$$(a_1, a_2, a_3, b_1, b_2, b_3, c_1, c_2, c_3) = (x_{21}, x_{22}, x_{23}, x_{11}, x_{12}, x_{13}, y_{11}, y_{12}, y_{13})$$

for the first call, and input

$$(a_1, a_2, a_3, b_1, b_2, b_3, c_1, c_2, c_3) = (y_{21}, y_{22}, y_{23}, x_{11}, x_{12}, x_{13}, y_{11}, y_{12}, y_{13})$$

for the second call. By means of the first call are computed the values $x = a_{11}$, $y = a_{12}$ and $z = a_{13}$, and $x = a_{21}$, $y = a_{22}$ and $z = a_{23}$ are computed by the second call.

```
procedure LES (a1, a2, a3, b1, b2, b3, c1, c2, c3: integer;
                    x, y, z : real);
begin
        det := (b1 – b2)(c1 – c3) – (b1 – b3)(c1 – c2);
```

```
        if ( det = 0 ) then
            the fitted points are not usable (the three points in f1 are collinear)
        else
            begin
                x : = [(a1 – a2)(c1 – c3) – (a1 – a3)(c1 – c2)] / det;
                y : = [(a1 – a3)(b1 – b2) – (a1 – a2)(b1 – b3)] / det;
                z := a1 – b1 · x  – c1 · y
            end {else}
    end {LES}
```

(5) Situations with more than three pairs of fitted points have been treated by Haberäcker, P.: *Digitale Bildverarbeitung*, Carl Hanser Verlag, München, 3rd ed., 1989.

4.3.3 Affine Image Transformations

(1) An image f has to be transformed into an image h according to a given affine coordinate transformation K, and following the backward transformation scheme "$h(x,y) = f(K(x,y))$"

Attributes:

Images: gray value images
Operator: geometrical with image value modification, window operator

Inputs:

transformation parameters $a_{11}, a_{12}, a_{13}, a_{21}, a_{22}, a_{23}$ of a non-singular transformation matrix,
variant VAR = 1, 2 or 3 for specifying a resampling method, and
parameters $b_{11}, b_{12}, b_{13}, b_{21}, b_{22}, b_{23}, b_{31}, b_{32}, b_{33}$ of a window function if variant VAR = 3 is selected.

(2) The following equations describe an ideal case of an one-to-one coordinate transformation from grid points to grid points,

$$h(x, y) = \begin{cases} f(a_{11}\cdot x + a_{12}\cdot y + a_{13},\ a_{21}\cdot x + a_{22}\cdot y + a_{23}) \\ \qquad , \text{if } 1 \le a_{11}\cdot x + a_{12}\cdot y + a_{13} \le M \quad \text{and} \\ \qquad\qquad 1 \le a_{21}\cdot x + a_{22}\cdot y + a_{23} \le N \\ 0 \qquad \text{otherwise} \end{cases}$$

with $(x, y) = (1, 1),..., (M, N)$, cp. Section 4.1 for examples. Unfortunately, such an ideal mapping onto grid points is a very rare event. Practically, problems arise because of the discrete nature of the image points, and because of the limitations of the image domain **R**:

(i) Starting with point (x, y) in image h it can happen that the point

$$K(x, y) = (a_{11} \cdot x + a_{12} \cdot y + a_{13}, a_{21} \cdot x + a_{22} \cdot y + a_{23})$$

is within the rectangle $[0.5, M + 0.5] \times [0.5, N + 0.5]$ of the real plane (i.e., within the image domain in some sense) but not a grid point, i.e. has no integer coordinates. Which f-value should be used?

(ii) Starting with point (x, y) in image h it can happen that the point

$$(a_{11} \cdot x + a_{12} \cdot y + a_{13}, a_{21} \cdot x + a_{22} \cdot y + a_{23})$$

is outside of the image domain **R**, i.e. not in the real rectangle mentioned under point (i). Thus, there exists no value in f "close to this location". Which value can be used for $h(x, y)$?

(iii) For points (r, s) in the image f it can happen that there exists no point in the image domain **R** of h such that $K(x, y) \approx (r, s)$, i.e. these pixels in f are not represented in the image h, and they "get lost".

The problems (ii) and (iii) have been already raised in Section 4.1.2 in a more simple way for image shifts.

A default value, e.g. the constant gray value 0, can be assigned to $h(x, y)$ in case the problem (ii) arises. The existence of problem (iii) has to be accepted, but it does not interfere with the computation of image h. The interesting case is problem (i). Here, the point

$$(a_{11} \cdot x + a_{12} \cdot y + a_{13}, a_{21} \cdot x + a_{22} \cdot y + a_{23})$$

is in the rectangle $[0.5, M + 0.5] \times [0.5, N + 0.5]$ of the real plane, but not a grid point. Therefore, a resampling can be chosen to compute $h(x, y)$:

$VAR = 1$: *Nearest neighbor resampling* considers the nearest grid point to $K(x, y)$, and the f-value in this grid point is used as h-value,

$$h(x, y) = f(\textbf{integer}(a_{11} \cdot x + a_{12} \cdot y + a_{13} + 0.5), \textbf{integer}(a_{21} \cdot x + a_{22} \cdot y + a_{23} + 0.5)).$$

$VAR = 2$: This *low-pass resampling method* (with a smoothing effect) uses the arithmetic average of f-values in those four grid points which are the closest to $(r, s) = (a_{11} \cdot x + a_{12} \cdot y + a_{13}, a_{21} \cdot x + a_{22} \cdot y + a_{23})$, i.e. in the points

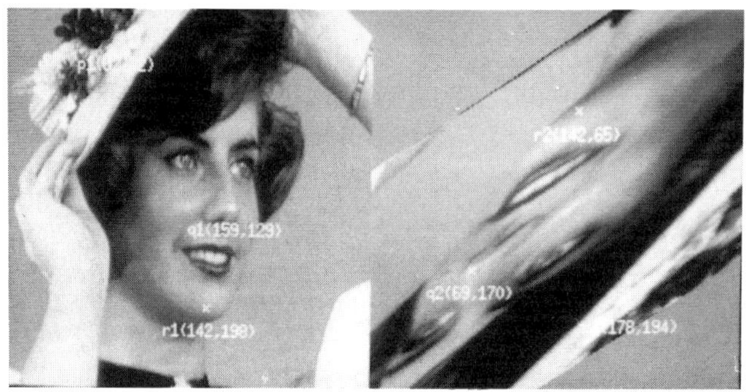

Figure 4.6: Three points are labeled in the right image as well as in the left image, defining three pairs of corresponding points. The affine transformation was defined (just for illustrating purposes) by the task of mapping the selected points in the left image onto the corresponding points in the right image. The result of this affine transformation is illustrated in the right image, with *VAR* = 3, and

1	2	1
2	4	2
1	2	1

was used as the window function kernel.

$$(\textbf{integer}(r), \textbf{integer}(s)), \quad (\textbf{integer}(r), \textbf{integer}(s) + 1),$$
$$(\textbf{integer}(r) + 1, \textbf{integer}(s)), \quad \text{and} \quad (\textbf{integer}(r) + 1, \textbf{integer}(s) + 1).$$

VAR = 3: This *linear convolution resampling method* selects the nearest grid point

$$(\textbf{integer}(a_{11} \cdot x + a_{12} \cdot y + a_{13} + 0.5), \textbf{integer}(a_{21} \cdot x + a_{22} \cdot y + a_{23} + 0.5))$$

as reference point of a 3×3 window. The convolution coefficients can be specified by the kernel:

b_{13}	b_{23}	b_{33}
b_{12}	b_{22}	b_{32}
b_{11}	b_{21}	b_{31}

It is suggested that the integer parameters b_{ij} should not be identical (otherwise, this would define a box-filter, cp. Section 6.1.2) to avoid a strong low-pass effect. An example to *VAR* = 3 is given in Figure 4.6.

(3) Many modifications of the resampling variants 2 and 3, as discussed under point **(2)**, are possible. For example, variant 2 can be refined by a linear interpolation in both coordinate directions (*bilinear interpolation resampling*), using the inverse Euclidean distances to point

$$(a_{11} \cdot x + a_{12} \cdot y + a_{13}, a_{21} \cdot x + a_{22} \cdot y + a_{23})$$

as weights of the *f*-values. A 5 × 5 window, e.g., can also be used in Variant 3.

(4) *Control structure*: here explicitly given

```
var r, s : real;
begin
        for x := 1  to M   do
                for y := 1  to N  do begin
                        r := a11 · x + a12 · y + a13;
                        s := a21 · x + a22 · y + a23;
   if (r ≤ 0.5 or r ≥ M + 0.5 or s ≤  0.5 or s ≥ N + 0.5) then
        v := a certain default value, e.g. equal to 0
   else                                        {(r, s) is in the image domain of f}
        if ( VAR = 1 ) then
                v := f( integer(r + 0.5),  integer(s + 0.5) )
        else
                if (VAR = 2) then begin
                        xl := integer(r);        xr := xl +1;
                        yd := integer(s);        yu := yd + 1;
                        if (1 ≤ xl and xr ≤ M and 1 ≤ yd and yu ≤ N) then
                        begin
                           v := f(xl, yd) + f(xr, yd) + f(xl, yu) + f(xr, yu) ;
                           v := v / 4
                           end {then}
                        else v := f( integer(r + 0.5),  integer(s + 0.5) )
                        end {then}
                else    begin                                        {VAR = 3}
                        bsum : = 0;
                        for i = 1 to 3  do
                                for j = 1 to 3  do
```

```
                        bsum := bsum + b_ij ;
            v : = 0;
            xc := integer(r + 0.5);        yc := integer(s + 0.5);
            if (1 < xc < M and 1 < yc < N) then begin
                for i = xc – 1 to xc + 1 do
                    for j = yc – 1 to yc + 1 do
                        v := v + b_ij · f(i, j);
              {ADJUST in procedure TEXTURE in Section 2.1}
                ADJUST(v/bsum; v)
                end {then}
            else v := f( integer(r + 0.5), integer(s + 0.5) )
            end {else};
            h(x, y) := v
        end {for}
    end
```

(5) Haberäcker, P.: *Digitale Bildverarbeitung*. Carl Hanser Verlag, München, 3rd ed., 1989.

Russ, J.C.: *The Image Processing Handbook*. CRC Press, Boca Raton, 1992.

5 GRAY SCALE TRANSFORMATIONS AND POINT OPERATORS

Point operators have been characterized in Section 1.4.2. In general these are image transforms which can be realized with high time efficiency for point-wise modifications of image values. In the position-independent case, a straightforward definition of a point operator T is given by a *gray scale transformation t*. An operator T maps an input image f into a resultant image h. It is

$$T(f)\ (x, y) = t(f(x, y)),$$

for image points (x, y) of the raster $\mathbf{R} = \{(x, y): 1 \leq x \leq M \wedge 1 \leq y \leq N\}$. By means of this operator the gray value histogram of the input image is modified into a gray value histogram of the resultant image. The gray scale transformation t as well as the *histogram modification* can be used for interactive specifications of operators T. These specifications can be realized for image editing by a very general user surface.

Point operators are special local operators. In this sense the gray scale transformation or *gradation function t* can also be understood as operator kernel. Operator kernels t of point operators are always order independent because the relevant "window" consists of a single point. In principle a position dependence from the current image point $p = (x, y)$ can exist. In this general case, the function t is a mapping of image values $f(x, y)$ and of image coordinates (x, y), i.e.

$$T(f)\ (x, y) = t(x, y, f(x, y)).$$

At first, a general simple control structure can be provided for the algorithmic realization of point operators, cp. Sections 3.3.3 or Figure 3.9. This pixel access is highly recommended if images have to be processed row-wise, cp. **(A)** in Section 3.2.2. In case of direct access to all image data this control structure can be simplified by adopting a complete and repetition-free run through all image points of the raster **R**.

5.1 GRAY SCALE TRANSFORMATIONS

In this Section several point operators for gray value images are considered which are defined by a position dependent, or by a position independent gray scale transformation t. They are based on the standard control structure "point operator" of Figure 3.9. See also examples (1) to (3) in Section 1.3.3 for this type of point operators.

5.1.1 Gray Value Scaling in a Selected Region

(1) For this operator a rectangular region of interest can be selected where all gray values should be modified by a constant factor. The gray values of the remaining image can maintain their previous values or can be replaced by a constant gray value.

Attributes:

Image:	$M \times N$ gray value images
Operator:	point operator, homogenous
Kernel:	position dependent, logically structured

Inputs:

scaling factor $c \geq 0$,
Variant $VAR = 1$ for the whole image raster, i.e. position-independence, or $VAR = 2$ for a rectangular region of interest,
coordinates XA, XE, YA, YE of the rectangular region of interest,
gray value outside the rectangular region of interest: $HINT = 0$ for black, i.e. gray value zero, $HINT = 1$ for original gray value, $HINT = 2$ for white, i.e. gray value $G - 1$.

(2) For $XA \leq x \leq XE$ and $YA \leq y \leq YE$ let

$$h(x, y) = \begin{cases} c \cdot f(x, y) & \text{, if } 0 \leq c \cdot f(x, y) \leq G - 1 \\ G - 1 & \text{otherwise .} \end{cases}$$

Outside the specified rectangular region of interest it holds $h(p) = 0$, $h(p) = f(p)$, or $h(p) = G - 1$ as defined by the parameter $HINT$.

(3) The chosen definition equation ensures that only values less or equal $G - 1$ are generated. Further variants are possible, e.g. setting to "white" (gray value $G - 1$) the field outside the region of interest by assigning a negative value to c.

(4) *Control structure*: point operator in Figure 3.9

Inputs:

```
parameter input of c and VAR;
if (VAR = 1) then begin
        XA := 1;        XE := M;        YA := 1;        YE := N
        end {then}
else    begin
                input of XA, XE, YA, YE ;
                input of HINT := 0, 1 or 2;
        end {else}
```

Operator kernel:

```
                         {transform u = f(x, y) into v = h (x, y) , cp. Figure 3.9 }
if ((x ≥ XA) ∧ (x ≤ XE) ∧ (y ≥ YA) ∧ (y ≤ YE)) then
        if (c ≥ 0) then
                             { ADJUST in procedure TEXTURE in Section 2.1 }
                call ADJUST(c · u, v)
        else    v := G – 1
else begin
        if (HINT = 0) then v := 0;
        if (HINT = 2) then v := G – 1
        end {else}
```

5.1.2. Linear Stretching to the Full Gray Value Range

(1) A selected gray value interval $[u_1, u_2]$ is linearly stretched to the whole gray scale $[0, G-1]$. A piece-wise linear scaling results for all gray values.

Attributes:

Images:	$M \times N$ gray value image
Operator:	point operator, homogenous
Kernel:	position independent, logically structured

Inputs:

Gray values u_1 and u_2, with $u_1 < u_2$ and $0 \le u_1, u_2 \le G - 1$.

(2) For all image points $p = (x, y)$ of the image raster it is

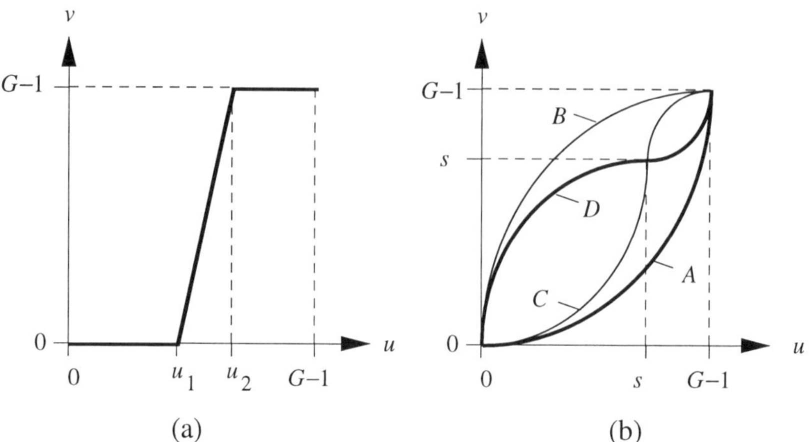

Figure 5.1: (a) Piece-wise linear gradation function. (b) Variant 1 with curves A $(r > 1)$ and B $(r < 1)$; and Variant 2 with S-shaped gradation functions C $(r > 1)$ and D $(r < 1)$ with inflexion point at gray value s, cp. Section 5.1.3.

$$h(p) = \begin{cases} 0 & , \text{ if } f(p) < u_1 \\ \dfrac{f(p) - u_1}{u_2 - u_1} \, (G - 1) & , \text{ if } u_1 \leq f(p) \leq u_2 \\ G - 1 & , \text{ if } f(p) > u_2 \, . \end{cases} .$$

This equation defines a *piece-wise linear gradation function* as shown in Figure 5.1 (a).

(3) A gray value stretching is very useful for improving *low contrast images*, in order to exploit the whole gray scale. Even under favorable image acquisition conditions, it results often that only a subrange of the gray value scale, say between u_1 and u_2, is non-empty. These two gray values can be derived from the global gray value histogram. For camera-acquired images, this effect can be due to the scene illumination or to the cameras automatic gain control. Low contrast is sometimes a systematic error for images taken under constant conditions. Then, the values u_1 and u_2 need to be estimated only once.
Gray value equalization, see Section 5.1.4, as well as stretching result in an extension of the exploited gray value domain to the full gray value scale. Unlike with equalization, after stretching the gray value histograms shape is similar to that of the original histogram.

(4) *Control structure*: point operator in Figure 3.9

Inputs:

```
Input of u1 and u2
```

Operator kernel:

```
                        {transform u = f(x, y)  in  v = h (x, y) , cp. Figure 3.9 }
if (u < u1)      then   v := 0
else   if (u > u2)    then    v := G – 1
       else                                    { ADJUST in Section 2.1 }
                 call ADJUST( [(u – u1) / (u2 – u1) ] · (G – 1), v)
```

(5) Zamperoni, P.: *Methoden der digitalen Bildsignalverarbeitung*. Vieweg Verlag, Wiesbaden, 2nd ed., 1991.

5.1.3 Variations of Gradation Functions

(1) This operator performs different gray value transformations, specified by different gradation functions. The Variant 1 is defined by a power function with arbitrary exponent, and the Variant 2 by an S-shaped characteristic curve with inflexion point at gray value s. The Variant 2 causes, in case (a) a stretching of the gray values close to s, and a squeezing of the gray values at both ends of the scale. In case (b) it causes a squeezing of the gray values close to s, and a stretching of the gray values at both ends of the scale.

Attributes:

Images:	$M \times N$ gray value image
Operator:	point operator
Kernel:	analytical (exponential), position independent

Inputs:

Exponent $r \geq 0$ (real number) of the power function, for both variants
Variant $VAR = 1$ or $VAR = 2$,
inflexion point s in case of Variant 2, with$0 \leq s \leq G - 1$.

(2) Assume normalized gray values in the interval [0, 1],

$$U = \frac{f(p)}{G-1} \qquad \text{and} \qquad S = \frac{s}{G-1}\,.$$

In case of Variant 1 let

$$h(p) = (G - 1) \cdot \left[\frac{f(p)}{G - 1}\right]^{r}$$

and in case of Variant 2 let $h(p)$ be either equal to $v_1(U)$ or equal to $v_2(U)$, with

$$h(p) = \begin{cases} v_1(U) = (G - 1)\dfrac{U^r}{S^{r-1}} & \text{, if } 0 \leq U \leq S \\ v_2(U) = (G - 1)\left(1 - \dfrac{(1 - U)^r}{(1 - S)^{r-1}}\right) & \text{, if } S \leq U \leq 1 . \end{cases}$$

(3) In case of Variant 1, the quotient is a number $U \leq 1$, representing the gray value $f(p)$ after normalization to the interval [0, 1]. Depending upon whether r is larger or smaller than 1, the power function is associated with a convex or a concave characteristic curve. As shown in Figure 5.1 (b), curves A and B, in the first case high gray values (bright image elements) are stretched and low gray values (dark pixels) are squeezed, and vice-versa in the second case.

By the use of the Variant 2, the transformation curves are generated as shown in Figure 5.1 (b), curves C and D. In contrast to the operator of Section 5.1.2 for $r > 1$, a smooth stretching occurs for gray values near to s, and a squeezing at both ends of the gray value scale. The inverse effect follows from $r < 1$. For obtaining a continuous shape of the transformation curve, the functions v_1 and v_2 have been chosen such as to meet the following conditions:

$$v_1(S) = v_2(S) = s \qquad \text{and} \qquad v'_1(S) = v'_2(S) = r \cdot (G - 1)$$

for the first derivatives. The steepness of the transformation curve at the inflexion point may be selected by choosing the input exponent r.

(4) *Control structure*: point operator in Figure 3.9

Inputs:

```
Parameter input of exponent r : real;
input of variant VAR ;
if (VAR = 2) then begin
    input of the inflexion point s, 0 ≤ s ≤ G – 1;
    S := s / (G – 1)
    end {if}
```

Operator kernel:

{transform $u = f(x, y)$ into $v = h(x, y)$, cp. Figure 3.9 }

if $(VAR = 1)$ **then**

$$w := (G-1) \cdot \left[\frac{u}{G-1}\right]^r$$

else begin

$U := u/(G-1);$

if $(u \leq s)$ **then** $w := (G-1) \cdot \frac{U^r}{S^{r-1}}$

else $w := (G-1) \cdot \left(1 - \frac{(1-U)^r}{(1-S)^{r-1}}\right)$

end {*else*};

{ *ADJUST* in procedure *TEXTURE* in Section 2.1 }

call $ADJUST(w, v)$

(5) Haberäcker, P.: *Digitale Bildverarbeitung*. Carl Hanser Verlag, München, 3rd ed., 1989.

Zamperoni, P.: *Methoden der digitalen Bildsignalverarbeitung*. Vieweg Verlag, Wiesbaden, 2nd ed., 1991.

5.1.4 Gray Value Histogram Equalization

(1) This operator modifies the gray values of an image so that they are about uniformly distributed after the transformation. The transformation of the original gray values $u = f(x, y)$ into the resultant gray values $v = h(x, y)$ is carried out by means of a gradation function $v = t(u)$. The function t is computed on the base of the given input image, cp. Section 1.3.3.

It is also possible to grade the equalization intensity by means of an additional parameter r.

Attributes:

Images:	$M \times N$ gray value image
Operator:	point operator, parameters defined by input image
Kernel:	logically structured

Inputs:

equalization exponent r.

(2) *HIST*(u) denotes the gray value histogram which has to be computed in a first run through the whole image f, for $u = 0, ..., G - 1$. *HIST* has been defined in Section 1.3.2. In a second run through f, the gray value transformation $v = t(u)$ is carried out, where

$$v = \frac{G-1}{Q} \sum_{w=0}^{u} HIST(w)^{r} \qquad \text{with} \qquad Q = \sum_{w=0}^{G-1} HIST(w)^{r} .$$

(3) The gray value transformation needed for the equalization can be derived from the gray value distribution function of the original image. In image processing, there is only a finite number of $M \times N$ image points and of G values. The gray value histogram is an estimate of the density function of the *global gray value distribution*. The sum histogram is an estimate of the distribution function, cp. Section 1.3.2. Since there is only a discrete number of gray levels and a finite number of samples, the gray value histogram of the transformed image has no ideal uniform distribution, but only in average. There can be gaps and peaks with different heights.

Two runs through the image data are necessary for the gray value equalization. In the first run, the histogram *HIST*(u) of the gray values u is computed. Then, following the equation given above, a *look-up-table* is generated for mapping the original gray value u onto the transformed gray value v, $0 \leq u, v \leq G - 1$. In a second run, this look-up-table is used for gray value transformation.

A stronger or weaker equalization can be obtained by adjusting the exponent r. The resultant histogram is uniformly distributed for $r = 1$. For $r > 1$, sparse gray values of the original image will occur more often than in the equalized image. The identical transformation follows from $r = 0$. A weaker equalization in comparison to $r = 1$ is obtained for $r < 1$.

The gray value range equalization may be used for improving the image quality if the original image covers only a part of the full gray scale. An insufficient exploitation of the full gray scale is mostly due to image acquisition circumstances, as e.g. low scene illumination or automatic gain control of the camera. In case of good gray value dynamics in the input image, an equalization can lead even to quality losses in form of unsharp edges.

(4) *Control structure*: point operator in Figure 3.9, two runs
Required data array: $q(1...G)$: *real* for the frequencies of the gray values $u = 0, ..., G - 1$. After the first run this data array contains the absolute (*integer*) value *hist*(u) in the position u+1. The gray value scale starts with gray value 0, but the array indexing starts at 1. Before the second run, a look-up-table is generated and it overwrites this data array. This look-up-table is used in the second run for the gray value transformation.

Input and initialization:

```
input of exponent r > 0 ;
for u := 1 to G do   q(u) := 0
```

Operator kernel (first run):

```
                          {read u = f(x, y) , write no value v in h, cp. Figure 3.9 }
q(u + 1) := q(u + 1) + 1
```

Preprocessing before the second run:

```
A := M · N;
for u := 1 to G  do q(u) := ( q(u) / A ) ^r ;
for u := 2 to G  do q(u) := q(u) + q(u – 1) ;
for u := 1 to G  do q(u) := (G – 1) · (q(u) / q(G))
```

Operator kernel (second run):

```
                          {transform u = f(x, y) in v = h (x, y) , cp. Figure 3.9 }
                                                        { ADJUST in Section 2.1 }
call ADJUST( q(u + 1), v)
```

(5) Fundamentals of gray value equalization are described, e.g., in

Ballard, D.H., Brown, C.M.: *Computer Vision*. Prentice-Hall, Englewood Cliffs, 1982.

Gonzalez, R.C., Wintz, P.: *Digital Image Processing*. Addison-Wesley, Reading, USA, 1987.

Zamperoni, P.: *Methoden der digitalen Bildsignalverarbeitung*. Vieweg Verlag, Wiesbaden, 2nd ed., 1991.

The role of the exponent r is extensively discussed in

Yaroslavsky, L.P.: *Digital Picture Processing*. Springer Series in Information Science, Springer-Verlag, Berlin, 1985.

in relation to strong or weak equalization.

5.2 GENERATION OF NOISY IMAGES

This Section presents some operators for the impairment of given input images by means of different types of *synthetic noise*. Sometimes this is a desired operation for image editing. Images with controlled impairments can also be used as test images for investigating noise-reducing operators, as e.g. smoothing or rank-order operators.

5.2.1 Generation of Spike Noise

(1) This operator adds to a given image negative or positive *spike noise interferences* in about the same ratio. The *average noise ratio* can be selected.

Attributes:

Images:	$M \times N$ gray value image
Operator:	point operator
Kernel:	position dependent, logically structured with dependency upon a random number

Inputs:

noise occurence ratio Q in average, each Q-th image point is noisy,
impulse height H, $0 \le H \le G-1$.

(2) Let z be a real random number, uniformly distributed between 0 and 1. A *random number generator* produces the next random number $z(p)$ for each consecutive image point $p = (x, y)$. Then, let

$$h(p) = \begin{cases} \mathbf{max}\{\, 0, f(p) - H\} & \text{, if } z(p) < q_{neg} \\ f(p) & \text{, if } q_{neg} \le z(p) \le q_{pos} \\ \mathbf{min}\{G-1, f(p) + H\} & \text{, if } z(p) > q_{pos} \end{cases}$$

for $q_{neg} = \dfrac{1}{2Q}$ and $q_{pos} = 1 - q_{neg}$.

(3) A software function *RANDOM* or *RND_EQU* is used in the program as random number generator. Such a function, generating a real, uniformly distributed random number z between 0 and 1 at each call, is available in most programming languages. Cp. Section 3.4.1 for a specific realization. At each image point a new z-value is generated and compared with the thresholds q_{neg} and q_{pos} as defined above. Since z is uniformly distributed, in average each $2Q$-th image point is affected by a negative noise impulse (original gray value minus impulse height H), and each $2Q$-th image point is affected by a positive noise impulse (original gray value plus impulse height H).

(4) *Control structure*: point operator in Figure 3.9

Inputs and initialization:

```
input of Q , for example 1 ≤ Q ≤ 100;
input of H;
q_neg := 0.5 / Q ;          q_pos := 1 − q_neg
```

Operator kernel:

```
                      {transform u = f(x, y)  in v = h (x, y), cp. Figure 3.9 }
v := u;          z := RND_EQU();
if (z < q_neg) then     v := max{ 0, u − H };
if (z > q_pos) then     v := min{ G − 1, u + H }
```

5.2.2 Generation of Images with Additive Random Noise

(1) Here the task is to generate additive, uniformly or normally distributed noise for impairing the original image.

Attributes:

Images:	$M \times N$ gray value image
Operator:	point operator
Kernel:	position dependent, logically structured with dependency upon a random number

Inputs:

Type of distribution: $VAR = 1$ for uniform distribution and $VAR = 2$ for normal distribution

Standard deviation s for the normal distribution, or range s for the uniform distribution.

(2) For the generated image h it holds

$$h(p) = \begin{cases} f(p) - \frac{s}{2} + s \cdot r_{equ} & \text{, if } VAR = 1 \\ f(p) + s \cdot r_{norm} & \text{, if } VAR = 2. \end{cases}$$

In this equation, $0 \leq r_{equ} \leq 1$ denotes a uniformly distributed random number and r_{norm} is a normally distributed random number with expected value 0 and variance 1.

(3) For developing and testing image enhancement methods for noise suppression, the generation of impaired test images with different noise distributions can be of interest. With this operator, original images can be impaired by additive uniformly or normally distributed noise.

For realizing uniformly distributed random numbers between 0 and 1, a suitable standard function *RANDOM* or *RND_EQU* , cp. Section 3.4.1, is available in most programming languages.

If a normal distribution is needed, uniformly distributed numbers can be used as input for generating the desired distribution. In **(4)** the procedure *RND_NORM* of Section 3.4.2 is used, which delivers pairs (z_1, z_2) of normally distributed numbers. Utilizing z_1 or z_2 leads to statistically equivalent normally distributed noise patterns.

The resultant real numbers in the equation in **(2)** are rounded up to integers or set uniformly to 0 or $G - 1$ by the procedure *ADJUST* of Section 2.1.

(4) *Control structure*: point operator in Figure 3.9

Inputs and initialization:

```
input of variant VAR = 1 or VAR = 2;
input of standard deviation s, e.g. 0 ≤ s ≤ 40;
initialization of RANDOM or of the global variable rand of
                                               procedure RND_EQU
```

Operator kernel:

```
                    {transform u = f(x, y) in v = h (x, y) , cp. Figure 3.9 }
if (VAR = 1) then
        w := u - s/2 + s · RND_EQU()
else    begin
                call RND_NORM(z1, z2);
                w := u + s · z1
                end {else};
call ADJUST(w, v)
```

(5) In the literature there exist different proposals for the considered transformation, cp. for example

Ahrens, J.H., Dieter, U.: *Extension of Forsythe's method for random sampling from the normal distribution*. Mathematics of Computation **27** (1973), pp. 927 - 937.

Knuth, D.E.: *The Art of Computer Programming. Vol. 2: Seminumerical Algorithms*, Addison-Wesley, Reading, USA, 1969.

The method described in

Press, W.H., Flannery, B.P., Teukolsky, S.A., Vetterling, W.T.: *Numerical Recipes*. Cambridge University Press, Cambridge, USA, 1988.

has been selected here because of its clarity.

5.3 BINARIZATION OF GRAY VALUE IMAGES

This Section deals with mappings of gray value images into binary or bilevel images. Such mappings can be of interest for image output (*halftone image*) or for binary representations of image structures. Furthermore, *binarization* can be understood as image improvement if the images consist of intrinsically binary patterns as text or drawings, and if these binary patterns are degraded by assuming intermediate gray values during the process of image acquisition.

5.3.1 Binarization with Hysteresis

(1) This operator realizes a binarization of gray value images with (in principle) constant threshold. The threshold is provided with hysteresis for suppression of digitization noise.

Attributes:

Images:	$M \times N$ gray value image
Operator:	window operator, sequential[1]
Kernel:	logically structured

Inputs:

binarization threshold s, $0 \leq s \leq G-1$,
hysteresis parameter L, for example $0 \leq L \leq 10$.

(2) Bilevel images h, with image values 0 and $G-1$, are generated by means of binarization. For a placed window $\mathbf{F}(p) = \{p, p_1, p_2, p_3\}$ with $p = (x, y)$, $p_1 = (x-1, y)$, $p_2 = (x-1, y-1)$ and $p_3 = (x, y-1)$, and for sequential processing, previously processed neighbors p_1, p_2 and p_3 of the current point $p = (x, y)$ can only have gray values 0 or $G-1$. Let

$$h(p) = \begin{cases} 0 & \text{, if } f(x, y) < s + \delta \cdot L \\ G-1 & \text{otherwise} \end{cases} ,$$

where

---

[1] This operator is not a point operator for a hysteresis parameter $L \neq 0$, because the binarization result of previously transformed picture points influences the result of the current picture point. Thus, the attribute "sequential" is justified. Since the common binarization (without hysteresis) is a point operator, this operator is dealt with in this Chapter 5 for reasons of clearness.

$$\delta = \begin{cases} +1 \text{ , if } h(x-1, y) + h(x-1, y-1) + h(x, y-1) < 2 \cdot (G-1) \\ -1 \quad \text{otherwise .} \end{cases}$$

(3) The introduction of hysteresis into the common binarization operator aims at suppressing full-scale ($G - 1$) inconsistent oscillations of the binarized output, occurring for gray values close to the threshold s. Such variations can occur if an image is raster-scanned line-by-line, especially at horizontal edges. The hysteresis reduces or increases the threshold depending on whether the majority of the current point's previously processed neighbors has been assigned the value $G - 1$ or 0. In other words: at binarization, each transition from 0 to $G - 1$ or vice-versa is affected by a certain degree of inertia. Figure 5.2 illustrates the effect of this operator upon an one-dimensional function $f(x)$ where the current threshold depends only upon the last processed value $h(x - 1)$.

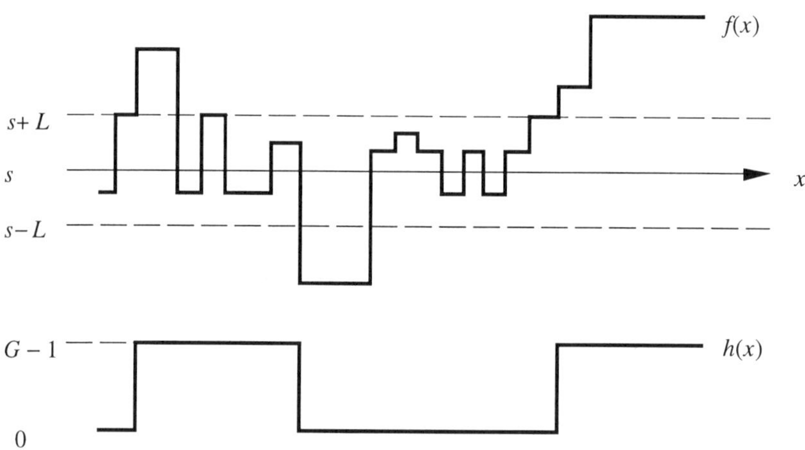

Figure 5.2: Illustration of the hysteresis effect in the binarization of the one-dimensional signal $f(x)$ shown above. Below the resultant binary signal $h(x)$ is given for the threshold s and the hysteresis interval $[s - L, s + L]$.

The realization of this operator requires the use of a data array for storing the already binarized gray values $h(x - 1, y - 1)$ and $h(x, y - 1)$ of the previous line. A Boolean array would be sufficient, because the binarization process represents a choice between only two values. In practice, an integer data array $g(1...M)$ can be used for the binarized previous line. Some binarization results, with and without hysteresis, are shown in the lower half of Figure 5.3.

(4) *Control structure*: special window operator, given explicitly
Required data array: line store array $g(1...M)$ for the binarized gray values of the previous line,
line store arrays $BUF(1...M)$ and $BUFOUT(1...M)$ for input and output.

```
begin {binarization with hysteresis}
    input of threshold s, 0 ≤ s ≤ G – 1;
    input of hysteresis parameter L ;
    for x := 1 to M do g(x) := 0 ;
    for y := 1 to N do begin
        read line y of input image into BUF(1...M);

        for x := 2 to M do begin
            SC := s + L ;
            if ((g(x) + g(x – 1) + BUF(x – 1)) > G – 1)
                then SC := s – L;
            if (BUF(x) > SC) then begin
                BUFOUT(x) := G – 1 ;        g(x) := G – 1
                end {then}
            else begin
                BUFOUT(x) := 0 ;            g(x) := 0
                end {else}
            end {for}

        write BUFOUT(1...M) into line y of output image
        end {for}
end {binarization with hysteresis}
```

(5) Zamperoni, P.: *Methoden der digitalen Bildsignalverarbeitung*. Vieweg Verlag, Wiesbaden, 2nd ed., 1991.

5.3.2 Recursive Binarization

(1) This operator realizes a binarization of gray value images by iterative computation of the binarization threshold. The determination of the binarization threshold can be performed with better time efficiency by using gray value histograms instead of image data.

Attributes:

Images:	$M \times N$ gray value image
Operator:	point operator, process controlled, in several runs
Kernel:	logically structured

Inputs:

convergence parameter $\varepsilon \geq 0$.

(2) The image *f* is binarized using an iteratively determined threshold *s*. Let

$$h(p) = \begin{cases} G-1 & \text{, if } f(p) > s \\ 0 & \text{otherwise .} \end{cases}$$

A constant threshold *s* is calculated by an iterative search based on the global gray value histogram *HIST*(*u*), cp. Section 1.3.2, with $0 \leq u \leq G{-}1$ and

$$0 \leq HIST(u) \leq 1 \qquad \text{and} \qquad \sum_{u=0}^{G-1} HIST(u) = 1 .$$

Let $s_0 = G/2$ be the initial threshold value. Let s_k be the value after *k*-th iteration. The search process stops if $|s_k - s_{k-1}| < \varepsilon$ is satisfied for a given value of the parameter ε, e.g. $\varepsilon = 3$ gray levels. Then, set $s := s_k$. In each iteration the average gray values μ_0 and μ_1 are computed from those image points having gray values less or equal s_{k-1}, or greater than s_{k-1}, respectively. Then, let

$$\begin{aligned} s_k &= \frac{\mu_0 + \mu_1}{2} \\ &= \frac{1}{2} \Bigg(\left[\sum_{u=0}^{s_{k-1}} u \cdot HIST(u) \right] \cdot \left[\sum_{u=0}^{s_{k-1}} HIST(u) \right]^{-1} \\ &\qquad + \left[\sum_{u=s_{k-1}+1}^{G-1} u \cdot HIST(u) \right] \cdot \left[\sum_{u=s_{k-1}+1}^{G-1} HIST(u) \right]^{-1} \Bigg) \end{aligned}$$

The binarization with this iteratively determined threshold *s* can also be provided with *hysteresis*, cp. Figure 5.3 top, right.

(3) The basic idea is that of positioning the threshold in the middle of the gray values μ_0 and μ_1. These are the average gray values of the image points assigned to the background and to the object region after binarization. Since the values of μ_0 and μ_1 depend upon a previous threshold, a number of iterations is necessary for accomplishing this search process. Typically the process converges after a few iterations.

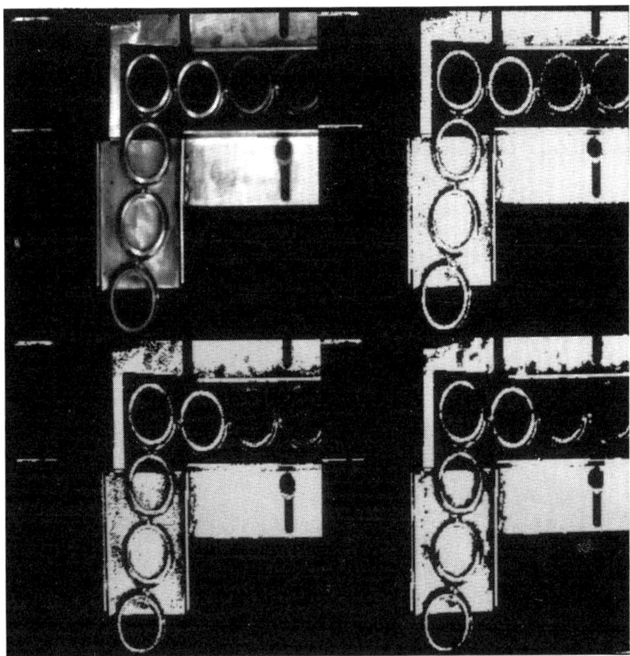

Figure 5.3: Results of different binarization methods of a gray value image shown top left. Recursive binarization of Section 5.3.2 leads to the threshold $s = 75$ for this image. By discriminant analysis, cp. Section 5.3.3, the threshold s = 74 follows for this image. Top right: recursive method. Below left: constant threshold $s = 95$ as determined by best visual subjective impression, without hysteresis. Below right: threshold $s = 95$ with hysteresis, $L = 15$.

This method leads to a good separation between the objects and the background in several applications, especially if both form regions of nearly the same size. However it is disadvantageous that no parameter adjustment is possible for matching with specific images.

For speeding up the computation, the search of the binarization threshold can be carried out on the gray value histogram. This histogram is computed in a first run. After determining a first threshold based on this histogram, the same histogram can be used for the further calculations of μ_0 and μ_1.

(4) *Control structure*: point operator in Figure 3.9

Required data array: $HIST(1...G)$ for absolute and relative frequencies of gray values $0, ..., G-1$.

Input and initialization:

```
input of ε, for example ε = 3;
for u := 1  to G  do HIST(u) := 0
```

Operator kernel (first run- histogram calculation):

```
                          { u = f(x, y) , no new value v for h, cp. Figure 3.9 }
HIST(1 + u) := HIST(1 + u) + 1
```

Determination of the binarization threshold:

```
for u := 1  to G  do HIST(u) := HIST(u) / (M · N);
s := G/2 ;                                {initial value of the threshold}
repeat
        s_a := s;          μ_0 := 0;        Σ_0 := 0;
        for u := 1  to s  do begin
                μ_0 := μ_0 +  u · HIST(u) ;      Σ_0 := Σ_0 + HIST(u)
                end {for};
        μ_1 := 0;         Σ_1 := 0;
        for u := s + 1  to G  do
                μ_1 := μ_1 + u · HIST(u) ;       Σ_1 := Σ_1 + HIST(u)
                end {for};
        if ( Σ_0 ≠ 0 and Σ_1 ≠ 0 ) then
                s := 0.5 · ( μ_0 / Σ_0  +  μ_1 / Σ_1 )
        else if ( Σ_0 ≠ 0 )     then   s := μ_0 / Σ_0
                                else   s := μ_1 / Σ_1
until ( | s − s_a | ≤ ε )
```

Operator kernel (second run - binarization):

```
                          { u = f(x, y)  and h (x, y) = v , cp. Figure 3.9 }
v := 0;
if (u > s )  then v  := G − 1
```

(5) The basic idea of this method was published by

Ridler, T.W., Calvard, S.: *Picture thresholding using an iterative selection method.* IEEE Trans. SMC-**8** (1978), pp. 630-632.

In

Magid A., Rotman, S.R., Weiss, A.M.: *Comment on "Picture thresholding using an iterative selection method".* IEEE Trans. SMC-**20** (1990), pp. 1238-1239.

it is shown that the resultant threshold s minimizes the average square deviation e,

$$e = \sum_{u=0}^{s} (u - \mu_0)^2 \, HIST(u) + \sum_{u=s+1}^{G-1} (u - \mu_1)^2 \, HIST(u) \,,$$

where the deviations are considered from both the gray values of the resultant image. The fast search of a binarization threshold in the gray value histogram was proposed by

Trussell, H.J.: *Comments on "Picture thresholding using an iterative selection method".* IEEE Trans. SMC-**9** (1979), p. 311.

5.3.3 Binarization Based on Discriminant Analysis

(1) This method considers the binarization as a problem of non-supervised *classification of pixels* into two classes, ω_0 for the (dark) background and ω_1 for the (bright) objects. The classification is based on the global gray value histogram. In this one-dimensional gray value space the calculated *discriminant function* is a constant s which is used as the binarization threshold. Due to the lack of a-priori information, s is determined only on the base of the gray value histogram by means of an optimality criterion known from the field of discriminant analysis.

Attributes:

Images: $M \times N$ - gray value images
Operator: point operator, process controlled, in several runs

(2) The computed threshold s is used for binarization,

$$h(p) = \begin{cases} G - 1 & , \text{ if } f(p) > s \\ 0 & \text{otherwise} . \end{cases}$$

The gray value s is used as a constant threshold for separating the two pixel classes. With respect to the predefined optimality criterion and to the global histogram of the input image, these two classes are maximally separated.

(3) The optimality criterion is as follows: the gray values of classes ω_0 and ω_1 should form separated clusters in the histogram. For a formal representation of this criterion, let

$$q(w) = SUMHIST(w) = \sum_{i=0}^{w} HIST(w) \quad \text{and} \quad \mu(w) = \sum_{i=0}^{w} w \cdot HIST(w) \,,$$

cp. Section 1.3.2. Thus $q(w)$ is an estimate of the gray value distribution, and $\mu(w)$ is the gray value averaged over the gray value domain 0, ..., w. Let $q_0(w)$ and $q_1(w) = 1 - q_0(w)$ be the absolute gray value probabilities of ω_0 and ω_1, respectively. Then,

$$\mu_0(w) = \mu(w) / q_0(w) \qquad \text{and} \qquad \mu_1(w) = \mu(w) / q_1(w)$$

denote the average gray values taking account of the conditional gray value probabilities. It is

$$\mu_1(w) = \frac{\mu_T - \mu(w)}{1 - q_0(w)} \quad \text{with} \quad \mu_T = q_0(w)\mu_0(w) + q_1(w)\mu_1(w),$$

where μ_T is the average gray value of $\omega_0 \cup \omega_1$. All these values, as well as the variances $\sigma_0^2(w)$ and $\sigma_1^2(w)$ of both classes, depend upon the parameter w. Only μ_T and the variance σ_T^2 of the total distribution are independent of the parameter w.

An optimality criterion, as used in discriminant analysis, is defined by maximizing of the quotient of $\sigma_B^2(w)$ (the variance between the two classes) and σ_T^2. This criterion is used for the calculation the binarization threshold. It holds

$$\begin{aligned} \sigma_B^2(w) &= q_0(w)\,(\mu_0(w) - \mu_T)^2 + q_1(w)\,(\mu_1(w) - \mu_T)^2 \\ &= \frac{[\mu_T q_0(w) - \mu(w)]^2}{q_0(w)\,[1 - q_0(w)]} \;. \end{aligned}$$

This criterion is equivalent to maximizing $\sigma_B^2(w)$, i.e.

$$s: \quad \sigma_B^2(s) = \max_{w=0...G-1} \quad \sigma_B^2(w) \quad .$$

To realize this operator, the relative sum histogram $SUMHIST(0...G-1)$ is computed and stored in an array $q(1...G)$. Then the binarization threshold s is determined by maximizing $\sigma_B^2(w)$, for $w = 0, ..., G-1$. The binarization is performed in a second run through the image.

(4) *Control structure*: point operator in Figure 3.9
Required data array: $q(1...G)$ for relative histograms

Initialization:

```
for w := 1 to G  do  q(w) := 0
```

Operator kernel (first run - histogram computation):

```
                                 { u = f(x, y) , no new value v in h, cp. Figure 3.9 }
q(1 + u) := q(1 + u) + 1
```

Computation of the binarization threshold

```
μ_T := 0;
for w := 1 to G do begin
      q(w) := q(w) / (M · N);        μ_T := μ_T + w · q(w)
      end {for}
Q_a := 0;        μ_a := 0;        MAX := 0;
for w := 1 to G do begin
      Q := Q_a + q(w);        μ := μ_a + w · q(w);
      T := [(μ_T · Q  –  μ)^2] / [Q · (1 – Q)] ;
      if ( T > MAX ) then begin
            s := w;            MAX := T
            end {if}
      Q_a := Q;        μ_a := μ
      end {for}
```

Operator kernel (second run - binarization):

```
                                 { u = f(x, y)  and h (x, y) = v , cp. Figure 3.9 }
v := 0;
if (u  > s )  then v  := G – 1
```

(5) The method has been described in

Otsu, N.: A *threshold selection method from gray-level histograms*. IEEE Trans. SMC-**9** (1979), pp. 62 - 66.

For the used optimality criterion see

Fukunaga, K.: *Introduction to Statistical Pattern Recognition*, 2nd ed. Academic Press, New York, 1990.

5.3.4 Halftoning by Means of a Threshold Matrix

(1) Let us consider gray value images with a high spatial resolution. The aim of *halftoning* is to produce about the same visual subjective impression as the original gray shade image, after performing a convenient position-dependent thresholding, cp. Figure 5.4. This operator is relevant for binary image output.

Attributes:

Images:	$M \times N$ gray value images with $G = 256$
Operator:	point operator, position dependent

(2) Ordered dithering was described in Section 2.2. Here thresholds are given by a 4×4 *threshold matrix*. This matrix is used in periodic, non-overlapping repetition on the image raster **R**. The following threshold matrix applies for $G = 256$,

$$\mathbf{T} = \begin{bmatrix} 0 & 128 & 32 & 160 \\ 192 & 64 & 224 & 96 \\ 48 & 176 & 16 & 144 \\ 240 & 122 & 208 & 80 \end{bmatrix}.$$

For the given spatial coordinates x, y of the considered image point, at first the matrix coordinates

$$i = x \ \mathbf{mod}\ n \quad \text{and} \quad j = y \ \mathbf{mod}\ n$$

are computed, and then the resultant gray value of a bilevel image h is calculated by

$$h(x, y) = \begin{cases} 0 & \text{, if } f(x, y) \le \mathbf{T}(i, j) \\ G - 1 & \text{, if } f(x, y) > \mathbf{T}(i, j) \end{cases}$$

(3) The selected threshold matrix can be modified in its size or in its numerical values for different values of G. Some general fundamentals have been given in Section 2.2. It can be advantageous to perform a gray value scaling and an image sharpening (contrast enhancement) before carrying out the halftoning.

(4) *Control structure*: point operator in Figure 3.9

Inputs:

```
DATA T(0, 128, 32, 160, 192, 64, 224, 96,
       48, 176, 16, 144, 240, 122, 208, 80)
                                    { indexed from 1 to 16 }
          { values are scaled for the input gray values G = 0, ..., 255 }
```

Operator kernel:

```
               { transform u = f(x, y) into v = h (x, y) , cp. Figure 3.9 }
i = x mod 4;          j = y mod 4;          s := i + 1 + (3 - j) · 4;
```

{matrix **T** in xy-coordinate system, see Figure 1.1}

if ($u \leq$ **T**(s)) **then** $v := 0$ **else** $v := 255$

(5) Ulichney, R. A.: *Digital Halftoning*. MIT Press, Cambridge, MA, 1987.

Figure 5.4: Halftone representation of a gray value image (top left). Top right: with the threshold matrix method (5.3.4). Below left: binarization with a constant threshold 130 (5.3.1). Below right: with the error distribution method (6.1.8).

5.4 POINT-TO-POINT OPERATIONS BETWEEN IMAGES

In this Section, a background function is first considered in Sections 5.4.1 and 5.4.2 as a second image. This background function can be, e.g., the result of *shading*. After computing such a backround function it can be, e.g., subtracted from the original image. Then, Section 5.4.3 deals with several combinations of two (arbitrary) input images.

5.4.1 Synthetic Background Compensation

(1) The spurious background function can be modeled by a slowly-varying gray value function in some cases. This operator compensates such a background function. For this aim, it subtracts a gray value pattern from the original image. This gray value pattern is synthetically generated and it is modeled by means of a Gaussian function depending on input parameters.

Attributes:

Images:	$M \times N$ gray value images or bilevel images
Operator:	point operator
Kernel:	analytical

Inputs:

parameters of the background function: coordinates XC and YC of the maximum, amplitude A of the maximum, exponent K of the Gaussian function,

variant $VAR = 1$ for the original image minus the background, or $VAR = 2$ for computing just the background function

(2) For the resultant image h it is

$$h(p) = \begin{cases} f(p) - g(p) & \text{, if } VAR = 1 \\ g(p) & \text{, if } VAR = 2 \end{cases}$$

where

$$g(p) = A \cdot \mathbf{exp}\left\{-\frac{K[(x - XC)^2 + (y - YC)^2]}{E^2}\right\}.$$

The parameter E^2 represents the *variance of the Gaussian function* and is defined as follows,

$$E^2 = [\mathbf{max}(XC, M - XC)]^2 + [\mathbf{max}(YC, N - YC)]^2 \ .$$

(3) This operator realizes a compensation of the background gray values. Nonhomogenous scene illumination or different sensitivity between the image middle and the image border of the imaging system (optics, CCD array etc.) can be reasons for inhomogeneous background functions. In fact, the disturbances are often multiplicative. However in such cases additive correction signals can also lead to satisfactory background compensations. The experiments have shown that the distribution of the background gray values can simply and efficiently be modeled by a Gaussian function in some cases, where height, variance and positioning in the image raster have to be specified by parameters.

The *contrast* of the correcting gray value function can be adjusted by the parameter A. The parameters XC and YC are the coordinates of the maximum of $g(p)$ which can be also displaced with respect to the image raster, e.g. in case of inhomogenous scene illumination. The parameter K determines the decreasing rate of the Gaussian function. It is related to the strength of the background variation which has to be compensated.

The value E is the maximum distance of the point (XC, YC) to the image border. For $M = N = 512$ and $K = 4.5$, the correction function is practically close to zero in that image corner which has a maximum distance to (XC, YC). Smaller values of K result into a more regular shape of the synthetic background.

(4) *Control structure*: point operator in Figure 3.9

Inputs and initialization:

```
parameter input of XC, YC, A and K;
input of variant VAR = 1 or VAR = 2;
MX := max(XC, M – XC);          MY := max(YC, N – YC);
EQ := MX² + MY²
```

Operator kernel:

```
                        { transform u = f(x, y) in v = h (x, y) , cp. Figure 3.9 }
```

$$w := A \cdot \mathbf{exp}\left\{-\frac{K[(x - XC)^2 + (y - YC)^2]}{EQ}\right\};$$

```
                                              { ADJUST from Section 2.1 }
call ADJUST(w, v);
if ( VAR = 1 ) then v := u – v;
if ( v < 0 )     then v := 0
```

(5) Ballard, D.H., Brown, C.M.: *Computer Vision*. Prentice-Hall, Englewood Cliffs, 1982.

Voß, K.: Shadingkorrektur in der automatischen Bildverarbeitung. Bild und Ton **36** (1983), S. 170-172.

5.4.2 Piece-wise Linear Background Subtraction

(1) The goal of this operator consists of extracting a background component of the gray value function. This component is estimated by a piece-wise linear approximation. By subtracting this background function from the input image in a certain ratio, the background compensation can be realized with a variable *contrast enhancement*.

Attributes:

Images:	$N \times N$ gray value image or two-channel images
Operator:	point operator, data driven, two runs
Kernel:	position dependent, analytical

Inputs:

square size L of the regular image raster partition in $L \times L$-squares for piece-wise linear approximation,

variant $VAR = 1$ for computing the weighted difference between original image and background, or $VAR = 2$ for computing just the background function

in case of $VAR = 1$: background weighting coefficient α with $0 \leq \alpha < 1$.

(2) After a regular partition of the image raster into squares of $L \times L$ image points, see Figure 5.5, there are Q^2 squares Δ_{cr} with $1 \leq c, r \leq Q = N/L$. For example, in the case of $N = 512$, the values $L = 512$, 256, 128 etc. are admissible. The central point of each square has the coordinates $XC(c, r)$ and $YC(c, r)$. The background image g is modeled by the plane

$$g(p) = g(x, y) = A(c, r) \cdot [x - XC(c, r)] + B(c, r) \cdot [y - YC(c, r)] + C(c, r) \,,$$

where $A(c, r)$, $B(c, r)$ and $C(c, r)$ are the coefficients of the regression plane $g = A\xi + B\eta + C$ for the square Δ_{cr}. This plane representation utilizes the relative coordinates ξ and η with the origin $(XC(c, r), YC(c, r))$. Assume N to be a power of two, then L is even and the geometrical center of square Δ_{cr} defined by

$$XC(c, r) = L \cdot (c - 1) + L/2 \qquad \text{and} \qquad YC(c, r) = L \cdot (r - 1) + L/2$$

is shifted of a half grid unit to the left and downwards with respect to the ideal center $(XC(c, r), YC(c, r))$. This shift has to be considered in the following formulas for calculating the coefficients of all the regression planes:

$$A(c, r) = \frac{1}{U} \sum_{\Delta_{cr}} f(x, y) \cdot \left(x - \frac{1}{2} - XC(c, r)\right) ,$$

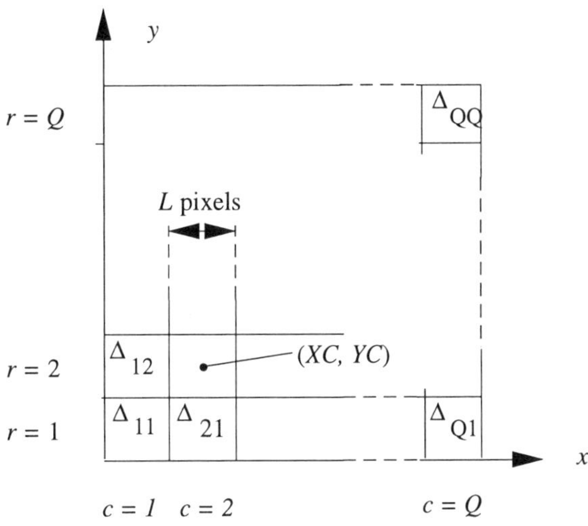

Figure 5.5: Schematic illustration of the division of the image raster **R** into identical $L \times L$ squares for computing piece-wise image approximations by regression planes.

$$B(c, r) = \frac{1}{U} \sum_{\Delta_{cr}} f(x, y) \cdot \left(y - \frac{1}{2} - YC(c, r)\right),$$

$$C(c, r) = \frac{1}{L^2} \sum_{\Delta_{cr}} f(x, y) .$$

Since all the squares Δ_{cr} have the same size, the used normalization value U is always given by the same formula:

$$\begin{aligned} U &= \sum_{\Delta_{cr}} \left(x - \frac{1}{2} - XC(c, r)\right)^2 \\ &= \sum_{\Delta_{cr}} \left(y - \frac{1}{2} - YC(c, r)\right)^2 \\ &= 2L \cdot \sum_{z=0}^{L/2-1} \left(z + \frac{1}{2} \right)^2 . \end{aligned}$$

The resultant gray value is

$$h(p) = \begin{cases} \dfrac{f(p) - \alpha\, g(p)}{1 - \alpha} & , \text{ if } VAR = 1 \\ g(p) & , \text{ if } VAR = 2 . \end{cases}$$

(3) The method as realized by this operator represents a simple solution to the general problem of piece-wise approximations of gray value functions by means of parametrical functions. For keeping the computing time in admissible limits, the domains of the different approximation functions have been chosen as squares of identical size, i.e. the shape of these domains is not correlated with the image content. By decreasing the values of L, the approximation error can be reduced.

The approximation is performed by means of linear functions (regression planes), determined by using the least square error criterion for each square Δ_{cr}. As an advantage of this method, the denominator U in the equations given above for A and B has to be calculated only once. $A(c, r)$ and $B(c, r)$ are the sums of the gray values in Δ_{cr} weighted by their relative coordinates with respect to the origin $(XC(c, r), YC(c, r))$. $C(c, r)$ is the average gray value in Δ_{cr}. For carrying out the operator all Q^2 coefficient triples $A(c, r)$, $B(c, r)$ and $C(c, r)$ are determined in a first run through the image. Then, in a second run, the gray values of the approximation functions are computed according to the above give equations for g and h.

In principle, the gray value range of the original image should remain unchanged after the background subtraction. For this reason, the weighted image difference is divided also by $1 - \alpha$.

(4) *Control structure*: point operator in Figure 3.9
Required data arrays: altogether five two-dimensional arrays, each array for $Q \times Q$ data, for $Q = N/L$:
$A(,)$, $B(,)$ and $C(,)$ for the coefficients,
$XC(,)$ and $YC(,)$ for the coordinates of the origins of Δ_{cr} .

Inputs and initializations:

```
input of the parameter L, an integer divisor of N;
input of VAR = 1 or VAR = 2;
if (VAR = 1 ) then input of the weighting coefficient α;
L2 := L/2 ;                S := 0 ;
for z := 0 to L2 – 1 do     S := S + (z + 0.5)² ;
U := 2 · L · S ;
for c := 1 to Q do
        for r := 1 to Q do begin
                A(c, r) := 0 ;          B(c, r) := 0 ;          C(c, r) := 0;
                XC(c, r) := L·(c – 1) + L2;   YC(c, r) := L·(r – 1) + L2
                end {for}
```

Operator kernel of the first run:

```
                    {read u = f(x, y), no new value h(x, y) = v, cp. Figure 3.9 }
c := 1 + integer[(x – 1) / L] ;
r := 1 + integer[(y – 1) / L] ;
A(c, r) := A(c, r) + u · ( x – 0.5 – XC(c, r) ) ;
B(c, r) := B(c, r) + u · ( y – 0.5 – YC(c, r) ) ;
C(c, r) := C(c, r) + u
```

Preprocessing for the second run:

```
for c := 1 to Q do
     for r := 1 to Q do begin
          A(c, r) := A(c, r) / U ;          B(c, r) := B(c, r) / U ;
          C(c, r) := C(c, r) / L² ;
          end {for}
```

Operator kernel of the second run:

```
                    {transform u = f(x, y) into h(x, y) = v, cp. Figure 3.9 }
c := 1 + integer[(x – 1) / L] ;
r := 1 + integer[(y – 1) / L] ;
w := A(c, r)·(x – 0.5 – XC(c, r)) + B(c, r)·(y – 0.5 – YC(c, r)) + C(c, r ) ;
if (VAR = 1) then w := (u – α · w) / (1 – α ) ;
call ADJUST(w, v)                              { ADJUST in Section 2.1 }
```

(5) The method has been proposed in

Pavlidis, T.: *Structural Pattern Recognition.* Springer, Berlin, 1977.

The computations of the parameters $A(c, r)$, $B(c, r)$ and $C(c, r)$ with the least square error criterion have been exposed in

Haralick, R.M.: *Edge and region analysis for digital image data.* Computer Graphics and Image Processing **12** (1980), pp. 60-73.

Here are also shown the advantages of a regular image field partition.

5.4.3 Some Operations with Two Images

(1) This program gathers several operators for obtaining a resultant gray value $h(x,y)$ by combining gray values $f_1(x, y)$ and $f_2(x, y)$ of two different input images. The program comprises the following operators: image addition, weighted image subtraction, image-to-image maximum, image-to-image minimum, image subtraction shifted of $G/2$, and a special inverse operation for the latter type of image subtraction.

Attributes:

Images:	pairs of $M \times N$ gray value images, two-channel scalar images
Operator:	point operator
Kernel :	analytical, order statistical or logically structured

Inputs:

selection of operation, $VAR = 1$ for *image addition*, $VAR = 2$ for *weighted image subtraction*, $VAR = 3$ for *image-to-image maximum*, $VAR = 4$ for *image-to-image minimum*, $VAR = 5$ for *image subtraction shifted of G/2*, and $VAR = 6$ for approximate *inversion of the latter image subtraction*,

weighting coefficient α, for $0 \le \alpha < 1$, in case of $VAR = 2$.

(2) For $VAR = 1$ it is

$$h(x, y) = \begin{cases} f_1(x, y) + f_2(x, y) & \text{, if } f_1(x, y) + f_2(x, y) \le G - 1 \\ G - 1 & \text{otherwise} \end{cases}$$

for $VAR = 2$ it is

$$h(x, y) = \begin{cases} 0 & \text{, if } f_1(x, y) - \alpha f_2(x, y) < 0\text{, otherwise} \\ [f_1(x, y) - \alpha f_2(x, y)]/(1 - \alpha) & \text{, if } \alpha \ne 0 \\ f_1(x, y) - f_2(x, y) & \text{, if } \alpha = 0 \end{cases}$$

for $VAR = 3$ it is

$$h(x, y) = \mathbf{max}\ \{f_1(x, y), f_2(x, y)\}\ ,$$

for $VAR = 4$ it is

$$h(x, y) = \mathbf{min}\ \{f_1(x, y), f_2(x, y)\}$$

and for $VAR = 5$ it is

$$h(x, y) = G/2 + [f_1(x, y) - f_2(x, y)]/2\ .$$

Inverting the equation of Variant 5, let be

$$h(x, y) = 2 \cdot f_1(x, y) + f_2(x, y) - G$$

the operation for the Variant 6. If h was computed by Variant 5 from f_1 and f_2, then the operation

$$2h(x, y) + f_2(x, y) - G$$

returns the original image $f_1(x, y)$. Because of the rounding in the Variant 5, there might be a difference in the lowest bits.

(3) Image addition, image-to-image maximum and minimum ($VAR = 1, 3, 4$) are simple operations which do not need to be explained. The result of an image addition must always be clipped at the upper gray scale bound $G - 1$ in order to be represented by a gray value.

For image subtraction ($VAR = 2$) the resultant value is clipped at the lower gray scale bound 0 for avoiding negative results. The subtraction can also be performed as a weighted subtraction. For example, this can be useful for subtracting a background image (shading correction) if a task demands an enhancement of high spatial frequencies (image details) of the image function. In the equation for Variant 2, the coefficients have been chosen so that the gray value range of the resultant image is about the same as in the original image. This is true in case that f_1 and f_2 feature about the same gray value range, as it happens in the case of background extraction by low pass filtering. An unweighted subtraction is performed under the convention of taking $\alpha = 0$.

The Variant 5 is an image subtraction shifted by $G/2$. This enables the realization of a subtraction even if the condition $f_1(x, y) \geq f_2(x, y)$ is not satisfied for all image points. Assume the default case of $G = 256$. For representing "negative image values" with only 8 bit per gray value, first the gray scale is reduced to 4 bit per gray value, and then the gray value 0 (black) is shifted to $G/2$ in the resultant image.

Thus the maximum negative (or positive) gray value difference is represented by the gray value 0 (or $G - 1$) in the resultant image. By neglecting the gray value resolution loss from 8 to 4 bit per gray value, the Variant-6 operation allows to invert this type of image subtraction. A certain discretization error has to be tolerated in the resultant image, as mentioned in **(2)**.

The operations of Variants 5 and 6 are useful for tasks in image approximation, especially for successive image approximations as e.g. by means of the Gauss-Laplace pyramid. In Variant 5, the input image $f_1(x, y)$ plays the role of the original image, and the input image $f_2(x, y)$ that of the approximated image. The result of Variant 5, or the image $f_1(x, y)$ of Variant 6, represents the approximation error. By use of Variant 6, the approximated image and the error image are combined into the original image. After the application of both operations on natural images, the reconstruction error, which is due to a loss in gray value resolution, remains acceptable over several reconstruction steps, as experienced in many practical application cases.

(4) *Control structure*: point operator, but special control structure (as specified here) because the data for f_1 and f_2 are read in two image files
Required data arrays: line store arrays $BUF1(1...M)$ and $BUF2(1...M)$ for inputs of lines from the input images $f_1(x, y)$ and $f_2(x, y)$, and line store array $BUFOUT(1...M)$ for the output lines.

```
begin
    for y := 1 to N do begin
        read line y from f1(x, y) into line store array BUF1(1...M);
        read line y from f2(x, y) into line store array BUF2(1...M);

        for x := 1 to M do begin
            u1 := BUF1(x);          u2 := BUF2(x);
            if (VAR = 1) then
                w := u1 + u2
            else  if (VAR = 2) then
                    if (α = 0) then
                        w := u1 − fu2
                    else
                        w := [u1 − α · u2] / (1 − α)
                  else  if (VAR = 3) then
                        w := max{u1, u2}
                    else  if (VAR = 4) then w := min{u1, u2}
                        else  if (VAR = 5) then
                                w := [G + u1 − u2]/2
                            else
                                w := 2 · u1 + u2 − G ;
            call ADJUST(w, v);              {ADJUST in Section 2.1}
            BUFOUT(x) := v;
            end {for};
        write BUFOUT(1...M) in line y of the resultant image h
    end {for}
end
```

(5) For the *Gauss-Laplace pyramid* as mentioned in **(3)**, see e.g.

Jähne, B.: *Digitale Bildverarbeitung*. Springer, Berlin, 1989.

Zamperoni, P.: *Methoden der digitalen Bildsignalverarbeitung*. Vieweg Verlag, Wiesbaden, 2nd ed., 1991.

Burt, P.J., Adelson, E.H.: *The Laplacian pyramid as a compact image code*. IEEE Trans. COM-**31** (1983), pp. 532 - 540.

5.5 IMAGE SEGMENTATION BY MULTILEVEL THRESHOLDING

Image segmentation aims at the clustering of groups of image points into connected sets. The resultant segments can either be represented by regions, e.g. labeled by a constant (symbolic) gray value, or by contours, e.g. by closed digital curves. *Multilevel thresholding methods* perform image segmentation on the base of information contained in the global gray value histogram.

5.5.1 Extraction of Constant Gray Value Lines

(1) This program's scope is the generation of bilevel images of lines or curves. Each one of the computed closed curves encloses a region of the original image, whose gray values are all in a predefined gray value domain.

Attributes:

Images:	$M \times N$ gray value image
Operator:	3×3 window operator[2]
Kernel:	logically structured, data driven

Inputs:

number 2^c of uniformly sized gray value intervals in which the full gray scale is divided, i.e. $1 \leq c \leq log_2 G - 1$.

(2) The gray value number G is assumed to be a power of two, $G = 2^b$. For two gray values u, v let **AND**(u, v) be the *logical bit-AND* in all bit positions. For example, for a gray value u and a number $B = G - 2^{b-c}$, for $0 \leq c \leq b$, it follows that **AND**(u, B) is a new gray value where all $b - c$ least significant bits of u are mapped

[2] Because of its conceptual affinity to simple point operators, this operator is described here in Chapter 5.

into 0. Let **F** be a 3×3 window, and let **F**(p) be the placed window with reference point p. Then, let

$$h(p) = \begin{cases} 0 & \text{, if } \mathbf{AND}(f(p), B) = \mathbf{AND}(f(q), B), \text{ for all } q \in \mathbf{F}(p) \\ G - 1 & \text{otherwise} \end{cases}$$

for $B = G - 2^{b-c}$ and $b = log_2 G$.

(3) Through this definition, the gray value scale is divided in 2^c intervals of equal size of 2^{b-c} gray levels. The resultant image is a bilevel one and its borders are those between neighboring gray value domains. These border lines are closed and they have a thickness of a single image point. For obtaining this graph of border lines, the gray value $f(p)$ of the current image point p is compared with the gray values of all 8-neighbors of p. The comparison is carried out between masked gray values, i.e. all the $b - c$ least significant bits are masked out. If there exists at least one neighbor q with a masked gray value not equal to **AND**($f(p)$, B), then $f(q)$ is in a different gray value interval than $f(p)$. In this case, the image point p is classified as a "white element" of the resultant line image and it is assigned the symbolic gray value $G - 1$. Otherwise it is $h(p)$:= 0 in the resultant image. In Figure 2.11 a resultant line image is shown for $c = 2$.

(4) *Control structure*: window operator in Figure 3.6 with $n = 3$, $a = 9$

Input and initialization:

```
input of parameter c  with 1 ≤ c ≤ b – 1;
B := G – 2^(b–c)
```

Operator kernel:

```
            {transform F(z) with 1 ≤ z ≤ a into v = h(x, y), cp. Figure 3.6 }
w_a := AND(F(a), B) ;         MIN := G ;           v := 0;
for z := 1 to a – 1 do begin
        w := AND(F(z), B);
        if (w < MIN)  then  MIN := w
        end {for};
if (w_a > MIN)  then v  := G – 1
```

(5) Haberäcker, P.: *Digitale Bildverarbeitung*. Carl Hanser Verlag, München, 3rd ed., 1989.

5.5.2 Thresholding by Extraction of the Histogram Extrema

(1) This procedure realizes a method for *image segmentation*. The gray value scale is divided into intervals by several thresholds, and each interval defines homogenous image regions. The determination of thresholds is based on the evaluation of the global gray value histogram. In principle, the minima of the histogram are used as thresholds. All gray values between two consecutive minima are replaced by the gray value of the maximum between these two minima. The histogram is smoothed before searching for extrema. The strength of smoothing determines the degree of segmentation coarseness.

Attributes:

Images:	$M \times N$ gray value images
Operator:	point operator
Kernel:	logically structured, data driven

Input:

width B, in gray values, of the one-dimensional window for histogram smoothing, B is assumed to be odd with $B = 2b + 1$, e.g. $b = 1, ..., 20$ for $G = 256$.

(2) The function $R(u)$ of a discrete variable u, for $0 \le u \le G-1$ defined below is derived from the global absolute sum histogram *sumhist*(u), cp. Section 1.3.2 for the definition of the function *sumhist*. The negative and the positive *zero crossings* of this function $R(u)$ are denoted by $NC(s)$, $1 \le s \le T$, and by $PC(s)$, for $1 \le s < T$. It holds

$$NC(s) < PC(s) < NC(s+1), \text{ for } \quad 1 \le s < T$$

and

$$1 \le NC(s),\ PC(s) \le G-1, \text{ for } \quad 1 \le s \le T\,.$$

The *gradation function* is defined by these zero crossings,

$$v = PC(r)\,, \text{ if } \quad NC(r) \le u < NC(r+1),$$

and it is used for this position dependent point operator. The function $R(u)$ is obtained by the difference between *sumhist*(u) and a smoothed version of *sumhist*(u) generated by aid of a one-dimensional low-pass with width of $B = 2b + 1$ gray levels:

$$R(u) = sumhist(u) - \frac{1}{B} \sum_{t=-b}^{b} sumhist(u+t) \qquad \text{for } b \le u \le G-1-b\,.$$

Figure 5.6: Segmentation of an image of a natural scene (top left) by multilevel thresholding with extraction of the histogram extrema, for different width B of the smoothing window. Top right: $B = 9$ and 22 regions. Below left: $B = 15$ and 14 regions. Below right: $B = 35$ and 7 regions.

(3) Multilevel thresholding for image segmentation can be considered as a clustering process in the one-dimensional feature space of the global gray value histogram. By cluster analysis, local maxima (*modes*) of the histogram can be recognized. The approach is based on the assumption that maxima correspond to relevant image regions. Each maximum is between two minima. Thus it is straightforward that all gray values between these two minima $NC(r)$ and $NC(r + 1)$ of the r-th mode are considered to define homogenous regions of the r-th mode. All image points of these regions are labeled by the gray value $PC(r)$ of the maximum of the r-th mode.

After segmentation, the image contains only the T gray values of the maxima, where each maximum characterizes a certain "type of regions". Often histograms feature fine-structured variations of small amplitude, for which the extrema can be considered to be irrelevant. Thus these irrelevant extrema can be suppressed by smoothing. The strength of smoothing is determined by the width parameter of the smoothing window.

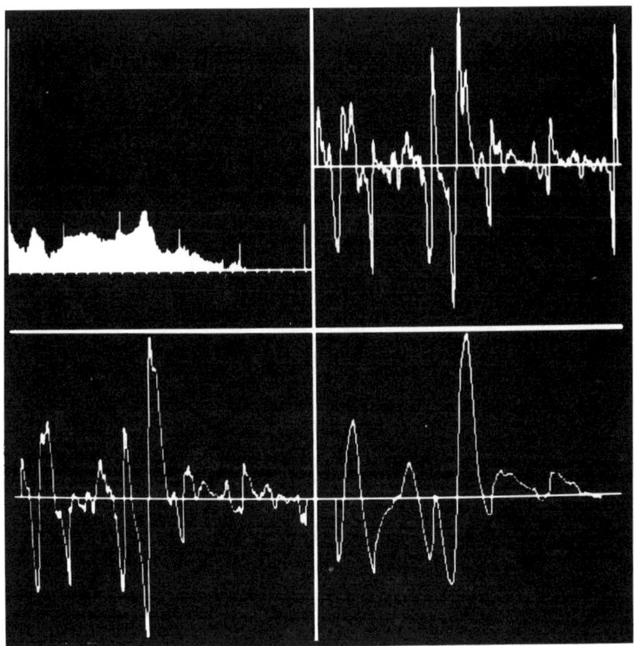

Figure 5.7: Gray value histogram of the input image (top left) and the graphs of the peak detection functions for the processed images of Figure 5.6. These graphs are at the same locations as the corresponding images of Figure 5.6.

The basic idea of this method consists in detecting inflexion points of decreasing or increasing curvature of the gray value distribution function, by analyzing minima and maxima of the histogram. The curvature can be computed by means of the first discrete derivative *sumhist'*(u) and of the second discrete derivative *sumhist''*(u). The computation of these derivatives has to be combined with the smoothing process mentioned above. Low-pass filtering is performed by convolving the histogram with a set of weighting coefficients which linearly decrease from the middle of the window, for $t = 0$ in the above definition of R, to the border of the window, i.e. for $t = \pm b$. The combination of this linear filter, which computes the curvature function, with a smoothing filter is realized by a difference between *sumhist*(u) and a smoothed version of *sumhist*(u), where a box-filter is used for smoothing. The result of this difference is a *peak detection function,* whose zero crossings are considered.

An example of the application of this multilevel thresholding is shown in Figure 5.6 and Figure 5.7. An image of a natural scene is segmented with different strength of *histogram smoothing*. The strength of the smoothing determines the coarseness of the segmentation, i.e. the number T of recognized modes $PC(r)$, $r = 1, ..., T$ and thus the number of resultant image regions.

(4) *Control structure*: point operator in Figure 3.9, two runs

Required data arrays: $NC(1...1 + T_{max})$ and $PC(1...T_{max})$ for storing the negative and positive zero crossings of the peak detection function, where T_{max} denotes the maximum allowable number of different region gray values, e.g. about 40 for $G = 256$, or at most $T_{max} = G$; T_{max} characterizes the maximum possible segmentation fineness;

$C(1...G)$ at first for the absolute histogram, and further on for the absolute sum histogram,

$R(1...G)$ for the peak detection function.

Inputs and initialization:

```
input of odd B ;        b := (B – 1)/2 ;
for z := 1 to Tmax do begin
                            NC(z) := 0 ;  PC(z) := 0
                     end {for};
NC(1 + Tmax) := G – 1;
for z := 1 to G do    C(z) := 0
```

Operator kernel, first run - histogram computation:

```
                     { u = f(x, y) , no new v = h (x, y), cp. Figure 3.9 }
C(1 + u) := C(1 + u) + 1
```

Preprocessing for the second run:

```
for z := 2  to G   do  C(z) := C(z – 1) + C(z) ;
for z := 1  to G   do
      if ((z ≤ b)  or (z ≥ G – b))  then   R(z) := 0
      else  begin
            CB := 0;
            for t := z – b  to z + b  do  CB := CB + C(t);
            R(z) := C(z)  –  CB / B
            end {else};
NC(1) := 0;            s := 2;              t := 1;
for z := b + 2  to G – b  do
      if ((R(z) < 0)  and (R(z – 1) ≥ 0))  then begin
            NC(s) := z – 1;             s := s + 1
```

```
                    end {then}
        else        if ((R(z) > 0) and (R(z – 1) ≤ 0)) then begin
                          PC(t) := z – 1;                    t := t + 1;
                          if (t > T_max) then
                                 program stop               { to many maxima }
                          end {if};
    NC(s) := G;           T := t – 1
```

Operator kernel, second run - image transform:

```
                              {transform u = f(x, y)  into v = h (x, y), cp. Figure 3.9 }
for s := 1 to T do
      if ((u ≥ NC(s)) and (u < NC(s + 1))) then v := PC(s)
```

(5) This method was proposed by

Sezan, M.I.: *A peak detection algorithm and its application to histogram-based image data reduction.* Computer Vision, Graphics and Image Processing **49** (1990), pp. 36 - 51.

based on results given by

Boukharouba, S., Rebordao, J.M., Wendel, P.L.: *An amplitude segmentation method based on the distribution function of an image.* Computer Vision, Graphics and Image Processing **29** (1985), pp. 47 - 59.

5.5.3 Multilevel Thresholding for Unimodal Histograms

(1) As in Section 5.5.2, the computation of gray value thresholds for image segmentation is only based on the global gray value histogram. The histogram intervals defining the homogenous regions are different from the previous Section, because an other image model is used. Here the goal consists in recognizing histogram sections with uniformly distributed gray values. It is assumed that such sections correspond to the image regions with two-dimensional linear shape (ramps, noisy gray value plateaus etc.). In order to recognize the intervals more easily, at first a non-linear histogram smoothing is performed. The interval limits of the recognized gray scale segments are used as thresholds for segmentation. All gray values between two consecutive thresholds are mapped onto the middle value of this interval. Besides uniformly distributed histogram domains, also "sharp modes", if they exist, are considered for threshold definition.

Attributes:

Images:	$M \times N$ gray value images
Operator:	point operator, two runs
Kernel:	logically structured, data driven

Inputs:

length L_1 of the one-dimensional window for histogram smoothing

(2) The thresholds T_z and the symbolic gray values S_z, which are defined between thresholds and which label the resultant gray value segments, satisfy the conditions

$$T_z < S_z < T_{z+1} \quad \text{and} \quad 1 \le T_z, S_z \le G - 1, \text{ for all } z \text{ with } 1 \le z \le Z_M \le Z_{max} .$$

These values are extracted, as explained later, from the smoothed gray value histogram. For the current input value $u = f(p)$, the resultant gray value $v = h(p)$ is equal to S_r if $T_r \le u < T_{r+1}$. The gradation function t,

$$t(u) = S_r \text{, for } T_r \le u < T_{r+1} \text{ ,}$$

defines the point operator.

(3) In previous work on segmentation by multilevel thresholding the *determination of histogram extrema* is the most popular approach. Sometimes the gray value distribution within homogenous image regions does not correspond to the image model assumed by this approach, e.g. that there should be an alternation of strong minima and strong maxima in the global gray value histogram. The segmentation method discussed in this Section represents an alternative solution based on a different model, and it can lead to improved results if the "minima-maxima-model" of Section 5.5.2 fails.

It can be observed experimentally that in many images the gray value distribution within homogenous regions can be approximated by an equal distribution. Non-continuous behavior of the histogram points to the beginning or ending of such an uniformly distribution section of the histogram.

In Figure 5.8, a schematic example is given for such a histogram. For obtaining such a histogram pattern it is assumed that a certain edge-preserving smoothing has been performed upon the original histogram (computed in a first run through the input image). This smoothing is realized by means of a one-dimensional closing, i.e. by cascading a one-dimensional maximum and a minimum filter in a histogram window of length of L_1 gray levels.

After smoothing the absolute gray value histogram $hist(u)$ is scanned from $u = 0$ to $u = G - 1$ for determining whether the current gray value u is either non-critical (label 0), a positive or negative edge value (labels 1 and 2), or an isolated maximum or minimum (labels 3 and 4). These typical configurations of three consecutive histogram values $hist(u-1)$, $hist(u)$ and $hist(u+1)$ are illustrated in Figure 5.8, below. The non-critical values (label 0) need no further treatment. The values $V(u)$ of the assigned labels are stored in the array $V(1...G)$.

pairs of labels $V(u_1)$, $V(u_2)$	recognized critical gray value T_z or. S_z
1, 1	$S_z := 0.5 \cdot (u_1 + u_2)$
1, 2	$S_z := 0.5 \cdot (u_1 + u_2)$
1, 4	$S_z := 0.5 \cdot (u_1 + u_2)$
2, 2	$S_z := 0.5 \cdot (u_1 + u_2)$
4, 2	$S_z := 0.5 \cdot (u_1 + u_2)$
arbitrary, 3	$S_z := u_2$
arbitrary, 4	$T_z := u_2$
2, 1	$T_z := 0.5 \cdot (u_1 + u_2)$

Table 5.1: Relation between pairs of labels and critical gray values.

After this process of label computation, the critical points T_z and S_z of the histogram, for $1 \le z \le Z_M$, are computed in two steps. In the first step, the array $V(1...G)$ is scanned, from $u = 0$ to $u = G - 1$, for detecting pairs of gray values u_1 and u_2 satisfying the following conditions:

$$u_1 < u_2,\ V(u_1) \neq 0,\ V(u_2) \neq 0 \text{ and } V(u) = 0, \text{ for all } u \text{ with } u_1 < u < u_2.$$

The critical gray values follow from the values $V(u_1)$ and $V(u_2)$, and Table 5.1 can be used for fast look-up. The critical gray values are stored in the lists T_z and S_z.

In this first step some thresholds T_z cannot be determined. Also here it can happen that S_z-values are computed without a T_z-value between. In the second step it has to be ensured that T_z-values (thresholds) and S_z-values (symbolic gray values for characterizing segmented regions) alternate in the gray value scale. This is necessary in the second run through the input image for generating symbolic gray values according to the gradation function given in **(2)**.

In the second step for all consecutive S_z-values S_{z-1} and S_z which are not satisfying the condition

$$\text{there exist a } t \text{ such that } S_{z-1} < T_t < S_z \, ,$$

an additional threshold $T_{new} = 0.5 \cdot (S_{z-1} + S_z)$ is inserted into the T_z-list such that this list remains sorted in increasing order.

In a second run through the input image, the gray values are transformed based on the lists T_z and S_z. The number Z_M of the elements in these two lists, i.e. the number of the region gray values corresponding to different segments, depends upon the strength of smoothing the the original histogram, i.e. upon the window length L_1. Thus the segmentation coarseness can be controlled by the parameter L_1.

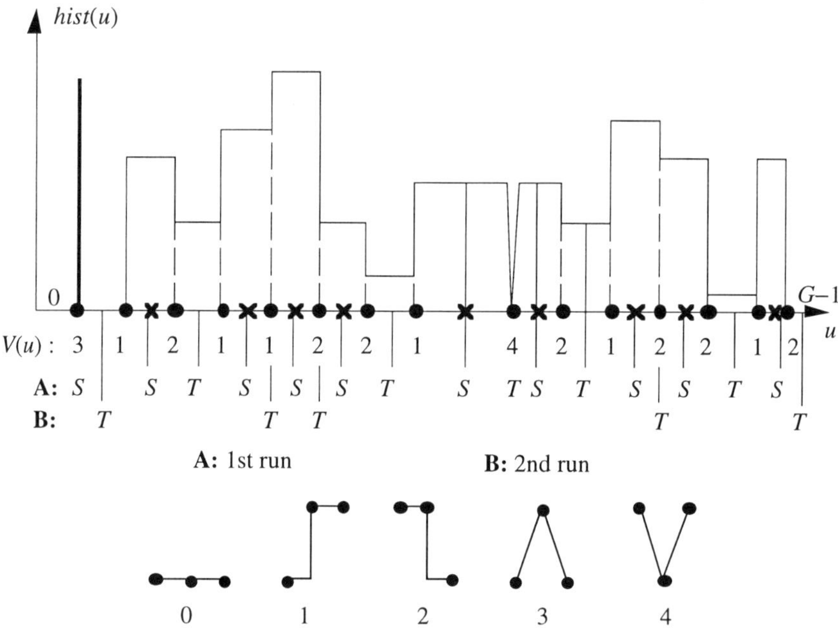

Figure 5.8: Gray value histogram for the method described in Section 5.5.3, assumed after smoothing by one-dimensional closing. Top: an example of a histogram with putting into evidence the gray values with labels 1, ..., 4. From these are derived the thresholds T and the symbolic gray values S for region labeling. Below: schematic representation of possible configurations of consecutive histogram values, and labels 0, ..., 4, for the classification of these configurations.

(4) *Control structure*: point operator in Figure 3.9, two runs

Required data arrays: $hist(1...G)$ for the original histogram, $Q(1...G)$ for the smoothed histogram, and $Q_0(1...G)$ as temporary memory,

$V(1...G)$ for labels 0 to 4 characterizing different gray value triplets in the histogram, see Figure 5.8,
$T(1...Z_{max})$ and $S(1...Z_{max})$ for the computed thresholds or for the new symbolic gray values; Z_{max} should be sufficiently large for a fine segmentation, e.g. for G= 256 about 50.

The operation of one-dimensional closing, performed on a gray value histogram, consists of succesive computations of the maximum and of the minimum in a 1-D window of L_1 gray levels. A standard routine can be used for this aim, see Section 3.4.3. Before performing this maximum-minimum operation, the histogram values are stored in a temporary memory $U(1...\ L_{1max})$. The number of arguments of this operator is smaller than L_1 at both ends of the gray scale, because the smoothing window cannot move behind the gray values 0 or $G-1$. The value L_{1max} restricts the maximum size of the 1-D smoothing window. For elements of the lists S_z and T_z the notation $S(z1)$ or $T(z2)$ is used in the following program.

Input and initialization:

```
input of the (odd) window width L1 for histogram smoothing;
L := (L1 – 1)/2 ;
for c := 1 to G do begin
        hist(c) := 0 ;    Q(c) := 0
        end {for};
for c := 1 to Zmax do begin
        S(c) := 0 ;       T(c) := 0
        end {for}
```

Operator kernel, first run - histogram computation:

```
                        { u = f(x, y) , no new v = h (x, y), cp. Figure 3.9 }
hist(1 + u) := hist(1 + u) + 1
```

Preprocessing for second run:

```
                    { application of the maximum operator to the histogram }
for u := 1 to G do begin
              NX := 1;    {NX counts arguments of maximum operator }
        for w := max{1, u – L} to min{G, u + L} do begin
              U(NX) := hist(w),     NX := NX + 1
              end {for}
        NX := NX – 1;
```

```
        Q0(u) := max{U(1), U(2),..., U(NX)}
        end {for};
                              { application of the minimum operator to the result }
for u := 1 to G do begin
        NX := 1;
        for w := max{1, u – L} to min{G, u + L} do begin
                U(NX) := Q0(w);        NX := NX + 1
                end {for}
        NX := NX – 1;
        Q(u) := min{U(1), U(2),..., U(NX)}
        end {for};
                              { classification of the gray levels into classes o to 4 }
V(1) := 0 ;      V(G) := 0;
for u := 2 to G – 1 do
        if ((Q(u) > Q(u – 1)) ∧ (Q(u) = Q(u + 1))) then V(u) := 1
        else if ((Q(u) = Q(u – 1)) ∧ (Q(u) > Q(u + 1))) then V(u) := 2
        else if ((Q(u) > Q(u – 1)) ∧ (Q(u) > Q(u + 1))) then V(u) := 3
        else if ((Q(u) < Q(u – 1)) ∧ (Q(u) < Q(u + 1))) then V(u) := 4
        else V(u) := 0;
                              { determination of the critical points - first step }
ZS := 1;            { ZS: is the counter for inputs into the list of S(z)-values }
ZT := 2;            { ZT: is the counter for inputs into the list of T(z)-values }
T(1) := 0;      V1 := 4;      u1 := 0;                 { initialization values }
for u := 2 to G – 1 do
        if (V(u) ≠ 0) then begin
                V2 := V(u);     u2 := u;
                if (V2 = 3) then begin   S(ZS) := u2;     ZS := ZS + 1
                                end {then}
                else if (V2 = 4) then begin  T(ZT) := u2;     ZT := ZT + 1
                                        end {then};
                W := 10 · V1 + V2 ;
                if ((W = 11) ∨ (W = 12) ∨ (W = 14) ∨ (W = 22) ∨ (W = 42))
                        then begin

                                S(ZS) := 0.5 · (u1 + u2);      ZS := ZS + 1

                                end {then}
                else if (W = 21) then begin
```

```
                              T(ZT) := 0.5 · (u1 + u2);      ZT := ZT +1
                              end {then};
                  V1 := V2;      u1 := u2
                  end {then};
      T(ZT) := G;   ZS := ZS – 1;
                        { determination of the critical points - second step }
      for z1 := 2 to ZS do begin
            z2 := 1;        flag := 0;
            while ((z2 ≤ ZT) ∧ (flag = 0)) do begin
                  if ((T(z2) > S(z1 – 1)) ∧ (T(z2) < S(z1))) then flag := 1;
                  z2 := z2 + 1
                  end {while};
            if (flag = 0) then begin
                        Tnew := 0.5 · (S(z1 – 1) + S(z1));
                          { insertion of a new value into the ordered list T(z) }
                        zone := 1;       z:= 2;
                        while ( (z ≤ ZT) ∧ (flag = 0) ) do begin
                              if ((Tnew ≥ T(z – 1)) ∧ (Tnew ≤ T(z)))
                                    then begin   zone := z ;    flag := 1
                                          end {then};
                              z := z + 1
                              end {while};
                        ZT := ZT + 1;
                        if (flag = 0) then T(ZT) := Tnew
                        else begin
                              for z := ZT – 1 to zone  step – 1 do
                                                            T(z + 1) := T(z);
                            T(zone) := Tnew
                            end {else}
                        end {then}
            end {for}
```

Operator kernel of the second run - gray value transform:

```
                        {transform u = f(x, y) into v = h (x, y), cp. Figure 3.9 }
      flag := 0;      z2 := 1;
      while ((z2 ≤ ZT – 1) ∧ (flag = 0)) do begin
```

```
if ((u ≥ T(z2)) ∧ (u < T(z2 + 1))) then begin
        z1 := 1;
        while ((z1 ≤ ZS) ∧ (flag = 0)) do begin
                if ((S(z1) ≥ T(z2)) ∧ (S(z1) < T(z2 + 1))) then begin
                        v := S(z1);     flag := 1
                        end {if}
                z1 := z1 + 1
                end {while}
        end {if};
z2 := z2 + 1
end {while}
```

(5) Zamperoni, P.: *An automatic low-level segmentation procedure for remote sensing images.* Multidimensional Systems and Signal Processing **3** (1992), pp. 29 - 44.

5.6. MULTI-CHANNEL IMAGES

In general, for a *multi-channel image* $f(x, y) = (f_1(x, y), f_2(x, y), ..., f_n(x, y)) = (u_1, u_2, ..., u_n)$ a tuple is considered as an image value at an image point (x, y). For such a vectorial image value and $n \geq 2$, certain operations can be defined between the different components, where each scalar image can be considered to be a component. In Section 5.6.1, a multi-channel image is transformed into a scalar image by different operations. In Section 5.6.2 a special multi-channel image, a *RGB color image*, is transformed into another multi-channel image, a *HSI color image.* Finally in Section 5.6.3 the generation from a scalar image of a multi-channel image, i.e. of a RGB color image is also discussed.

5.6.1 Basic Arithmetic Operations

(1) This program allows the selection of different *point operators for multi-channel images*. The generated image value is scalar. Thus a multi-channel image is transformed into a scalar image. The program contains the following operations: overlay, channel difference, ratio of channels, averaging and difference of subsequences, cp. also Section 5.4.3 for further operations.

Attributes:

Images: multi-channel $M \times N$ Images

Operator: point operator
Kernel: analytical, different variants

Inputs:

selection of operation, $VAR = 1$ for superposing (overlay), $VAR = 2$ for the channel difference, $VAR = 3$ for the ratio of channels, $VAR = 4$ for averaging, and $VAR = 5$ for the difference of subsequences,
parameter $c_1, c_2, \ldots, c_n$ for $VAR = 1$,
Parameter i, j and c for $VAR = 2$,
Parameter i, j, c_1, c_2 for $VAR = 3$, and
Parameter i, j, k for $VAR = 5$.

(2) For a multi-channel image f let $f(x, y, t) = f_t(x, y) = u_t$ be the gray value of channel t at image point (x, y), for $1 \leq t \leq n$. In the case of $VAR = 1$, $h(x, y)$ is equal to

$$\frac{1}{n} \cdot \sum_{t=1}^{n} c_t \cdot f(x, y, t) \qquad \text{with} \qquad \sum_{t=1}^{n} c_t = 1 \ ,$$

for $VAR = 2$ it is equal to

$$f(x, y, i) \ - \ f(x, y, j) \ + \ c \qquad \text{for } 1 \leq i, \ j \leq n \ ,$$

for $VAR = 3$ it is equal to

$$\frac{f(x, y, i) - f(x, y, j)}{f(x, y, i) + f(x, y, j)} \cdot c_1 \ + \ c_2 \qquad \text{for } 1 \leq i, j \leq n \ ,$$

for $VAR = 4$ it is equal to

$$\frac{1}{n} \cdot \sum_{t=1}^{n} f(x, y, t),$$

and for $VAR = 5$ it is equal to

$$\frac{1}{j - i + 1} \cdot \sum_{t=i}^{j} f(x, y, t) \ - \ \frac{1}{k - j} \cdot \sum_{t=j+1}^{k} f(x, y, t) \qquad \text{for } \ 1 \leq i \leq j < k \leq n \ .$$

(3) The *image superposition* operation is useful already for $n = 2$. For example, the graphical information can be given in the channel 2 as a bilevel image which can be used as an overlay on the channel 1. In the case of the *channel differences*

($VAR = 2$), a constant c is needed for avoiding negative gray values, e.g. $c = G/2 - 1$ can be used. Here it can also be suggested to divide the values $f(x, y, i)$ and $f(x, y, j)$ by 2 because otherwise the difference can assume values between $-(G-1)$ and $+(G-1)$. However if both channels have "nearly" the same gray value content, this additional division by 2 compensation may become redundant. In the Variant **(4)** the division by 2 is in practice superfluous. The channel difference operation can be used for the deletion of constant additive noise or for shading compensation. In the case of $VAR = 3$ (the *ratio of channels*) the resultant image h is invariant with respect to the multiplication of the gray values in channels i and j with the same constant. Such a constant can be used for compensating a different intensity of illuminations. Here the constants c_1 and c_2 should be adjusted in dependence of the nature of the multi-channel images. *Averaging* ($VAR = 4$) can be of interest for image sequences, e.g. to obtain a noise reduction integrating several images of a static scene. *Differences of subsequences* ($VAR = 5$) can be of interest for detecting "essential differences" in image sequences.

A possible aim of operations as superposition, channel differences or channel ratio, can also be the reduction of the number n of channels. Meaningful operations for the multi-channel images can also be defined by non-arithmetical point operations, e.g. by maximum or by minimum operations, which can be used for the so-called *sandwich effect*. Logical operations between the channels can be used for compositions ("If gray value at (x, y) in channel 2 is larger then s, then value at (x, y) in channel 1 plus value at (x, y) in channel 3, otherwise gray value 0" etc.). Such a logical-arithmetic combination will depend upon specific image contents and the aims of an image processing.

(4) *Control structure*: point operator for multi-channel images, here specified.

Required data arrays: line store arrays $BUF(1,1...M)$,..., $BUF(n,1...M)$ for lines of the input images $f_1(x, y)$,..., $f_n(x, y)$,
line store array $BUFOUT(1...M)$ for the resultant image h..

begin
 input of the parameter VAR for the selection of an operation;
 if ($VAR = 1$) **then** input of parameters $c_1, c_2, ..., c_n$ satisfying $\sum_{t=1}^{n} c_t = 1$;
 if ($VAR = 2$) **then begin**
  input of the channel indices i, j with $i \neq j$; $c = G/2 - 1$
  end $\{if\}$;
 if ($VAR = 3$) **then** input of the parameters i, j with $i \neq j$, input ofc_1, c_2;
 if ($VAR = 5$) **then** input of the parameters i, j, k with $i \leq j < k$;

```
for y := 1 to N do begin
        for t := 1 to n do
                read line y of the image f_t into line store array BUF(t, 1...M);
        for x := 1 to M do begin
                for t := 1 to n  do u(t) := BUF(t, x);
                w  := 0;
                if (VAR = 1) then begin
                        for t := 1 to n  do w := w + c_t · u(t);
                        w := w / n
                        end {then}
                else    if (VAR < 4) then begin
                                D := u(i)  –  u(j);
                                if (VAR = 2) then     w := D + c
                                else  begin                          { VAR = 3 }
                                        S := u(i) + u(j);
                                        if (S = 0) then S := 1;
                                        w := (D · c1) / S   +   c2
                                        end {else}
                        else    if (VAR = 4) then begin
                                        for t := 1 to n do w := w + u(t);
                                        w := w / n
                                        end {if}
                                else  begin                          { VAR = 5 }
                                        L := 0;          R := 0
                                        for t := i to j  do L := L + u(t);
                                        L := L / (j – i + 1);
                                        for t := j + 1 to k do R := R + u(t);
                                        R := R / (k – j);         w := L – R
                                        end {else};
                call ADJUST(w, v);  BUFOUT(x) := v;
                end {for};
        write BUFOUT(1...M) into line y of resultant image h
    end {for}
end
```

(**5**) Haberäcker, P.: *Digitale Bildverarbeitung*. Carl Hanser Verlag, München, 3rd ed., 1989.

5.6.2 Color Model Conversion

(1) A three-channel RGB image has to be transformed into an HSI image, cp. Section 1.1.3.

Attributes:

Images:	$M \times N$ RGB images
Operator:	point operator
Kernel:	analytical

(2) For a color $q = (r, g, b)$ in the RGB model, the task consists in computing the color representation $q = (h, s, i)$ in the HSI model. The hue H of color q is given by

$$H = \begin{cases} \delta & \text{, if } b \leq g \\ 360° - \delta & \text{, if } b > g \end{cases}$$

where

$$\delta = \textbf{arc cos} \frac{\dfrac{(r-g)+(r-b)}{2}}{\sqrt{(r-g)^2 + (r-b)(g-b)}} \; ;$$

and h is the hue H scaled into the range 0, 1, ..., G–1. The saturation S is given by

$$S = 1 - 3 \cdot \frac{\textbf{min}\,\{r, g, b\}}{r + g + b} \; ,$$

and s is the saturation S scaled into the range 0, 1, ..., G–1. The intensity i of the color q is equal to

$$i = \frac{r + g + b}{3}$$

and thus in the range 0 to G–1.

(3) The computer representation of color images is normally given in the RGB model. The conversion into the HSI model can be of graphical interest for the representation of these channels. In general, the HSI model has some benefits for image segmentation of color images.

(4) *Control structure*: point operator, but special control structure (here specified) because the input data come from three different images f_1 (channel for Red), f_2 (channel for Green) and f_3 (channel for Blue)

Required data arrays: line store arrays $BUF(1, 1...M)$, $BUF(2, 1...M)$ and $BUF(3, 1...M)$ for the input of lines from the images $f_1(x, y)$, $f_2(x, y)$ and $f_3(x, y)$, respectively, and
$BUFOUT(1, 1...M)$ for the hue h (the resultant image h_1), $BUFOUT(2, 1...M)$ for the saturation s (the resultant image h_2) and $BUFOUT(3, 1...M)$ for the intensity i (the resultant image h_3).

```
begin
   for y := 1 to N do begin
         read line y from f1(x, y) into line store array BUF(1, 1...M);
         read line y from f2(x, y) into line store array BUF(2, 1...M);
         read line y from f3(x, y) into line store array BUF(3, 1...M);
         for x := 1 to M do begin
               u1 := BUF(1, x);       u2 := BUF(2, x);       u3 := BUF(3, x);
               Z := ((u1 – u2) + (u1 – u3)) / 2;
               Q := root( (u1 – u2)^2 + (u1 – u3)(u2 – u3) );
               if (Q ≠ 0)     then delta := arc cos( Z / Q )
                                          { delta in radiants, 0 ≤ delta < π }
                              else delta not defined;     {u1 = u2 = u3 }
               if (u3 ≤ u2)   then H := delta else H:= 2π – delta;
               w := H · G / 2π;        { G is the number of gray levels }
               call ADJUST(w, v);  BUFOUT(1, x) := v;
               SUM := u1 + u2 + u3;         MIN := min{u1, u2, u3};
               if (SUM ≠ 0)  then   S := 1 – 3 · (MIN / SUM)
                             else   S is undefined;  {u1 = u2 = u3 = 0 }
               w := S · G;
               call ADJUST(w, v);  BUFOUT(2, x) := v;
               w := SUM / 3;
               call ADJUST(w, v);  BUFOUT(3, x) := v
               end {for};
         write BUFOUT(1, 1...M) into line y of the resultant image h1;
         write BUFOUT(2, 1...M) into line y of the resultant image h2;
         write BUFOUT(3, 1...M) into line y of the resultant image h3
   end {for}
end
```

(5) Fellner, W.D.: *Computer Grafik.* Wissenschaftsverlag, Mannheim, 1988.

5.6.3 Pseudo-Coloring

(1) The task consists in transforming a gray value image into a RGB color image. A unique color value has to be assigned to each gray value. Alternatively, the interval method or the specification of the transformation functions can be chosen.

Attributes:

Images:	$M \times N$ gray value images
Operator:	point operator
Kernel:	analytical, logically structured

Inputs:

$VAR = 1$ for the interval method and $VAR = 2$ for the specification of transformation functions,

interval limits l_1 , ..., l_{n-1} with $0 < l_1 < ... < l_{n-1} < G$ and colors $c_1, c_2, ..., c_n$ with $c_k = (r_k, g_k, b_k)$ for $VAR = 1$,

two transformation functions $i_R(u)$ and $i_G(u)$ for $VAR = 2$.

(2) A gray value image f is transformed into a RGB image h. For the *interval method* assume that a number n, $n > 1$, is predefined as number of the different gray value intervals with corresponding color values,

$$\text{intervals:} \quad 0 = l_0 < l_1 < ... < l_{n-1} < l_n = G$$
$$\text{colors:} \quad c_1, c_2, ..., c_{n-1} \quad \text{with } c_k = (r_k, g_k, b_k) .$$

The image transformation is given by

$$h(x, y) = c_k , \quad \text{if} \quad l_{k-1} \leq f(x, y) < l_k, \quad \text{for} \quad k = 1, 2, ..., n .$$

For the *method based on transformation functions*, three special color transformations $i_R(u)$, $i_G(u)$, $i_B(u)$ with integer values are used, satisfying the conditions

$$0 \leq i_R(u), i_G(u), i_B(u) \leq G-1, \quad \text{for} \quad u = 0,1,...,G-1$$

and

$$i_R(u) + i_G(u) + i_B(u) = G-1, \quad \text{for} \quad u = 0, 1, ..., G-1 .$$

In this case, the image transformation is given by

$$h(x, y) = (\ i_R(f(x, y)),\ i_G(f(x, y)),\ i_B(f(x, y))\ ).$$

In Figure 5.9 is displayed an example of the definition of such color transformations.

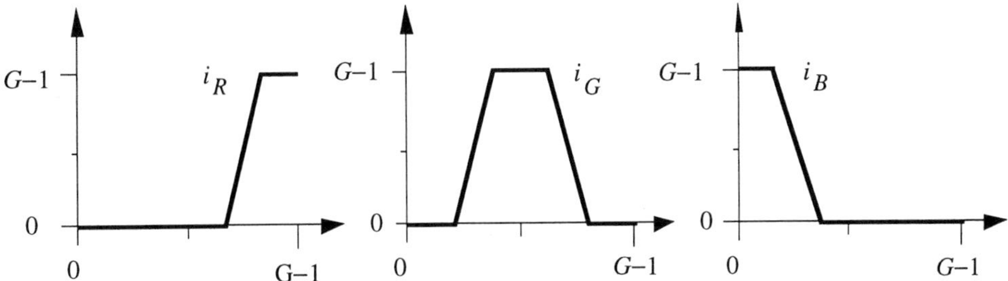

Figure 5.9: Example of color transformation functions.

(3) *Look-up-tables of computer displays* can be used for efficient pseudo coloring. The representation of images on a screen is modified by altering the look-up-tables instead of computing new image values. Pseudo-coloring can be realized with this technique in real-time since about beginning of the 70's.

The following is a program for computing pseudo-coloring explicitly. For the input of colors with the interval method, the use of a program for interactive color specification (see some computer graphics textbook) is suggested. For defining color transformation functions, it is sufficient to define two functions instead of three because of the assumed normalization in **(2)**. Such functions can be generated by an interactive drawing of gradation curves, e.g. by the selection of points on the curve and by moving these points to new locations. Such interactive solutions for the input of graphical data are described in the computer graphics literature.

(4) *Control structure*: point operator, but special control structure (here specified) because data are generated in three images h_1 (channel for Red), h_2 (channel for Green) and h_3 (channel for Blue).

Required data arrays: line store arrays $BUF(1...M)$ for input of lines from input image $f(x, y)$,
and $BUFOUT(1, 1...M)$, $BUFOUT(2, 1...M)$, $BUFOUT(3, 1...M)$ for output into resultant images h_1,h_2 and h_3.

```
begin
    input of variant VAR = 1 or VAR = 2;
    if (VAR = 1) then begin
        input of n > 1;
        input of the interval limits l_1 , ..., l_{n-1} with 0 < l_1 < ... < l_{n-1} < G;
        l_0 := 0;          l_n := G;
        input of the colors c_1, c_2, ..., c_n with c_k = (r_k, g_k, b_k)
                                  { interactive input is suggested, see (3) }
    end {then}
```

```
else begin
        input of the color transformation functions i_R(u) and i_G(u);
        for u = 0 to G–1 do i_B(u) := max{0, G – i_R(u) – i_G(u) – 1};
                                { interactive input is suggested, see (3) }
        end {else};
    for y := 1 to N do begin
        read line y from f into the line store array BUF(1...M);
        for x := 1 to M do begin
            u := BUF(x);
            if (VAR = 1) then begin
                k := 0;
                repeat k:= k+1 until (u < l_k);
                v1 := r_k ;      v2 := g_k ;      v3 := b_k
                end {then}
            else begin
                v1 := i_R(u);    v2 := i_G(u);    v3 := i_B(u)
                end {else}
            BUFOUT(1, x) := v1;          BUFOUT(2, x) := v2;
            BUFOUT(3, x) := v3
            end {for};
        write BUFOUT(1, 1...M) into line y of the resultant image h1;
        write BUFOUT(2, 1...M) into line y of the resultant image h2;
        write BUFOUT(3, 1...M) into line y of the resultant image h3
    end {for}
end
```

(5) For the design of interactive programs for the input of graphical data see, e.g.,

Fellner, W.D.: *Computer Grafik*. Wissenschaftsverlag, Mannheim, 1988.

Foley, J.D., van Dam, A., Feiner, S.K., Hughes, J.F.: *Computer Graphics*. 2nd ed., Addison-Wesley, Reading, 1990.

6 WINDOW FUNCTIONS AND LOCAL OPERATORS

This Chapter deals with *local operators* where every computed gray value depends upon all available gray values inside the placed picture window $\mathbf{F}(f, p)$, and upon their spatial distribution inside this window. The operators are presented in Sections according to the type of task for which they might be applied. Most of the current tasks of low-level vision are addressed by the different Sections, with aims such as image improvement, contour finding or image segmentation. In this Chapter, operators are presented for the smoothing of gray values for noise suppression, for the edge detection, for the enhancement of image details, of texture or of curves, and for the region growing.

The operators of Section 6.5 are characterized by the same methodical principles but applied for different tasks. In this Section the extraordinary flexibility of rank order operators becomes clear as well as the manifold possibilities for realizing specific signal dependent image transformations.

6.1 SMOOTHING AND NOISE REDUCTION

The operators of this Section serve for *smoothing* the gray value functions. They are based on the assumption that small local gray value fluctuations are impairments. The desired image enhancement should be realized without loss in image sharpness. In an extended sense, smoothing can also be used for flattening of texture patterns of a noise-free image. In such a case, smoothing is directed on the gray value agglomeration for image segmentation instead of the image restoration.

6.1.1 Linear Convolution with User-defined Kernel

(1) This operator realizes a *linear convolution* with a $n \times n$ convolution kernel. Typical values are $n = 3, 5$ or 7. All $a = n^2$ coefficient values can be specified by the user

at program start. A convolution kernel is *symmetric* if there are two diagonal symmetry axes and two axes-parallel symmetry axes. For symmetric convolution kernels it is sufficient to specify $Z = (n^2 + 4n + 3) / 8$ coefficients at program start. Two principal cases are considered in the program:

(A) The sum SK of all coefficients of the convolution kernel is not equal to 0. The discrete impulse answer of such a filter has *low-pass character*. Thus it contains a background component as a constant part. For obtaining this component all coefficient values are normalized to SK in the program.

(B) It holds $SK = 0$. In this case the filter has *high-pass character* and emphasizes higher spatial frequencies (*contrast enhancement*, *edge detection* etc.). The impulse answer of the operator contains negative parts which can only be represented in the resultant image by non-negative gray values. The user has two choices for the definition of the resultant image. The output is computed either by

($B1$) taking the absolute value or by

($B2$) shifting the value corresponding to black to the gray level $G / 2$, and by adding the convolution result afterwards. In general a reduction of the gray value dynamics of the convolution result is not necessary for ensuring output values in the range of $0,1,...,G-1$.

Attributes:

Images:	$M \times N$ gray value images
Operator:	window operator
Kernel:	linear (convolution with defined kernel)

Inputs:

Variant $SYM = 1$ for symmetric convolution kernel, and $SYM = 0$ for non-symmetric kernel,

coefficients $K(z)$ of the convolution kernel, for $z = 1, 2, ..., a$ or for $z = 1, 2, ..., Z$,

if $SK = 0$ then Variant $VAR = 1$ for the gray value representation by absolute value computation, or $VAR = 2$ for shifting the gray value scale,

parameter $PARSEQ$ for selecting either *parallel* or *sequential processing*.

(2) The $n \times n$ coefficients of the convolution kernel are applied in parallel or sequentially to a placed window $\mathbf{F}(p)$ of size $n \times n$. If

$$SK = \sum_{z=1}^{a} K(z) \neq 0$$

then

$$h(p) = \frac{1}{SK} \sum_{z=1}^{a} K(z) \cdot F(z) \ .$$

Otherwise, if $SK = 0$ then

$$h(p) = \begin{cases} \left| \sum_{z=1}^{a} K(z) \cdot F(z) \right| & , \text{if } VAR = 1 \\ \frac{G}{2} + \sum_{z=1}^{a} K(z) \cdot F(z) & , \text{if } VAR = 2 \ . \end{cases}$$

In both cases, a mapping of the results into the gray value range 0 to $G - 1$ can be obtained by subsequent application of the procedure *ADJUST* in Section 2.1.

(3) This operator combines basic ideas of several, often used *linear filters*. The convolution kernel has to be defined by the user at program start. Often the convolution kernel can be assumed to be symmetric. In this case, the input of a (i.e. for $n = 3, 5, 7$ up to 49 coefficients) coefficients is simplified for the user by restricting the input to $(n^2 + 4n + 3) / 8$ coefficients of half of a quadrant of the window. Because of the symmetry conditions, these coefficients can be extended to the remaining part of the window. The order of the coefficient input corresponds to the indices of the window gray values in the list $F(z)$, for $z = 1,..., a$, as shown in Figure 3.5. In the symmetric case, the coefficient indices correspond to the spatial ordering as shown in Figure 6.1. These indices are equal to the order of the coefficient input. For $p = (x, y)$, the coefficient $K(1)$ is assigned to the gray value $f(x, y)$, the coefficient $K(10)$ is assigned to the gray value $f(x + 3, y + 3)$ etc. Roughly it can be distinguished between low-pass or high-pass filters for all realizable filter types :

(a) *Low-pass filters* do not alter the original gray value if all gray values inside the picture window are identical. Therefore, in the case $SK \neq 0$ this sum SK is used for normalization. The coefficients $K(z)$ of the filter are the coefficients of a two-dimensional discrete impulse answer of the desired low-pass. In this way, different low-pass characteristics can be realized.

(b) *High-pass filters* produce output gray value zero for zones with constant gray value. Thus SK has to be zero. Therefore, some filter coefficients must be negative, i.e. the resultant "gray values" can be also negative. It has to be decided how negative results are mapped into the non-negative gray value scale of 0 to $G - 1$.

The first mentioned possibility ($B1$) in point **(1)** makes sense, e.g., for the representation of results of edge operators, or for the representation of extracted configurations which should be recognized independently with respect to the gradient direction. Such configurations can be lines, isolated points or others. Absolute value computation is a non-linear process, and it is used here only for representation of results of an intrinsically linear filter.

The second variant ($B2$) is a linear mapping. This variant is basically information preserving. The full gray value scale has to be scaled to the proper dimension for a univocal mapping of all the possible values of the filter output.

(c) Also *other filters* as typical low-pass or high-pass ones can be realized by means of the program. There is no restriction to the parameter setting of the possible coefficient sets. A specified coefficient set is classified into cases (a) or (b) depending upon $SK \neq 0$ or $SK = 0$.

For realizing recursive linear filters, the control structure contains also the option of sequential processing, cp. Section 3.3.1.

			10
		6	9
	3	5	8
1	2	4	7

Figure 6.1: The spatial index order of the filter coefficients in the case of a symmetric convolution kernel.

(4) *Control structure*: window operator in Figure 3.6, centered, with $n = 3, 5$ or 7 and $a = n^2$

Required data arrays: $K(1...a)$ and $KZ(1...Z)$ for intermediate storing of the coefficients of the convolution kernel,

$TAB(1 \; ... \; a - 1) = (2, 3, 2, 3, 2, 3, 2, 3, 4, 5, 6, 5, 4, 5, 6, 5, 4, 5, 6, 5, 4, 5, 6, 5, 7, 8, 9, 10, 9, 8, 7, 8, 9, 10, 9, 8, 7, 8, 9, 10, 9, 8, 7, 8, 9, 10, 9, 8)$. This array describes the mapping between the indices z, $1 \leq z < a$, of coefficients $K(z)$ and the indices $TAB(z)$, $1 < TAB(z) \leq Z$, of the reduced array $KZ(...)$ for the case of symmetric kernels.

Inputs and initialization:

```
input of SYM and PARSEQ;
if (SYM = 0) then
        input of a  coefficients K(1), ..., K(a)
```

```
else begin
        Z := (n^2 + 4n + 3) / 8;
        input of Z coefficients KZ(1), ..., KZ(Z);
        K(a) := KZ(1);
        for z := 1 to a - 1 do          K(z) := KZ(TAB(z))
        end {else};
SK := K(a);
for z := 1 to a - 1 do          SK := SK + K(z);
if (SK = 0) then input of parameter VAR
```

Operator kernel:

```
              {transform F(z), with 1 ≤ z ≤ a, into v = h(x, y), cp. Figure 3.6 }
S := K(a) · F(a);
for z := 1 to a - 1 do          S := S + K(z) · F(z);
if (SK = 0) then
        if (VAR = 1) then       v := | S |
                       else     v:= G / 2+ S
else    v := S / SK
```

(5) Haberäcker, P.: *Digitale Bildverarbeitung*. Carl Hanser Verlag, München, 3rd ed., 1989.

Wahl, F.M.: *Digital Image Processing*. Artech House, Norwood, 1987.

Zamperoni, P.: *Methoden der digitalen Bildsignalverarbeitung*. Vieweg Verlag, Wiesbaden, 2nd ed., 1991.

6.1.2 Smoothing with a Separable Unweighted Averaging Filter

(1) A *box filter* is a simple *smoothing* operator performing an unweighted arithmetic averaging, cp. *AVERAGE* in Section 1.3.3. A box filter transforms an image as if this image would be scanned through an ideal slot. The convolution kernel is separable if all coefficients are identical. This property is used for the reduction of the computing time, cp. Section 3.1.3.

Attributes:

Images:	$M \times N$ gray value images
Operator:	window operator, in two runs
Kernel:	linear (convolution with fixed kernel), separable

Inputs:

window size n .

Figure 6.2: Results of smoothing an image of a natural scene (top left) by applying different methods, in each case for 7×7 windows. Top right: separable *box filter*. Below left: separable *binomial filter*. Below right: *smoothing in a selected neighborhood.*

(2) For the values $F(1)$, ..., $F(a)$ as read into the placed window $\mathbf{F}(p)$ it is

$$h(p) = \frac{1}{a} \sum_{z=1}^{a} F(z),$$

where $a = n^2$

(3) In comparison to the *Gaussian low-pass*, the box filter is characterized by a smaller value of the product

$$\text{band width} \times \text{edge width}.$$

The occurrence of "ringing" near the edges, caused by the box filter, is disadvantageous. However for image processing the strength of the realized smoothing effect (e.g. for *noise suppression*) is often as important as the faithful reproduction of the gray value function in the transmission band of the filter. As far as this property is concerned, the box filter is superior to the Gaussian low-pass and to the *binomial filter* (cp. Section 6.1.3).

The box filter and the binomial filter are separable, cp. Section 3.1.3. They can be realized in two runs, using window sizes of $n \times 1$ or $1 \times n$ pixels. The successive convolution with the two one-dimensional kernels is equivalent to the convolution with the kernel of the desired two-dimensional low-pass.

The computing time reduction by separation is of great importance for smoothing filters, because smoothing with quite large windows is often necessary. Sometimes these windows can not be considered any more as "much smaller than the image grid". In the case of large operator windows, say of the magnitude of $n = N / 5$, the image border can not remain unprocessed. Here, the intersection between placed window and image grid is chosen for defining the current operator window in the image border. Thus for border points the operator window is smaller than $n \times n$, and the smoothing effect is weaker.

For a further reduction of computing time, averaging in the given window is realized by updating a running average: switching to the next image point, those gray values which are not contained any more in the new one-dimensional window are subtracted, and the new gray values are added, cp. Section 3.1.4.

In Figure 6.2 the result of a box filter is shown in comparison to results of two other smoothing filters.

(4) *Control structure*: special for one-dimensional windows, explicitly given here
Required data arrays: line store arrays $L1(z)$, $L2(z)$ and $L3(z)$, with $z = 1, ..., M$, for accumulation of intermediate results
Inputs and initialization:

```
        input of odd n;
        k = (n – 1) / 2;                   { half of the window size }
        XM := N – k;          YM := M – k
```

First run with n × 1-window

```
for y := 1 to N do begin
    read line y from input image f into array L1(1...M);
    Q := 0;
    for x := 1 to k do     Q := Q + L1(x);
    Q := Q / k;
    for x := 1 to M do begin
        if (x ≤ k + 1) then
                Q := [Q · (x + k – 1) + L1(x + k)] / (x + k)
        else if (x > XM) then
            Q := [Q · (n – x + XM + 1) – L1(x – k – 1)] / (n – x + XM)
        else
                Q := [Q · n + L1(x + k) – L1(x – k – 1)] / n;
```

```
        L2(x) := Q
        end {for};
        write L2(1...M) into line y of intermediate image g
    end {for}
```

Second run with n × 1-window

```
    for x := 1 to M do   L1(x) := 0;
    for y := 1 to k do begin
        read line y from intermediate image g into array L2(1...M);
        for x := 1 to M do   L1(x) := L1(x) + L2(x);
        end {for};
    for x := 1 to M do   L1(x) := L1(x) / k ;
    for y := 1 to N do begin
        if (y ≤ k + 1) then begin
            read line y + k from g into array L2;
            for x := 1 to M do
                L1(x) := [L1(x) · (y + k – 1) + L2(x)] / (y + k)
            end {then}
        else if (y > YM) then begin
            read line y – k – 1 from g into array L2;
            for x := 1 to M do
                L1(x) := [L1(x) · (n – y + YM + 1) – L2(x)] / (n – y + YM)
            end {then}
        else begin
            read line y + k from g into array L2;
            read line y – k – 1 from g into array L3;
            for x := 1 to M do
                L1(x) := [L1(x) · n + L2(x) – L3(x)] / n
            end {else};
        write array L1 into line y of resultant image h
        end {for}
```

(5) The separability of this convolution kernel has been treated in

Jähne, B.: *Digital Image Processing*. Springer, Berlin, 1991.

For comparisons between different linear smoothing filters, see

Wahl, F.M.: *Digital Image Processing*. Artech House, Norwood, 1987.

6.1.3 Smoothing with a Separable Binomial Filter

(1) This operator is a *linear low-pass filter* for smoothing by weighted averaging. For windows of $n \times n$ pixels, the multiplicative weights are derived from the binomial coefficients

$$\binom{n-1}{i}, \quad \text{with} \quad 0 \le i < n \ .$$

In comparison to the box filter in Section 6.1.2, the smoothing effect of this *binomial filter* is weaker. The spectrum of this convolution kernel is nearly isotropic and it is a-periodically decreasing. Separability of this filter allows low computing time even in case of large windows.

Attributes:

Images:	$M \times N$ gray value image
Operator:	window operator, in two runs
Kernel:	linear (convolution with fixed kernel), separable

Inputs:

window size n .

(2) It is suggested to use the centered two-dimensional *ij*-coordinate system for a clearer presentation of this operator. However for the realization by separation, a different one-dimensional pixel index is used. The resultant gray value is given by

$$h(x, y) = \frac{1}{S_n} \sum_{j=0}^{n-1} \sum_{i=0}^{n-1} C_{ij} \cdot f(x - \mathrm{k} + i, y - \mathrm{k} + j) \quad \text{with} \quad S_n = \sum_{j=0}^{n-1} \sum_{i=0}^{n-1} C_{ij} \ .$$

The weights C_{ij} are the elements of the following $n \times n$ matrix $\| C_{ij} \|$,

$$\| C_{ij} \| = \left[\binom{n-1}{0} \cdots \binom{n-1}{i} \cdots \binom{n-1}{n-1} \right] * \left[\binom{n-1}{0} \cdots \binom{n-1}{j} \cdots \binom{n-1}{n-1} \right]^T .$$

The *convolution* operation is denoted by $*$. The coefficient

$$C_{ij} = \binom{n-1}{i} \cdot \binom{n-1}{j} \quad , \text{for} \quad 0 \le i, j \le n-1,$$

is equal to the product of the i-th element of the line vector with the j-th element of the column vector. Note that both masks of the line and column vector are identical. In general it holds

$$S_n = \left[\sum_{i=0}^{n-1} \binom{n-1}{i} \right]^2 = (2^{n-1})^2 = 2^{2n-2} .$$

(3) The binomial filter can be applied as a smoothing filter for *noise suppression*. The smoothing effect of this filter corresponds to its low-pass characteristics. It possesses some of the good properties of the *Gaussian low-pass* (isotropy, steep and monotonously decreasing transfer function in the frequency domain). However it is easier to realize in comparison to the Gaussian low-pass. The *box filter* of Section 6.1.2 does not possess these good properties. This is due to the fact that the sharp zero-rise-time impulse answer of this filter in the spatial domain corresponds to an unbounded transfer function in the frequency domain, with sinus-type behavior and decreasing amplitudes. In comparison to the box filter, the smoothing effect of the binomial filter is reduced. In order to achieve a strong smoothing with the binomial filter, larger windows have to be used. This leads to higher demands in computing time. However the separable realization (cp. Section 3.1.3) leads to a time complexity of order $2n$, instead of n^2 for the non-separated window function. The separability of this operator follows immediately from the definition of the kernel as a convolution product of two (identical) coefficient masks, one for the horizontal and one for the vertical direction. The realization of this operator takes place in two runs, one for convoluting the image with the horizontal mask, and one for convoluting with the vertical mask. Therefore, an image array g is needed for saving the intermediate results of the first run. This array g is used as input for the second run. The values of the *binomial coefficients*

$$\binom{n-1}{i} \qquad \text{for } i = 0, ..., k$$

are computed before starting the first run. They are stored in a one-dimensional array $C(i)$. The remaining coefficients follow from the symmetry relation, i.e.

$$\binom{n-1}{k-r} = \binom{n-1}{k+r} \qquad \text{for } r = 0, ..., k. .$$

The computation of the coefficients for a selected value n can be realized iteratively, starting with $n_0 = 3$. Odd numbers n are preferably used for the window sizes. Thus a one-dimensional temporary array $CC(1 \dots n-1)$ is convenient for storing intermediate results for even values of n_t, i.e. for $t = 1, 3, \dots, n-4$.

Figure 6.2. shows the result of applying this smoothing filter in comparison to two other methods.

(4) *Control structure*: special, here specified

Required data arrays: $C(1 \dots n_M)$ for the computed binomial coefficients for window sizes $n \leq n_M$, where n_M denotes the maximum allowed window size, and for intermediate results of the t-th iteration with odd n_t,

$CC(1 \dots n_M)$ is the intermediate memory for the results of the t-th iteration with even n_t,

(By choosing the format *integer*4* for arrays C and CC, the resultant upper bound of n_M is 23, i.e. binomial coefficients for $n = 23$ can be uniquely represented in this format, but not for $n = 24$.)

$LE(1 \dots M)$, $SC(1 \dots M)$ are the line store arrays for reading lines from the original image, and for writing lines into the resultant image,

$W(1 \dots M)$ is the intermediate memory for storing one line in the second run (final computation of the resultant gray values in the current line).

Inputs and initialization:

```
input of n;
                              { initialization of arrays C and CC by value 0}
for z := 1 to n_M do begin
     C(z) := 0 ;     CC(z) := 0
     end {for};
                    { initialization of the binomial coefficients for n = 3 }

     C(1) := 1;      C(3) := 1;
     C(2) := 2;      CC(1) := 1;
     Z := (n – 3) / 2 ;
  { sum of all coefficients of the one-dimensional mask of length n }
     S := 2^(n–1) ;
                { computation of the one-dimensional binomial coefficients
                                              based on the Pascal triangle}
 for t_1 := 1 to Z do begin
     for t_2 := 2 to (2 + 2 t_1) do
          CC(t_2) := C(t_2) + C(t_2 – 1);
     for t_2 := 2 to (3 + 2 t_1) do
          C(t_2) := CC(t_2) + CC(t_2 – 1)
     end {for}
```

First run

```
for y := 1 to N do begin
        read line y from the original image into line store array LE(1...M);
        for x := k + 1 to M – k do begin
                Q := C(k + 1) · LE(x);
                for t3 := 1 to k do
                        Q := Q + C(k+1 – t3) · [ LE(x – t3) + LE(x + t3) ];
                SC(x) := Q / S
                end {for};
        write array LE(1...M) into line y of intermediate image g
        end {for}
```

Second run

```
for y := k + 1 to N – k do begin
        for x := 1 to M do    W(x) := 0 ;
        for t4 := 1 to n do begin
                read line (y – k – 1 + t4) from image g into line store array
                        LE(1...M);
                for x := 1 to M do    W(x) := W(x) + C(t4) · LE(x)
                end {for};
        for x := 1 to M do    SC(x) := W(x) / S ;
        write line SC(1...M) into line y of the resultant image h
        end {for}
```

(5) Jähne, B.: *Digital Image Processing*. Springer, Berlin, 1991.

6.1.4 Smoothing in a Selected Neighborhood

(1) This operator can be viewed as a weighted averaging. An appropriate weight is computed for each pixel in the placed window $\mathbf{F}(p)$. These weights depend upon the difference between the pixel gray value and the gray value $f(p)$ of the current point p. This aims for a *noise suppression* without loss in *edge sharpness*.

Attributes:

Images:	$M \times N$ gray value image
Operator:	window operator
Kernel:	data driven, order independent

Inputs:

window size n .

(2) The gray values of the input image relative to the current point p and with the centered ij-coordinate system are assumed to be stored in $F(1),...,F(a)$. For the resultant image h it is

$$h(p) = \frac{1}{S} \sum_{z=1}^{a} [\,(G - 1) - |F(a) - F(z)|] \cdot F(z)$$

with

$$S = \sum_{z=1}^{a} [(G - 1) - |F(a) - F(z)|]\ .$$

(3) The disadvantage of the *linear smoothing filters* is attenuating, besides the noise, also the high spatial frequencies. This leads to a loss of details and to unsharp edges. The aim of many non-linear filtering approaches is to reduce noise without blurring edges. This is also the aim of averaging in a selected neighborhood.

The basic idea of this approach is as follows: Neighbors of the current point p are examined to estimate whether they belong to the same image region as p, or not. In the positive case, these neighbors are assigned a higher weight than the neighbors estimated to belong to a different image region, e.g., because they are separated from p by a more or less pronounced edge. The task is to perform such an estimation as simply, and as efficiently as possible. Here one can choose between a "soft" and a "strong" decision criterion. A strong criterion classifies neighbors of p into two classes, i.e. elements of the same region or elements of another region. The filter as specified here realizes a soft weighting, where the degree of membership to the same region depends linearly upon the absolute gray value difference.

A comparison of this smoothing filter with two other smoothing filters is shown in Figure 6.2.

(4) *Control structure*: window operator in Figure 3.6, centered, with $a = n^2$
Required data array: $K(1 \ldots a - 1)$ for the weighting coefficients of all pixels of $\mathbf{F}(p)$ except of point p itself.

Inputs:

input of the odd parameter n

Operator kernel:

```
{transform F(z), with 1 ≤ z ≤ a, into v = h(x, y), cp. Figure 3.6 }

S := G – 1;     P := (G – 1) · F(a);
```

```
for z := 1 to a – 1 do begin
        K := (G – 1) – |F(a) – F(z)|;   P := P + K · F(z) ;   S := S + K
        end {for};
    v := P / S
```

(5) The described operator was basically proposed in

Nagao, M., Matsuyama, T.: *Edge preserving smoothing*. Computer Graphics and Image Processing **9** (1979), pp. 394-407.

Scher, A., Dias Velasco, F.R., Rosenfeld, A.: *Some new image smoothing techniques*. IEEE Trans. SMC-**10** (1980), pp. 153-158.

The specific form of the operator as given here is a simplified variant of the non-linear smoothing filter which was proposed in

Wang, D., Vagnucci, A.H., Li, C.C.: *Gradient inverse smoothing scheme and the evaluation of its performance*. Computer Graphics and Image Processing **15** (1981), pp. 167-181.

and which was improved in

Lee, J.S.: *Digital image smoothing and the sigma filter*. Computer Vision, Graphics, and Image Processing **24** (1983), pp. 255-269.

Wang, D., Wang, Q.: *A weighted averaging method for image smoothing*. Proc. 8*th* ICPR, Paris, 1988, pp. 981-983.

6.1.5 Adaptive Smoothing Based on Local Statistics

(1) This adaptive smoothing operator outputs a smoothed image ($h(p) \approx \mu$ = local average), where the observed *local variance* σ^2 is estimated to be characteristic of noise. It outputs no smoothed image ($h(p) \approx f(p)$) where the local variance is in the first approximation the same as that of the input image.

Attributes:

Images:	$M \times N$ gray value images
Operator:	window operator
Kernel:	data driven, order independent

Inputs:

window size n and estimated value of the noise variance σ_r^2.

(2) Assume that the local variance σ^2 and the average μ have been computed for the current point p of an image f. Then, let

$$h(p) = K \cdot f(p) + (1 - K)\mu \qquad \text{with} \qquad K = \frac{\sigma^2}{\sigma^2 + \sigma_r^2},$$

Figure 6.3: Upper half: original image (cp. Figure 2.7) with additive normally distributed noise (standard deviation of 20 gray levels), and the result of smoothing based on the local statistics with window size 7×7 (Section 6.1.5). Lower half: *radar remote sensing image* (SAR) and the result of *smoothing by adaptive quantile filtering* with window size 7×7 (Section 6.1.6)

where σ_r^2 is a value of the noise variance, estimated or given a-priori at program start.

(3) The adaptive behavior of the previous smoothing operator of Section 6.1.4 was obtained by fitting each filter coefficient to the local image data. Here, adaptation is controlled by the most important parameters of the local gray value first order statistics, namely by the average μ and by the variance σ^2. The resultant gray value is a weighted sum of the original gray value $f(p)$ and of μ, where the weighting coefficient K depends upon the local image data.

For theoretical considerations on this operator assume the following situation: The input image f is an additive superposition of a noise-free image f_u with variance σ_u^2, and a non-correlated noise with variance σ_r^2.

For the local variance inside the placed window $\mathbf{F}(p)$ one obtains

$$\sigma^2 = E[f(p) - \mu]^2 = \sigma_u^2 + \sigma_r^2 \ .$$

At edges or at other details it is $\sigma^2 >> \sigma_r^2$. Here no smoothing should take place; thus $K \approx 1$ is chosen. However, if the variance of the image is small inside the placed window $\mathbf{F}(p)$ then a noisy plateau is assumed in this window. In this case of $\sigma^2 \approx 0$, smoothing is realized by averaging by taking $K \approx 0$.

Assume that the average and the variance of the noise-free image $f_u(p)$ can be approximated by the observed values μ and σ^2. Then, the equation given in **(2)** specifies an optimal linear estimation of f_u in the sense of the *linear least mean square error* (*LLMSE*).

This smoothing method has the disadvantage that the value σ_r^2 should be known in advance, at least as an a-priori estimate. This can be achieved by measurements for similar images, generated under comparable conditions, or by knowledge about the image acquisition process. If a good estimation of σ_r^2 is ensured, then this fast method delivers good results with respect to noise suppression and preservation of edge sharpness. In Figure 6.3, an example is given for an application of this operator for smoothing an original image with additive normally distributed noise. This operator is here particularly efficient, because the value of $\sigma_r = 20$ gray levels was known in advance as a parameter used in generating the noisy image.

(4) *Control structure*: window operator in Figure 3.6, centered, with $a = n^2$

Inputs and initialization:

```
parameter inputs of n and σr ;                    { σr in gray levels}
Sr := σr²
```

Operator kernel:

```
              {transform F(z), with 1 ≤ z ≤ a, into v = h(x, y), cp. Figure 3.6}
σ2 := 0;          μ := 0;
for z:= 1 to a do begin
       μ := μ + F(z);        σ2 := σ2 + F(z)²          end {for};
μ := μ / a;       σ2 := σ2 / a – μ²;
K := σ2 / (σ2 + Sr);      w := K · F(a) + (1 – K) μ;
call ADJUST (w, v)
```

(5) Lee, J.S.: *Digital image enhancement and noise filtering by use of local statistics.* IEEE Trans. PAMI-**2** (1980), pp. 165-168.

Melsa, J.L., Cohn, D.L.: *Decision and Estimation Theory*. McGraw-Hill, New York, 1978.

6.1.6 Smoothing by Adaptive Quantile Filtering

(1) This operator is based on properties of the *median filter* (cp. Section 6.5.1). The strength of smoothing is controlled in dependence upon the local image data, more precisely, upon the value $L = \sigma / \mu$ where σ and μ denote the standard deviation and the average gray value of the given placed window, respectively. The adaptation should ensure the following behavior of this filter: Large values of L are typical for windows with large variance and/or low average gray value; here the smoothing should be weak. These regions are assumed to be rich in contrast and relatively dark. If L is small, i.e. the window has small variance and/or bright average gray value, then smoothing should be strong. These regions are assumed to be

(i) low in contrast, and a small value of σ is considered to be due to noise, cp. Section 6.1.5, and

(ii) bright and thus more affected by noise than dark regions.

The assumption (ii) holds for *multiplicative noise*, where the noise intensity is assumed to be proportional to the average gray value.

Attributes:

Images:	$M \times N$ gray value images
Operator:	window operator
Kernel:	data driven, order independent

Inputs:

window size n.

(2) For the analytical definition of the filter function, first the meaning of two thresholds T_0 and T_1, with $0 \leq T_0 \leq T_1 \leq G-1$, should be explained. Both thresholds are computed for each window. These thresholds T_0 and T_1 can be regarded as A_0- or A_1-quantiles of the local sum histogram $S(u) := SUMHIST(\mathbf{F}(f, p), u)$, cp. Section 1.3.2 and Figure 6.4 (a). Based on the histogram $S(u)$, these *quantiles* are defined as values

$$A_0 = S(T_0) \quad \text{and} \quad A_1 = S(T_1) \quad , \text{with} \quad 0 \leq A_0 \leq A_1 \leq 1.$$

In the program, at first A_0 and A_1 are computed. Then, by means of the inverse function $T = S^{-1}(A)$, the thresholds T_0 and T_1 are determined. In practice, $A_1 = 1 - A_0$ is assumed, i.e. it is taken symmetrically to A_0 with respect to the symmetry center 0.5. The operator result is defined as

$$h(p) = \begin{cases} T_0 & , \text{if } f(p) < T_0 \\ f(p) & , \text{if } T_0 \leq f(p) \leq T_1 \\ T_1 & , \text{if } f(p) > T_1 . \end{cases}$$

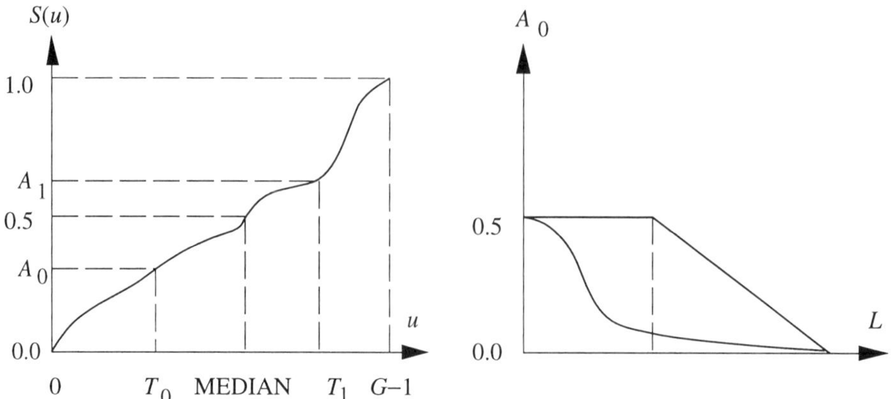

Figure 6.4: Left: a sum histogram $S(u)$ of the local gray values u. Right: functional relation between the control parameter $L = \sigma / \mu$ and the lower quantile A_0.

The values A_0 and A_1 are computed, as described below, adaptively in dependence of the local image data.

(3) This operator performs a smoothing by means of an *adaptive rank order filter* (cp. Section 6.5.3). Without adaptation, the original gray value $f(p)$ would always be replaced by the *median* $S^{-1}(0.5)$ of all gray values inside the placed window $\mathbf{F}(p)$. The same function would be realized for any local gray value distribution. Thus smoothing would always be strong. With adaptation, weaker smoothing effects can be obtained according to the control parameter L, see below. The possible differences between $f(p)$ and $h(p)$ are reduced in comparison with the fixed median operation. With respect to the transformation equation given in **(2)**, these differences are restricted to be zero ($h(p) = f(p)$), if $T_0 = 0$ and $T_1 = G - 1$.

The aim of the adaptation is to fit the filter function to the local image content in the sense of the typical cases as discussed in **(1)**. For this aim, the quotient

$$L = \frac{\sigma}{\mu}$$

between the *local standard deviation* and the local average is used as control parameter. This approach is similar to the one described in Section 6.1.5: large σ-values are assumed to characterize high-contrast image content (edges, detail in patterns), and small σ-values are assumed to characterize noise in flat regions. A_0 is close to 0 and A_1 is close to 1 for large L-values. Thus T_0 is close to 0 and T_1 is close to $G - 1$ in that case. The position of μ as denominator in the definition of L has also the effect that smoothing in bright image regions is more intense than in dark image regions. This corresponds to the assumption of multiplicative noise with stronger intensity in bright regions.

The function $A_0(L)$ is decreasing monotone. The analytic relation between A_0 and L can be given, for example, by a *Gaussian function* (as chosen here), or by a piece-wise linear curve. Figure 6.4 (b) illustrates both alternatives, but in the program a Gaussian function is assumed.

It has been observed that the density function $dS(u)/du$ normally is not symmetric, but rather similar to a *Rayleigh distribution*, e.g. for *SAR images*. For this reason the global average gray value is shifted with respect to the local mid-range gray value. Therefore, the normalized difference C between the average μ and the median,

$$C = \frac{\mu - MEDIAN}{G - 1},$$

is used as a *measure of asymmetry*. For compensating this gray value shift of the average, the following corrected values A_{0C} and A_{1C} are determined and used instead of A_0 and A_1,

$$A_{0C} = A_0\,(1 + C) \qquad \text{and} \qquad A_{1C} = (1 - A_0)(1 - C)\,.$$

These corrections are based on the locally measured value of C.

The lower half of Figure 6.3 gives an insight into the result obtained with this filter on a SAR image. The operator can be used for *segmentation preprocessing* because of its flattening effect upon a texture which is typical for SAR images. The *edge sharpness* is substantially preserved.

(4) *Control structure*: window operator in Figure 3.6, centered, with $a = n^2$
Required data array: $S(1...G)$ for the sum histogram of the current picture window $\mathbf{F}(f, p)$

Inputs and initialization:

input of parameter n

The inverse function $SM1(A) = S^{-1}(A)$, $0 \le A \le 1$, is defined by the function procedure given below.

Operator kernel:

{transform $F(z)$, with $1 \le z \le a$, into $v = h(x, y)$, cp. Figure 3.6 }

```
μ := 0 ;            σ2 := 0;
for u := 1 to G do    S(u) := 0;
for z:= 1 to a do begin
        μ := μ + F(z);        σ2 := σ2 + F(z)^2;
        S(F(z) + 1) := S(F(z) + 1) + 1
        end {for};
μ := μ / a;                              σ2 := σ2 / a – μ^2;
S(1) := S(1)/a;
for u := 2 to G do    S(u) := S(u–1) + S(u) / a;
L := √σ2 / μ;
A0 := 0.5 · exp(–3.912 · L^2);           A1 := 1 – A0;
C := (μ – SM1(0.5)) / (G – 1);
A0C := A0 (1 + C);                       A1C := A1 (1 – C);
T0 := SM1(A0C);                          T1 := SM1(A1C);
if (F(a) < T0) then                      { F(a) is the value of image f in point p}
                v := T0
        else    if (F(a) > T1) then   v := T1
                else    v := F(a)
```

Function procedure SM1:

```
function SM1(A: real): integer;
begin
        u := 1; flag := 0;
        while ((u ≤ G)  and  (flag = 0) do
                if (S(u) ≥ A) then begin
                        SM1 := u;      flag := 1
                        end {then}
                else u := u + 1
end {SM1}
```

(5) The *SAR-images* (synthetic aperture radar) are typical images affected by *multiplicative noise*. Therefore, the operator described in this Section originally was called *filter for smoothing of SAR images* by its developers

Alparone, L., Baronti, S., Carlà, R., Puglisi, C.: *A new adaptive digital filter for SAR images: test of performance for land and crop classification on Montespertoli area.* Proc. IGARSS (International Geographic and Remote Sensing Symposium), Houston, 1992, pp. 899-901.

and this filter was used for the suppression of *speckle noise* (that is the typical noise in SAR images), see Figure 6.3 below left. The behavior of function $S(T)$ has been also studied in this paper.

Besides this specific application, this operator can be suggested also for images with similarly structured noise. Some special and more complicated methods for smoothing multiplicative noise are described, e.g., in

Kuan, D.T., Sawchuk, A.A., Strand, T.C., Chavel, P.: *Adaptive noise smoothing filter for images with signal-dependent noise.* IEEE Trans. PAMI-**7** (1985), pp. 165-177.

6.1.7 Elimination of Small Objects in Binary Images

(1) For "cleaning" bilevel images, it can be useful to eliminate all objects which can be enclosed in a subwindow of $(n-2) \times (n-2)$ pixels. The window size used for this operator is n.

Attributes:

Images:	$M \times N$ bilevel images
Operator:	window operator, sequential
Kernel:	order dependent, logically structured

Inputs:

window size n

(2) Let **F** be a $n \times n$ window. Assume a *bilevel image* f where the gray value u_0 corresponds to the background regions, and the gray value u_1 to the object regions. For the resultant image h let $h(p) = u_0$, if $f(p) = u_0$ or if there exists an image point q as current point of a placed picture window $\mathbf{F}(q)$, such that

- $p \in \mathbf{F}(q)$,
- p is not on the border of $\mathbf{F}(q)$, and
- $f(r) = u_0$ for all pixels r on the border of $\mathbf{F}(q)$.

Otherwise, let $h(p) = u_1$, i.e. no gray value change occurs with respect to $f(p)$.

(3) In dealing with bilevel images sometimes it is desirable to eliminate small objects. The presence of these small objects, which can be considered as *artifacts*, can be due to the *binarization* or to the application of *morphological operators*. This elimination is important before *object counting* or before *component labeling* (cp. Section 7.1.1), for avoiding to count irrelevant components. Assume that the property "an object is completely contained inside a $(n-2) \times (n-2)$-window" is the criterion for the elimination.

This task could also be solved by several runs of the operation of *erosion* (cp. Section 6.5.2). The number of runs would depend upon the size of the objects which have to be eliminated. A different approach would be the use of a special sequential operator in two runs with opposite directions of scanning.

The given solution aims at a fast realization in a single run. A special sequential control structure is needed which can easily be implemented. During this single run, in a placed window **F**(p) not only the gray value of the current point p can be modified, but also the gray value of each point in the interior of **F**(p), by replacing u_1 with the value u_0.

The conventions for the control structure of the centered ij-coordinate system, as given in Section 3.3.1, are extended here by the possibility that all gray values in line store arrays $BUF(1...M,$ **ind**$(2...n-1))$ can be addressed and also modified.

The gray values from the interior of **F**(p) are stored in the locations

$$BUF(x-k+1...x+k-1, \mathbf{ind}(2...n-1)).$$

Assume an odd n and thus $k = (n-1)/2$. Whenever gray values in these line store arrays are changed from u_1 to u_0, also the corresponding gray values in the line store array $BUFOUT(1...M)$ must be changed into u_0. $BUFOUT$ is used for the intermediate storage of the processed gray values, before writing a processed line into the resultant image.

The operator examines whether all pixels on the window border have gray value u_0, i.e. whether they belong to the background. If so, then all pixels in the interior of the window are assigned the value u_0.

(4) *Control structure*: window operator in Figure 3.6, centered, with $a = n^2$, but with a few modifications (s. below)

Required data arrays (an example for $n \leq 7$):

$ZA(k) = (1, 9, 25)$ and $ZE(k) = (8, 24, 48)$ for the first and the last value of index z of the data arrays **xind**(z) and **yind**(z), cp. Section 3.3.1, for different window sizes $n = 2k + 1$. By varying z between $ZA(k)$ and $ZE(k)$, one obtains the coordinates of the pixels on the window border in the centered ij-coordinate system.

Inputs and initialization:

```
input of parameter n;                                    {odd n}
k := (n – 1) / 2
```

Operator kernel:

```
                                        { generate values in BUFOUT(x)}
      BUFOUT(x) := BUF(x + xind(a), ind(k + 1 + yind(a));
                                        { corresponds to h(p) := f(p)}
      z := ZA(k);     flag := 0;
      while ((z ≤ ZE(k) ∧ (flag = 0)) do
            if (BUF(x + xind(z), ind(k + 1 + yind(z))) ≠ u0)  then flag := 1
                                                          else z := z + 1;
      if (flag = 0) then begin
            for j := 1 – k to k – 1 do
                  for i := 1 – k to k – 1 do
                        BUF(x + i, ind(k + 1 + j)) := u0;
            for i := 1 – k to k – 1 do      BUFOUT(x + i) := u0
            end {if}
```

6.1.8 Halftoning by Means of Error Distribution

(1) A *binarization* of image gray values has to be realized for utilizing a bilevel image output, or for a representation on a b/w display. The visual impression of the original gray value image should be retained as well as possible. There should be no "sharp" transition between the two gray levels but a "smooth" binary representation, cp. Figure 5.4. The method of *error distribution* is exposed in this Section.

Attributes:

Images:	$M \times N$ gray value images
Operator:	window operator, sequential
Kernel:	analytical

Inputs:

binarization threshold S, $0 \le S \le G-1$

(2) A bilevel image h with values 0 and $G - 1$ is sequentially generated. The difference between the gray value $f(p)$ and the generated value $h(p)$ is considered to be the error. This error is distributed onto the next values of f in the 8-neighborhood of f.

The image points $p = (x, y)$ of the image grid **R** are processed in *raster scan order*. For the current point $p = (x, y)$ let

$$h(p) := \begin{cases} 0 & , \text{if } f(p) < S \\ G - 1 & , \text{if } f(p) \geq S \end{cases} .$$

The initial default threshold can be set to $S = G / 2$. The error $f(p) - h(p)$ is added to image f:

$$\begin{aligned} f(x+1, y) &:= f(x+1, y) &&+ 7 \cdot error / 16, \\ f(x-1, y+1) &:= f(x-1, y+1) &&+ 3 \cdot error / 16, \\ f(x, y+1) &:= f(x, y+1) &&+ 5 \cdot error / 16 \qquad \text{and} \\ f(x+1, y+1) &:= f(x+1, y+1) &&+ error / 16. \end{aligned}$$

The sum of the four added values should be exactly equal to *error*. Therefore, the fourth value *error* / 16 is replaced in the program by the difference of *error* to the sum of the first three summands.

(3) Generally the quality of the resultant image can be improved by adopting a less systematic scan order, e.g. alternating from left to right and from right to left.

(4) *Control structure*: window operator in Figure 3.6, centered, with $n = 3$ and $a = 9$, parallel (sequential processing takes place implicitly in the line store arrays, i.e. without actually modifying values of the input image f); a simple threshold operation is performed at border points (rows 1 and N, columns 1 and M) to ensure binarization also at these locations.

Inputs :

input of parameter S, $0 < S \leq G - 1$

Operator kernel:

```
                {transform F(z) , z = 1, ..., a, into v = h(x, y), cp. Figure 3.6}
if F(a) < S then   v := 0  else  v := G – 1;
error  := F(a) – v;
error1 := integer (7 · error / 16 + 0.5);
error2 := integer (3 · error / 16 + 0.5);
error3 := integer (5 · error / 16 + 0.5);
error4 := error – error1 – error2 – error3;
BUF(x + 1, ind(2))  :=  BUF(x + 1, ind(2))    + error1;
BUF(x – 1, ind(3))  :=  BUF(x – 1, ind(3))    + error2;
BUF(x, ind(3))      :=  BUF(x, ind(3))        + error3;
BUF(x + 1, ind(3))  :=  BUF(x + 1, ind(3))    + error4
```

(5) The method was originally proposed by

Floyd, R.W., Steinberg, L.: *An adaptive algorithm for spatial grey scale.* SED 75 Digest. Society for Information Display (1975), pp. 36-37.

Several variants of the method are described in

Knuth, D.E.: *Digital halftones by dot diffusion.* ACM Transact. on Graphics **6** (1987), pp. 245-273.

6.2 EDGE EXTRACTION

The operators described in this Section have the scope of generating a so-called "edge map" of an input image. An *edge map* contains all *contour lines* of objects, or of homogeneous image regions (e.g. as bright lines on a dark background), and contains only such lines. In the ideal case, an edge map should allow a transformation into a bilevel line image by simple binarization with a fixed threshold (cp. Section 5.3). Such an ideal bilevel line image should represent all relevant contours of the objects and of the regions in graphical form.

6.2.1 One-Pixel-Edge Operator

(1) The edge map extracted by this operator is defined by the difference between the current gray value and the minimum gray value inside a 3×3-window. The generated edges have a width of one pixel, in contrast to most of the other edge operators. Therefore, this operator is also called "one-sided". A binarization of the edge map is possible by means of a given constant threshold.

Attributes:

Images:	$M \times N$ gray value images
Operator:	window operator
Kernel:	order dependent, order statistical, analytical or logically structured

Inputs:

$VAR = 1$ for binarization, and $VAR = 2$ for an edge map without binarization, binarization threshold S, $0 \leq S \leq G - 1$ in case of $VAR = 1$.

(2) At first an intermediate image g is defined by

$$g(p) = f(p) - \min_{q \in \mathbf{F}(p)} f(q) .$$

Then, the final operator result is

$$h(p) = \begin{cases} g(p) & , \text{if } VAR = 2 \\ \text{binary otherwise, i.e.} \begin{cases} 0 & , \text{if } g(p) < S \\ G-1 & , \text{if } g(p) \geq S \end{cases} \end{cases} .$$

(3) This operator produces good *edge maps* if the gray value function consists substantially of "plateaus of constant gray value". The images segmented by *multi-threshold methods* are typical examples of such gray value functions (cp. also Section 5.5) where this operator can extract meaningful edge maps. Often edges are not recognized if they assume the shape of "gray value ramps". This operator's most-significant feature is that of generating edges with a single pixel width. This can be an important requirement in some applications. Assume bright objects on a dark background. Then, the detected edges are in the interior of the objects, i.e. they are their inner borders. Without binarization, the edge map is a gray value image where gray values have the meaning of edge strength. A *binarization* with a constant gray value threshold S can follow. By selecting S, only edges with a certain minimum height are represented.

(4) *Control structure*: window operator in Figure 3.6, centered, with $n = 3$ and $a = 9$

Inputs:

```
input of parameter S;                          { 0 < S ≤ G – 1}
```

Operator kernel:

```
            {transform F(z), with 1 ≤ z ≤ a, into v = h(x, y), cp. Figure 3.6 }
MIN := G – 1;
for z := 1 to 9 do
        if (F(z) < MIN) then MIN := F(z);
v:= F(9) – MIN;                                { here is F(9) = f(x, y) }
if (VAR = 1) then
        if (v < S) then        v := 0
                   else        v := G – 1
```

(5) Zamperoni, P.: *Methoden der digitalen Bildsignalverarbeitung*. Vieweg Verlag, Wiesbaden, 2nd ed., 1991.

6.2.2 Standard Edge Operators

(1) This program combines three "classical" edge operators (Sobel, Kirsch, pseudo-Laplace). Optionally a threshold can be specified for generating a bilevel edge map.

Attributes:

Images:	$M \times N$ gray value images
Operator:	window operator
Kernel:	order dependent, analytical or statistical

Inputs:

selection of edge operator, $VAR = 1$ for the *Sobel operator*, $VAR = 2$ for the *Kirsch operator*, or $VAR = 3$ for the *pseudo-Laplace operator*,
$BIN = 1$ for binarization, or $BIN = 0$ if no binarization is desired,
binarization threshold S, $0 \leq S \leq G - 1$, if $BIN = 1$.

(2) In case of the centered ij-coordinate system (cp. Figure 3.5 for $n = 3$) it was agreed that gray values $f(q)$, with $q \in \mathbf{F}(p)$, are stored in an array $F(z)$, $z = 1 \ldots 9$. For $n = 3$ it follows $a = 9$. $F(a)$ is the value at the current point p. The selected edge operators are defined as follows[1]:

SOBEL:
$$h(p) = |2\,(F(3) - F(7)) + F(2) + F(4) - F(6) - F(8)| + |2\,(F(5) - F(1)) + F(4) + F(6) - F(2) - F(8)|;$$

KIRSCH:
$$h(p) = [\max_{z=1,\ldots,8} \{|5\,(F(z) + F(z \oplus 1) + F(z \oplus 2)) - 3\,(F(z \oplus 3) + F(z \oplus 4) + F(z \oplus 5) + F(z \oplus 6) + F(z \oplus 7))|\}]/15$$

and

PSEUDO-LAPLACE: $h(p) = |F(1) + F(3) + F(5) + F(7) - 4 \cdot F(9)|$.

The first convolution kernel of the Sobel operator is shown below at left. The second convolution kernel follows from this by 90° rotation. Below at right one of the eight convolution kernels of the Kirsch operator is also shown, for $z = 1$. The seven remaining kernels follow by repeated 45° rotation:

+ 1	+ 2	+ 1
0	0	0
− 1	− 2	− 1

− 3	+ 5	+ 5
− 3	0	+ 5
− 3	− 3	− 3

[1] $\oplus$ denotes the modulo 8 addition.

(3) These three edge operators are described in many textbooks, cp. e.g. the references in **(5)**. Thus their properties are discussed here only very briefly.

The result of the Sobel operator is the sum (an other variant: the maximum) of two convolution results. One convolution has a kernel for horizontal edges, and the other for vertical edges. Each kernel corresponds to a derivation into a direction which is orthogonal to the edge direction. At the same time, each kernel performs also a certain amount of smoothing in edge direction for noise suppression.

The Kirsch operator utilizes a set of eight convolution kernels. Each convolution kernel can be viewed as a matched mask representing a model of an ideal edge in one of the eight basic directions (8-neighborhood of a regular square grid). The edge model of the Kirsch operator also contains smoothing in edge direction.

The pseudo-Laplace operator is the absolute value of the *Laplace operator* which is the two-dimensional second discrete derivative

$$\frac{\partial^2 f}{\partial x^2} + \frac{\partial^2 f}{\partial y^2}$$

of the gray value function f. The absolute value is used for dealing with negative (not representable) edge values which are treated as positive edge values. In contrast to the other two operators, no smoothing takes place. It follows that the pseudo-Laplace operator is very noise-sensitive. Furthermore, only the 4-neighborhood of the current point is used in the definition of this operator.

The coefficients of the convolution kernels are directly given above in analytical form. In all three cases, the sum of the coefficients of the convolution kernel is equal to 0. This is because the result of an edge operator should be 0 for an image zone with constant gray level.

Because of its noise sensitivity, the pseudo-Laplace operator is of limited practical relevance. It is given here only for completeness. Regarding noise suppression, Sobel and Kirsch operator demonstrate much better performance. Often, the Kirsch operator allows a better match to possible edge directions, because it contains a set of eight direction-dependent convolution kernels. However, because of the maximum computation for the eight convolution products, the Kirsch operator is slower than the Sobel operator.

In general, the edge width is of two pixels because in all of the three operators the absolute value is computed, i.e. edge pixels are detected on both sides of an edge. In Section 6.2.1 an "one-sided" edge operator for producing edges with one-pixel width has been described.

The generation of a gray-value *edge map* by means of the considered operators is often just the first step towards the computation of a bilevel contour image. The contours of objects should be represented by bright lines on a dark background, or vice-versa. This aim can be reached by binarization of a gray-value edge map by means of a constant gray value threshold. Thus the possibility of direct binarization is also included in the program as an optional thresholding.

Figure 6.5 illustrates the results of applying the Sobel or the Kirsch operator to a natural scene.

Figure 6.5: Examples for the application of three edge operators. Top left: original image, top right: *Sobel operator*, below left: *Kirsch operator*, below right: *morphological edge operator*.

(4) *Control structure*: window operator in Figure 3.6, centered, with $n = 3$ and $a = 9$

Inputs:

input of parameters *VAR*, *BIN* and, if $BIN = 1$, of S

Operator kernel:

{transform $F(z)$, with $1 \le z \le a$, into $v = h(x, y)$, cp. Figure 3.6 }

```
if (VAR = 1) then
        v := |2 (F(3) – F(7)) + F(2) + F(4) – F(6) – F(8)|
               + |2 ( F(5) – F(1)) + F(4) + F(6) – F(2) – F(8)|
else    if (VAR = 2) then begin
               v := 0;
               for z := 0 to 7 do begin
                      g = |5 (F(z) + F(z ⊕ 1) + F(z ⊕ 2))
               – 3 (F(z ⊕ 3) + F(z ⊕ 4)+ F(z ⊕ 5) + F(z ⊕ 6) + F(z ⊕ 7))| ;
                      if (g > v) then v := g
                      end {for};
               v := v / 15
               end {then}
        else
               v := |F(1) + F(3) + F(5) + F(7) – 4 F(9)|;
if (BIN = 1) then
        if (v < S) then       v := 0
        else                  v := G – 1
```


(5) Ballard, D.H., Brown, C.M.: *Computer Vision*. Prentice-Hall, Englewood Cliffs, 1982.

Haberäcker, P.: *Digitale Bildverarbeitung*. Carl Hanser Verlag, München, 3rd ed., 1989.

Jähne, B.: *Digital Image Processing*. Springer, Berlin, 1991.

Rosenfeld, A., Kak, A.C.: *Digital Image Processing*. Vol. 1 and 2, Academic Press, New York, 1982.

Wahl, F.M.: *Digital Image Processing*. Artech House, Norwood, 1987.

Zamperoni, P.: *Methoden der digitalen Bildsignalverarbeitung*. Vieweg Verlag, Wiesbaden, 2nd ed., 1991.

Davis, L.S.: *A survey of edge detection techniques*. Computer Graphics and Image Processing **4** (1975), pp. 248-270.

6.2.3. Morphological Edge Operator

(1) The edge detection is realized by a point-to-point combination of the resultant images of two different morphological operations on gray value images. This operator is characterized by a good detection of edge positions and by noise robustness. The second property is ensured by including a smoothing operation.

Attributes:

Images: $M \times N$ gray value images
Operator: window operator with 5×5 window
Kernel: order statistical, analytical

Inputs:

none, because the window size $n = 3$ is a constant (in practice, a 5×5 window operation is realized).

(2) Let $g(p)$ be the result of smoothing the original image $f(p)$ by a 3×3 *box filter* (cp. Section 6.1.2). Then, it is

$$h(p) = \mathbf{min} \left\{ \left(g(p) - \underset{q \in \mathbf{F}(p)}{\mathbf{min}} \{g(q)\} \right), \left(\underset{q \in \mathbf{F}(p)}{\mathbf{max}} \{g(q)\} - g(p) \right) \right\} .$$

(3) The principle of this edge detector is explained in Figure 6.6 by means of a one-dimensional gray value transition. In this operator, edge detection is based on the fact that erosion and dilation shift the gray value edges into opposite directions. The point-to-point difference images

$$dilation(g) - g \qquad \text{and} \qquad g - erosion(g)$$

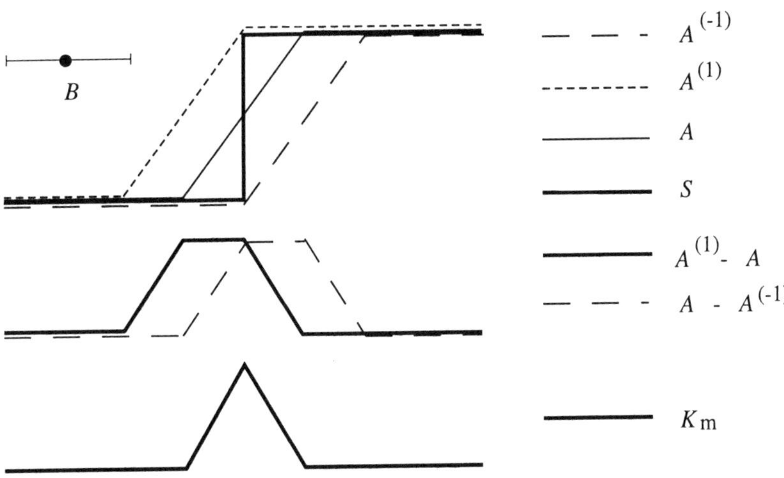

Figure 6.6: Explanation of the principle of the morphological edge operator by means of an one-dimensional edge S. A: smoothed edges, $A^{(-1)}$, $A^{(1)}$: erosion and dilation of A by the one-dimensional window B as structuring element; K_m: result of edge extraction.

are unsharp edge maps shifted into the direction of the bright or of the dark side the edge. The morphological operations of erosion and dilation are realized on gray value images by the operations

$$\min_{q \in \mathbf{F}(p)} \quad \text{or} \quad \max_{q \in \mathbf{F}(p)} .$$

The operator window $\mathbf{F}(p)$, here assumed to be consistently of size 3×3, plays the role of the structuring element. The point-by-point combination of these two edge maps by a minimum operation results into a sharp edge map. The maximum of this resultant image coincides with the position of the steepest edge slope.

The initial smoothing of the original image is necessary to avoid to start from ideal step edges. With such ideal edges, if no smoothing is performed, the final result would be zero. Furthermore, this initial smoothing realizes a noise suppression similar to the *Laplacian-of-Gaussian edge operator*, cp. Section 6.2.4.

The operator could be realized in two runs with 3×3 windows. *Smoothing* could be performed in the first run, and *erosion* or *dilation* could be performed in the second run simultaneously. However all these processes can be combined into a single run. For this aim, the required window size is 5×5 because of the convolution of a 3×3 window with itself. In this expanded window, at first 9 smoothed gray values of the elements inside the inner 3×3 window are calculated. Then, in the same run, the local extrema are computed. The gray value $f(p)$ of the current pixel, which is contained in all the smoothing windows, is not considered in the averaging operation.

Figure 6.5, below right, shows the result of an application of the morphological edge detector to the original image shown at top left.

(4) *Control structure*: window operator in Figure 3.6, centered, with $n = 5$
Required data array: $Q(1...9)$ for the smoothed gray value of the 3×3 window of the operator kernel.

Operator kernel:

{transform $F(i, j)$, with $-2 \le i, j \le 2$, into $v = h(x, y)$, cp. Figure 3.6 }

$Q(1) := F(0, 1) + F(1, 1) + F(2, 1) + F(1, 0) + F(2, 0)$
$\quad + F(0, -1) + F(1, -1) + F(2, -1);$
$Q(2) := F(0, 2) + F(1, 2) + F(2, 2) + F(0, 1) + F(1, 1)$
$\quad + F(2, 1) + F(1, 0) + F(2, 0);$
$Q(3) := F(-1, 2) + F(0, 2) + F(1, 2) + F(-1, 1) + F(0, 1)$
$\quad + F(1, 1) + F(-1, 0) + F(1, 0);$

```
Q(4) := F(- 2, 2)+F(- 1, 2) + F(0, 2) + F(- 2, 1) + F (- 1, 1)
        + F(0, 1)  + F(- 2, 0) + F(- 1, 0);
Q(5) := F(- 2, 1) + F(- 1, 1) + F(0, 1) + F(- 2, 0) + F(- 1, 0)
        + F(- 2, - 1) +  F(- 1, - 1) + F(0, - 1);
Q(6) := F(- 2, 0) + F(- 1, 0) + F(- 2, - 1) + F(- 1,  - 1) + F(0, - 1)
        + F(- 2, - 2) +  F(- 1, - 2) + F(0, - 2);
Q(7) := F(- 1, 0) + F(1, 0) + F(- 1, - 1) + F(0, - 1) + F(1, - 1)
        + F(- 1, - 2) +  F(0, - 2) + F(1, - 2);
Q(8) := F(1, 0) + F(2, 0) + F(0, - 1) + F(1, - 1) + F(2, - 1)
        + F(0, - 2) +  F(1, - 2) + F(2, - 2);
Q(9) := F(- 1, 1) + F(0, 1) + F(1, 1) + F(- 1, 0) + F(1, 0)
        + F(- 1, - 1) +  F(0, - 1) + F(1, - 1);
MIN := 8 · (G - 1);   MAX := 0;
for z := 1 to 9 do begin
        if   (Q(z) > MAX)    then MAX := Q(z);
        if   (Q(z) < MIN)    then MIN := Q(z)
        end {for};
v := min {(Q(9) - MIN)/8, (MAX - Q(9))/8}
```

(5) The morphological edge detector has been proposed by

Lee, J.S., Haralick, R.M., Shapiro, L.G.: *Morphologic edge detection*. Proc. 8th International Conference on Pattern Recognition, Paris, 1986, pp. 369-373.

and uses common morphological operations (erosion and dilation), cp.

Haralick, R.M., Sternberg, S.R., Zhuang, X.: *Image analysis using mathematical morphology*. IEEE Trans. PAMI-**9** (1987), pp. 532-550.

6.2.4 Edge Detection by Gaussian Filtering (LoG and DoG)

(1) The *Laplacian-of-Gaussian filter* is an edge operator also known as *LoG-filter*, as *Marr-Hildreth filter*, or as *Mexican-hat operator*. It combines a two-dimensional second derivation of the gray value function with a noise-reducing smoothing by a Gaussian low-pass filter. Edge pixels are detected at zero crossings of the final convolution result.

Attributes:

Images:	$M \times N$ gray value images
Operator:	window operator , in two runs
Kernel:	linear (convolution with a constant kernel), separable

Inputs:

window width n,

Variant $VAR = 1$ if the resultant image should be generated by shifting the middle gray value onto $G / 2$, or $VAR = 2$ if the resultant image should be generated by computing the absolute value,

parameter K for scaling has not to be inputted because it is implicitly determined by the program.

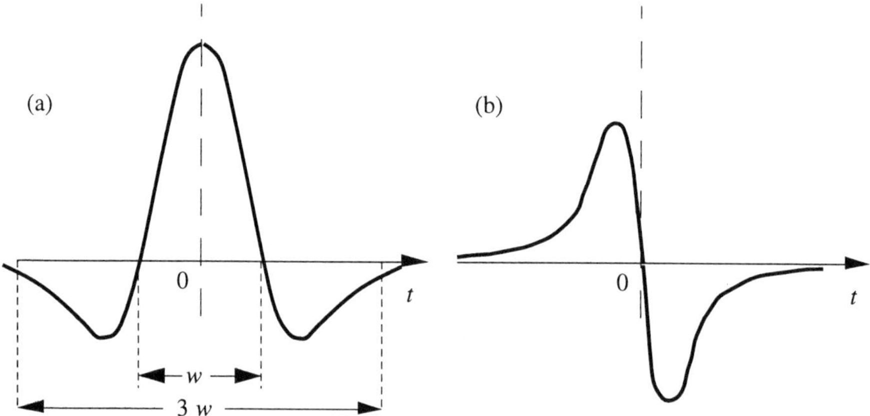

Figure 6.7: (a) One-dimensional cut of the impulse answer of the LoG-filter; the value $w = 2\sigma\sqrt{2}$ is the width of the central part with positive answer. (b) The answer of the LoG-filter for an one-dimensional ideal edge.

(2) With respect to the centered *ij*-coordinate system it is

$$h(p) = \sum_{j=-k}^{k} \sum_{i=-k}^{k} c(i, j) \cdot f(i, j)$$

with

$$c(i, j) = K \cdot \left(2 - \frac{i^2 + j^2}{\sigma^2} \right) \cdot \mathbf{exp}\left[- \frac{i^2 + j^2}{2\sigma^2} \right],$$

where K is a scaling constant and σ^2 denotes the *variance*. This formula for the convolution coefficients $c(i, j)$ holds only in principle, because the normalization of these coefficients is not yet considered. For $VAR = 1$ it is

$$h(p) := G / 2 \quad + \quad h(p) / 2,$$

and for $VAR = 2$ it is

$$h(p) := | h(p) | .$$

(3) The LoG-filter is a combination of the two-dimensional second derivative of the gray value function (*high-pass filter* for edge detection) and of a smoothing operation by means of a two-dimensional Gaussian filter (*low-pass filter* for noise suppression). Because of the linearity of both filters they can be applied in arbitrary order. The coefficients $c(i, j)$ of the discrete convolution kernel result from both filters. Assume for a moment that i, j are the continuous spatial coordinates. Then the coefficients represent the sum

$$\nabla^2 g(i, j) = \frac{\partial^2 g}{\partial i^2} + \frac{\partial^2 g}{\partial j^2} ,$$

where

$$g(i, j) = \mathbf{exp}\left[- \frac{(i^2 + j^2)}{2\sigma^2} \right]$$

is a *Gaussian function* with variance σ^2, and ∇^2 denotes the *Laplacian operator* (cp. Section 3.1.3 and 6.2.2).

The convolution of a gray value function with the LoG-operator kernel, as described in **(2)**, results into an edge map with expected gray value average at 0 because the LoG-filter has an overall band-pass characteristics. In Figure 6.7 (a), a cut of its rotation-symmetric impulse answer is illustrated qualitatively. The answer to an ideal edge (cp. Figure 6.7 (b)) shows two extrema of contrary signs. The *zero crossing* between both extrema corresponds to the position of the steepest edge slope. For representing the result of edge detection on a screen by a (non-negative) gray value function, one of the following three solutions can be chosen:

- The resultant values are divided by 2, and value $G / 2$ is subsequently added. Thus a negative extreme is darker than $G / 2$, and a positive extreme is brighter than $G / 2$. This solution has been chosen for the example shown in Figure 6.8.

- The resultant gray values are computed as absolute values. Then, in the resultant edge map edges appear as parallel bright lines with a thin dark line between them, and the background is dark.
- Only zero-crossings are labeled with a constant gray value. Thus it results a bilevel image with edges at exact positions.

The function $c(i, j)$ of the continuous spatial coordinates i, j is separable. It holds

$$c(i, j) = c_1(i) \cdot c_2(j) + c_2(i) \cdot c_1(j) \ ,$$

where $c_1(t)$ and $c_2(t)$ are suitable one-dimensional functions of the variable t, with

$$c_1(t) = K \cdot \left(1 - \frac{t^2}{\sigma^2}\right) \cdot \mathbf{exp}\left[- \frac{t^2}{2\sigma^2} \right] \quad \text{and} \quad c_2(t) = \mathbf{exp}\left[- \frac{t^2}{2\sigma^2} \right].$$

These functions are (so far assumed to be continuous) one-dimensional convolution kernels. The function $c_1(t)$ is also called the "small LoG" (cp. Figure 6.7 (a) for its shape), and the function $c_2(t)$ is a Gaussian function.

To realize the step from continuous to discrete convolution kernels, assume that the scanning interval τ, i.e. the image grid parameter, has been optimally chosen taking account of the *sampling theorem* of signal theory. The amplitude spectrum of the LoG-filter has band-pass characteristics. It is zero for the spatial frequency $\omega = 0$, it has its maximum at

$$\omega = \omega_M = \sqrt{2} \ / \ \sigma \ ,$$

and it decreases practically to zero at the *cutoff frequency* ω_g. This cutoff frequency ω_g can be situated empirically at $\omega_g = 3\omega_M$. Thus it follows that

$$\tau = \frac{2\pi}{2\omega_g} = \frac{\pi\sigma}{3\sqrt{2}} \approx 0.74 \cdot \sigma \ .$$

It holds $w = 2\sqrt{2}\,\sigma$ for the width w of the central part of the convolution kernel, shown in Figure 6.7 (a). A certain number W of discrete samples (at grid point positions of the image grid) falls inside this interval w. Since τ is the distance between the sampling points, altogether it follows that $W \approx 3.8$ based on the empirical estimate of ω_g, and on the assumption that τ satisfies the sampling theorem, i.e.

$$\tau = \frac{2\pi}{2\omega_g} \ .$$

Thus inside the interval w there should be at least $W \geq 4$ pixels. Altogether (cp. Figure 6.7 (a)), the convolution kernel should have a width of a least $n = 3W + 1 \geq 13$ pixels, i.e. should contain at least 13 filter coefficients, in order to avoid *aliasing*. However often in practical image processing LoG-filters are used with smaller windows which do not satisfy the condition derived above.

For realizing this operator, the coefficients of the convolution kernel in practice are not computed on the basis of the equation given in **(2)**, because the separability of the kernel allows a computing time reduction. This is of special importance because the LoG-filter is well suited for detecting thick edges which require large operator windows. Thus the coefficients of the two one-dimensional discrete convolution kernels $c_1(T)$ and $c_2(T)$ have to be determined for integer values of T in the range between $-1.5 \cdot W$ and $1.5 \cdot W$,

$$c_1(T) = K\left(1 - \frac{8T^2}{W^2}\right) \mathbf{exp}\left[-\frac{4T^2}{W^2}\right] \qquad \text{with } W = \frac{n}{3}\text{, and}$$

$$c_2(T) = \frac{1}{\sqrt{2\pi}\,\sigma}\, \mathbf{exp}\left[-\frac{4T^2}{W^2}\right] = \frac{2}{\sqrt{\pi}\,W}\, \mathbf{exp}\left[-\frac{4T^2}{W^2}\right].$$

The computed coefficients have to satisfy also the following two conditions,

$$C_1 = \sum c_1(T) = 0 \qquad \text{and} \qquad C_2 = \sum c_2(T) = 1\,.$$

About the first condition: The "small LoG" is a band-pass. Thus for a constant gray value in the operator window the resultant gray value should be equal to 0.

About the second condition: The result of smoothing a given picture window with constant gray value by means of the Gaussian low-pass should be the original gray value of the assumed image.

To satisfy these two conditions, the computed coefficient values are finally corrected by the value $-C_1 / n$ for the "small LoG" and by the value $(1 - C_2) / n$ for the Gaussian low-pass.

The algorithmic realization takes place in two runs, and for each run there are two parallel processes, as shown by the scheme

$$f(x,y) \rightarrow \begin{cases} \boxed{c_1(x)} \rightarrow u_1(x,y) \rightarrow \boxed{c_2(y)} \rightarrow h_1(x,y) \\ \boxed{c_2(x)} \rightarrow u_2(x,y) \rightarrow \boxed{c_1(y)} \rightarrow h_2(x,y) \end{cases} \oplus \rightarrow h(x,y)\,.$$

giving account of the separability of the convolution function.

After the first run, there are two resultant images $u_1(x, y)$ and $u_2(x, y)$. Again, there are two resultant images $h_1(x, y)$ and $h_2(x, y)$ after the second run. The sum of these two images h_1 and h_2 is the resultant image $h(x, y)$. But note that for practical realization, both runs can be integrated into a single run, as specified in the program description at point **(4)**.

Figure 6.8: Result of processing a natural scene with the LoG-filter and $n = 21$. The resultant gray value is shifted by $G / 2$ (scaled negative image).

Figure 6.8 shows the result of the LoG-filter with parameter $n = 21$. The input is a natural scene, and the output has been generated by shifting the middle gray value to $G / 2$.

Sometimes the *DoG-filter* (difference of Gaussians) is proposed alternatively to the LoG-filter. By applying the DoG-filter, the resultant image is the difference between two intermediate results obtained with different Gaussian low-pass filters. For these filters, the continuous convolution kernels (impulse answers) are Gaussian functions,

$$g_1(i, j) = \mathbf{exp}\left[-\frac{i^2 + j^2}{2\sigma_1^2}\right] \quad \text{and} \quad g_2(i, j) = \mathbf{exp}\left[-\frac{i^2 + j^2}{2\sigma_2^2}\right], \text{ with } \sigma_1 > \sigma_2 .$$

Several studies have investigated how to approximate a LoG-filter with a given value σ by means of a DoG-filter with values σ_1 and σ_2, where σ_1 and σ_2 have to be computed from σ. Assuming $\sigma = \sigma_1$ and $\sigma_1 = \gamma\sigma_2$ as a first approach, the best correspondence between transmission functions and between impulse answers of both filters has been observed for about $\gamma = 1.6$, where the DoG-filter has a stronger low-pass characteristics in comparison to the LoG-filter. This equivalence can be further improved by taking $\sigma_1 \neq \sigma$.

Also the DoG-filter allows separation, and thus a fast performance. The program follows the structure given for the LoG-filter. Then, the computing time and the memory requirements are about the same for both types of filters. Thus the LoG- and the DoG-filter can be realized so as to have an equivalent behavior in practical situations.

(4) *Control structure*: window operator in Figure 3.6, centered, here given explicitly

Required data arrays: $c_1(0...k_m)$ and $c_2(0...k_m)$ for the coefficients of the two one-dimensional convolution kernels, with $k_m = (n_m - 1) / 2$, where n_m is the maximum allowed window size (note that because of $c_1(-z) = c_1(z)$ and $c_2(-z) = c_2(z)$, for $z = 1,..., k$ only half of the coefficients have to be computed),

line store arrays $H_1(1...M)$ and $H_2(1...M)$ for the gray values of the current lines of the intermediate images $h_1(x, y)$ and $h_2(x, y)$,

line store array $H(1...M)$ for the finally resultant image h,

line store arrays $U_1(1...M, 1...n_m)$ and $U_2(1...M, 1...n_m)$ for the input gray values of the second run..

Because of the assumed row-wise exchange of the image data, the convolution with a $1 \times n$ window in the second run is realized for all image points of this line simultaneously.

Inputs and initialization:

input of window width n and parameter VAR;
for odd n it follows $k := (n - 1) / 2$;

{ computation of the filter coefficients}

$c_1(0) := 1$; $c_2(0) := 1$; $S_1 := 1$; $S_2 := 1$;
for $z := 1$ **to** k **do begin**
$c_1(z) := [1 - 8 (3z / n)^2]$ **exp** $[-4 (3z / n)^2]$;
$c_2(z) :=$ **exp** $[-4 (3z / n)^2]$;
$S_1 := S_1 + 2\, c_1(z)$; $S_2 := S_2 + (12 / n\sqrt{\pi})\, c_2(z)$
end {*for*};

```
{correction of the filter coefficients, parameter K is implicitly obtained by this
                                                                         correction}
        S1 := S1 / n;    S2 := (1 – S2) / n;
        for z := 0 to k do begin
                c1 (z) := c1 (z) – S1;                c2 (z) := c2 (z) + S2
                end {for}
```

Convolution in one single run

```
    for q := 1 to n do
            read line q into array BUF(1...M, q);
    for y := k + 1 to N – k do begin
            for q := 1 to n do
                    for x := k + 1 to M – k do begin
                            v1 := c1 (0)·BUF(x, q);        v2 := c2 (0)·BUF(x, q)
                            for i := 1 to k do begin
                                    v1 := v1 + [BUF(x – i, q) + BUF(x + i, q)]·c1(i);
                                    v2 := v2 + [BUF(x – i, q) + BUF(x + i, q)] · c2(i)
                                    end {for};
                            U1 (x, q) := v1;
                            U2 (x, q) := v2
                            end {for};
            for x := 1 to M do begin
                    H1 (x) := 0;    H2 (x) := 0
                    end {for};
            for x := 1 to M do
                    for q := 1 to n do begin
                            H1(x) := H1(x) + U1 (x, q) · c2 (|k + 1 – q|);
                            H2(x) := H2(x) + U2 (x, q) · c1 (|k + 1 – q|)
                            end {for};
            if (VAR = 1) then    H(x) := G / 2 + [H1 (x) + H2 (x)] / 2
                            else    H(x) := |H1 (x) + H2 (x)|;
            write the content of array H(1...M) into line y of the resultant image;
            read line y + k + 1 into line store array BUF(1...M, 1);
            permutation of indices q, 1 ≤ q ≤ n, of the line store arrays
                    BUF(1...M, q) according to the standard control-structure, cp.
                    Figure 3.8
            end {for}
```

(5) With respect to edge extraction, the LoG-filter has several favorable properties which are extensively described in textbooks, cp. e.g.

Jähne, B.: *Digital Image Processing*. Springer, Berlin, 1991.

These properties can be briefly listed as follows:

(a) the impulse answer corresponds to the *human perception of edges*, cp. publications by *David Marr*,

(b) low noise sensitivity,

(c) well detected edge locations, even for weak edge slopes.

For the theoretical background of the LoG-filter, see

Marr, D.: *Vision: A Computational Investigation into the Human Representation and Processing of Visual Information*. Freeman, San Francisco, 1982.

Marr, D., Hildreth, E.: *Theory of Edge Detection*. Proc. R. Soc. London, B207, 1980, pp. 187-217.

The realization of this filter by separation, its approximation by a difference of two low-pass filters, and the accuracy of the recognizable edge positions are dealt with, e.g., in

Siohan, P., Pelé, D., Ouvrard, V.: *Two design techniques for 2-D LoG-Filters*. in: Kunt, M. (ed.), Proc. SPIE Conf. on Visual Communications and Image Processing, Lausanne, 1990, pp. 970-981.

Sotak, G.E., Boyer, K.L.: *The Laplacian-of-Gaussian kernel: a formal analysis and design procedure for fast, accurate convolution and full-frame output*. Computer Vision, Graphics, and Image Processing **48** (1989), pp. 147-189.

The realization proposed in this Section follows

Chen, J.S., Huertas, A., Medioni, G.: *Fast convolution with Laplacian-of-Gaussian masks*. IEEE Trans. PAMI-**9** (1987), pp. 584-590.

Huertas, A., Medioni, G.: *Detection of intensity changes with subpixel accuracy using Laplacian-of-Gaussian masks*. IEEE Trans. PAMI-**8** (1986), pp. 651-664.

The suggested values for the filter parameters are based on experimental results in the comparative study

Della Giustina, D.: *Progetto nel dominio della frequenza di filtri LoG per l'estrazione dei contorni*. Master thesis No. 197/90, Dipartimento di Elettronica e Informatica, Università di Padova, 1990.

This extensive experimental research did prove that optimal approximations of the LoG-filter by the DoG-filter can be obtained for $\sigma_2 = 0.776\sigma$ and $\sigma_1 = 1.6 \cdot 0.776\sigma = 1.24\sigma$. For the approximation of the LoG-filter by the DoG-filter see also

Sommer, G., Meinel, G.: *The design of optimal Gaussian DOLP edge detectors*. in Yaroslavskii, L.P. et al. (eds.), Computer Analysis of Images and Patterns, Proc. 2nd. Int. Conf. CAIP'87, Wismar, 1987, pp. 82-89.

6.2.5. Deriche Edge Operator

(1) The *Deriche edge operator* was developed on the base of the *Canny edge operator*. The realization of this operator requires a rather complex program and large memory arrays. However the results of edge detection are often better than those obtained by simpler methods, and the program run needs a low computing time.

Attributes:

Images:	$N \times N$ gray value images
Operator:	window operator, eight runs
Kernel:	linear (convolution with constant kernel), separable, sequential

Inputs:

filter parameter α, positive real number

(2) This operator is characterized by some advantageous properties which are emphasized here:

- The operator realizes an optimal edge detection with respect to several quality criteria, as sensitivity to true edges, robustness against noise, and accuracy of edge positions, cp. **(5)**.
- It is separable and therefore can be realized by a fast algorithm.
- Because of its recursive structure, large areas of influence (cp. Section 3.1.3 and 3.2.3) can be realized although the observation windows is small and of fixed size. The size of the area of influence (corresponds to the window size in the case of parallel operators) can be adjusted by a single parameter α. The computing time is independent of α.
- The processing is realized in eight runs through the image, which are denoted by (i) ... (viii) in **(2)** and in **(4)**. There are four runs in horizontal direction, and four in vertical direction. The number eight results as follows:
  - The resultant image $h(x, y)$ is a combination of edge maps $H(x, y)$ and $V(x, y)$, where $H(x, y)$ represents the horizontal edges, and $V(x, y)$ represents the vertical edges,

$$h(x, y) = \sqrt{H(x, y)^2 + V(x, y)^2} \quad \text{or} \quad h(x, y) = \mathbf{max}\ \{|H(x, y)|, |V(x, y)|\}\ .$$

  - Both intermediate edge maps are the result of a convolution with a separable convolution kernel (cp. Section 3.1.),

$$K_h(x, y) = k_1(x) \cdot k_2(y) \quad \text{for} \quad H(x, y) \quad \text{and}$$
$$K_v(x, y) = k_2(x) \cdot k_1(y) \quad \text{for} \quad V(x, y)\ .$$

- The functions k_1 and k_2 are recursive and non-causal. However both functions can be separated into two recursive and causal convolution kernels with opposite scan orders, i.e. one from left to right, and one from right to left. The results of both runs are combined by addition. This includes also a certain compensation of the fact that the result depends upon the image processing scanning order, which is a typical effect of sequential operations.

- Two disadvantages of this operator are the complexity of the method and the necessity to realize a special control structure, requiring several image arrays and random pixel access.

In the following formulae, the intermediate images g_{v1}, g_{v2}, g_{h1}, g_{h2}, and g_{hv} are the outputs of eight one-dimensional causal recursive filters of size 1×3 or 3×1. The values a, a_0, a_1, a_2, a_3, b_1, and b_2 of their coefficients are functions of the input parameter α:

$$a = -(1 - e^{-\alpha})^2, \qquad b_1 = -2\,e^{-\alpha}, \qquad b_2 = e^{-2\alpha}, \qquad a_0 = \frac{-a}{1 - \alpha\, b_1 - b_2},$$

$$a_1 = a_0\,(\alpha - 1)\,e^{-\alpha}, \qquad a_2 = a_1 - a_0\,b_1, \qquad a_3 = -a_0\,b_2\,.$$

The resultant horizontal edge map $H(x, y) = g_{h1}(x, y) + g_{h2}(x, y)$ is the outcome of performing the following four runs:

(i) $\quad g_{v1}(x, y) = f(x, y-1) - b_1 \cdot g_{v1}(x, y-1) - b_2 \cdot g_{v1}(x, y-2),$

(ii) $\quad g_{v2}(x, y) = f(x, y+1) - b_1 \cdot g_{v2}(x, y+1) - b_2 \cdot g_{v2}(x, y+2)$

$\quad g_{hv}(x, y) = a \cdot (g_{v1}(x, y) - g_{v2}(x, y)),$

(iii) $\quad g_{h1}(x, y) = a_0 \cdot g_{hv}(x, y) + a_1 \cdot g_{hv}(x-1, y) - b_1 \cdot g_{h1}(x-1, y) - b_2 \cdot g_{h1}(x-2, y),$

(iv) $\quad g_{h2}(x, y) = a_2 \cdot g_{hv}(x+1, y) + a_3 \cdot g_{hv}(x+2, y) - b_1 \cdot g_{h2}(x+1, y) - b_2 \cdot g_{h2}(x+2, y).$

For the resultant image $V(x, y) = g_{h1}(x, y) + g_{h2}(x, y)$ for vertical edges, the same image arrays can be used for storing intermediate results as for computing the image $H(x, y)$. The image $V(x, y)$ is obtained after performing the following four runs:

(v) $\quad g_{v1}(x, y) = f(x-1, y) - b_1 \cdot g_{v1}(x-1, y) - b_2 \cdot g_{v1}(x-2, y),$

(vi) $\quad g_{v2}(x, y) = f(x+1, y) - b_1 \cdot g_{v2}(x+1, y) - b_2 \cdot g_{v2}(x+2, y),$

$$g_{hv}(x, y) = a \cdot (g_{v1}(x, y) - g_{v2}(x, y)),$$

(vii) $$g_{h1}(x, y) = a_0 \cdot g_{hv}(x, y) + a_1 \cdot g_{hv}(x, y-1) - b_1 \cdot g_{h1}(x, y-1) - b_2 \cdot g_{h1}(x, y-2),$$

(viii) $$g_{h2}(x, y) = a_2 \cdot g_{hv}(x, y+1) + a_3 \cdot g_{hv}(x, y+2) - b_1 \cdot g_{h2}(x, y+1) - b_2 \cdot g_{h2}(x, y+2).$$

Finally, the desired resultant image is computed by

$$h(x, y) = \sqrt{H(x, y)^2 + V(x, y)^2}\ .$$

(3) This operator has band-pass characteristics. The *cutoff frequencies* can be adjusted by specifying the filter parameter α. Like other edge operators (cp. Sections 6.2.2, 6.2.3 and 6.2.4), the band-pass characteristics follows from a combination of

- high-pass filtering (e.g. differentiation) for edge detection and
- low-pass filtering for the smoothing of statistically distributed noise.

The filter size is constantly equal to 3 x 1 pixels. However the area of influence of this filter increases in size with decreasing values of α. A large area of influence corresponds to a strong smoothing effect. This simple dependency allows edge detection for small as well as for large operator windows with the same computing effort. In comparison, the computing time would increase with the square of the window size for a two-dimensional non-recursive edge operator, and it would increase linearly for a separable edge operator.

The time efficiency of this operator is very important for a good edge detection. In some cases, optimum results can be achieved by combining resultant images of two edge detectors which feature "complementary" properties:

(1) One edge detector is realized with a small area of influence (high α). This operator produces thin edges in accurate positions. Because of the low smoothing effect, this operator is very noise sensitive. This noisy output is especially disturbing in "homogeneous" regions, far away from edges.

(2) The other edge detector is realized with a large area of influence (low α). The result is characterized by thick and unsharp edges, but it is relatively free of noise.

The combination of both resultant images can be performed, e.g., by point-wise minimum computation (cp. Section 5.4.3). The resultant image of operator (2) is used as a mask to "cut out" those regions in the resultant image of operator (1), which are far away from edges.

This technique of combining two different operators is illustrated in Figure 6.9. At the top left the original synthetic image is shown. It consists of a bright object on a dark background, and it has some additive normally distributed noise ($\sigma = 30$). In the lower half there are the results of the Deriche operator for $\alpha = 0.6$ (left) and $\alpha = 2.5$ (right). The image at top right has been obtained by an AND-combination (point-wise minimum) of both images shown below, and this image is characterized by low noise.

In the upper half of Figure 6.10 is shown the result of the Deriche operator for a natural scene, with $\alpha = 2$. For obtaining thin edges, it is suggested to make use of $\alpha = 2 \ldots \alpha = 3$ as typical values.

(4) *Control structure*: here specified, with direct pixel access
Required data arrays: line store arrays $Z1(1...N)$, $Z2(1...N)$, $Z3(1...N)$, $ZE(1...N)$, $ZA(1...N)$,
image arrays for intermediate results, $B1(1...N, 1...N)$ and $B2(1...N, 1...N)$.

Inputs and initialization:

input of filter parameter α;
computation of a, b_1, b_2, a_0, a_1, a_2, a_3 (in this order) as defined under point **(2)**

COMPUTATION OF $H(x, y)$

run (i): bottom-up

```
for y := 3 to N do begin
    read line y – 1 from the original image into line store array ZE(1...N);
    for x := 1 to N do
        B1(x, y) := ZE(x) – b1·B1(x, y – 1) – b2·B1(x, y – 2)
    end {for}
```

run(ii): top-down

```
for y := N – 2 to 1 step –1 do begin
    read line y + 1 from the original image into line store array ZE(1...N);
    for x := 1 to N do begin
        B2(x, y) := ZE(x) – b1·B2(x, y + 1) – b2·B2(x, y + 2);
        B1(x, y) := a(B1(x, y) – B2(x, y))
        end {for}
    end {for}
```

runs (iii) - (iv): left to right or right to left

```
for y := 1 to N do begin
    for x := 1 to N do    Z1(x) := B1(x, y);
    for x := 3 to N do
        Z2(x) := a0·Z1(x) + a1·Z1(x – 1)
                              – b1·Z2(x – 1) – b2·Z2(x – 2);
    for x := N – 2 to 1 step –1 do
        Z3(x) := a2·Z1(x + 1) + a3·Z1(x + 2)
                              – b1·Z3(x + 1) – b2·Z3( x + 2);
    for x:= 1 to N do    ZA(x) := Z2(x) + Z3(x);
    write line store array ZA(1...N) into line y of image h
    end {for}
```

COMPUTATION OF $V(x, y)$

run (v) - (vi): left to right or right to left

```
for y := 1 to N do begin
    read line y from the original image into line store array ZE(1...N);
    for x := 3 to N do
        Z2(x) := ZE(x – 1) – b1·Z2(x – 1) – b2·Z2(x – 2);
    for x := N – 2 to 1 step –1 do
        Z3(x) := ZE(x + 1) – b1·Z3(x + 1) – b2·Z3(x + 2);
    for x := 1 to N do    B1(x, y) := a·(Z2(x) – Z3(x))
    end {for}
```

run (vii): bottom-up

```
for y := 3 to N do
    for x := 1 to N do
        B2(x, y) := a0·B1(x, y) + a1·B1(x, y)
                              – b1·B2(x, y – 1)  – b2·B2(x, y – 2)
```

run (viii): top-down

```
for y := N – 2 to 1 step – 1 do begin
    for x := 1 to N do    Z1(x) := B2(x, y);
    for x := 1 to N do
        B2(x, y) := a2·B1(x, y + 1) + a3·B1(x, y + 2)
                              – b1·B2(x, y + 1)  – b2·B2(x, y + 2);
```

```
        for x := 1 to N do begin
            Z := B2(x, y) + Z1(x, y);      ZA(x) := sqrt(Z^2 + h(x, y)^2)
            end {for}
        write line store array ZA(1...N) into line y of image h
        end {for}
```

(**5**) The Deriche operator, published in

Deriche, R.: *Fast algorithms for low-level vision*. IEEE Trans. PAMI-**12** (1990), pp. 78-87.

is based on the operator proposed by J. Canny, see

Canny, J.: *A computational approach to edge detection*. IEEE Trans. PAMI-**8** (1986), pp. 679-698.

J. Canny investigated the optimum transfer function or impulse answer of a non-recursive linear edge detector satisfying a specific quality criterion covering the properties listed in (**2**). In other words, the main three advantages of this operator are: sensitive edge recognition, noise robustness, and accuracy of edge position. The Deriche operator in its turn realizes a good approximation of this optimal impulse answer by means of a recursive, non-causal and separable filter. A decomposition of this filter into two recursive, causal filters has been given in

Deriche, R.: *Optimal edge detection using recursive filtering*. Proc. First International Conference on Computer Vision (ICCV), London, June 1987.

where also the corresponding filter coefficients can be found.

6.2.6 Contra-harmonic Filter

(1) This edge detector computes an edge value as a difference between an estimate C_M of the local maximum and an estimate C_m of the local minimum. These estimates are obtained by non-linear computations of averages (contra-harmonic average) of all the gray values inside $\mathbf{F}(f, p)$. This method can be viewed as a generalization of the method described in Section 6.2.1, with improved noise suppression behavior.

Attributes:

Images:	$M \times N$ gray value images
Operator:	window operator
Kernel:	analytical, data driven, order independent

Inputs:

window width n,
exponent r (a positive real number).

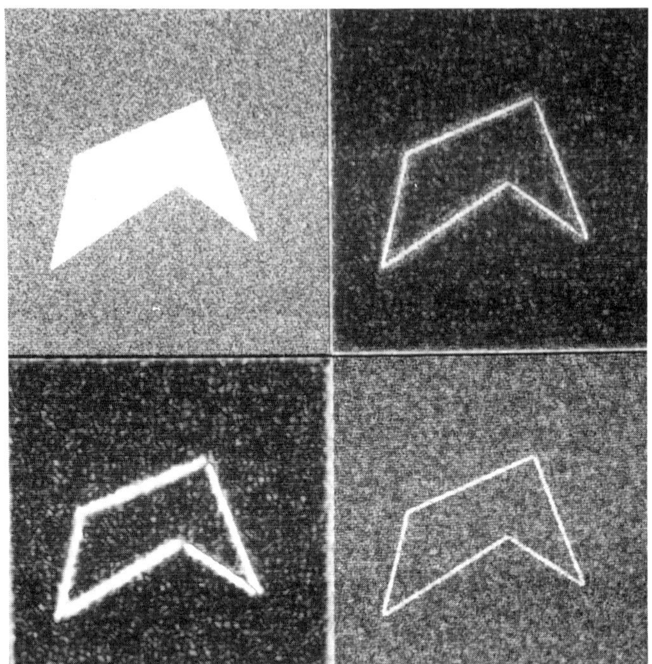

Figure 6.9: *Deriche edge operator*. Top left: synthetic image of a bright object on dark background with additive normally distributed noise (σ = 30 gray levels). Top right: image-to-image minimum of the two images shown below. Below: Results of the Deriche edge operator for $\alpha = 0.6$ (left) and $\alpha = 2.5$ (right).

(2) For the current point p and the corresponding window values C_M and C_m it is

$$h(p) = C_M - C_m\,.$$

Let $F(1), ..., F(a)$ be all gray values inside the window $\mathbf{F}(f, p)$. Then,

$$C_M = \frac{\sum_{z=1}^{a} F(z)^{r+1}}{\sum_{z=1}^{a} F(z)^{r}} \qquad \text{and} \qquad C_m = \frac{\sum_{z=1}^{a} F(z)^{-r+1}}{\sum_{z=1}^{a} F(z)^{-r}}\,.$$

(3) The basic idea of this *range edge detector* consists in identifying an edge value with the difference between the estimates of the local maximum and of the local minimum. The edge operator from Section 6.2.1 is a very simple example of such a type of an edge detector. Because of the presence of statistical and of impulse noise, the measured values of the local extrema can be bad estimates of the true gray value extrema. By weighted averaging it is possible to obtain improved estimates. The weights should be selected such that the local maximum as well as the local minimum are enhanced.

In this sense, the formulae in **(2)** can be considered as weighted sums of the local gray values $F(z)$. The coefficient normalization is represented by the sum in the denominator. For C_M, the weighting coefficients are equal to $F(z)^r$, and for C_m, they are equal to $F(z)^{-r}$. The number r was chosen as a positive real. Thus in the first case the higher gray values, and in the second case the lower gray values are enhanced.

This operator features good robustness against noise and outliers. This is due to the smoothing effect of averaging. This operator is less noise sensitive in comparison to the one-pixel-edge operator of Section 6.2.1, because the following inequality holds:

$$\text{local minimum} \leq C_m \leq C_M \leq \text{local maximum}.$$

Therefore it can happen that low contrast edges are suppressed.

Figure 6.10: Top left: Image of a natural scene. Top right: result of the *Deriche edge operator* for $\alpha = 2$. Below left: an aerial view. Below right: result of the *contra harmonic filter* with a 5×5window and $r = 1.2$.

The lower half of Figure 6.10 shows on the left an aerial view which is rich in texture, and on the right the resultant image obtained with the contra-harmonic filter and a 5×5 window with $r = 1.2$. Spurious contours arising in the textured zones and considered as false edges are substantially suppressed as a consequence of the smoothing effect of this filter.

(4) *Control structure*: window operator in Figure 3.6, centered, with $a = n^2$

Inputs:

```
        input of parameter n  (window size) and r > 0
```

Operator kernel:

```
                    {transform F(z), with 1 ≤ z ≤ a, into v = h(x, y), cp. Figure 3.6 }
        N_M := 0;       Z_M := 0;       N_m := 0;       Z_m := 0;
        for z := 1 to a do begin
              N_M := N_M + F(z)^(r+1);  Z_M := Z_M + F(z)^r;
              N_m := N_m + F(z)^(-r+1); Z_m := Z_m + F(z)^(-r)
              end {for};
        w := (N_M / Z_M) – (N_m / Z_m);
        call ADJUST(w, v)                            { ADJUST from Section 2.1 }
```

(5) Pitas, I., Venetsanopoulos, A.N.: *Nonlinear Digital Filters*. Kluwer Academic Publishers, Boston, 1990.

Pitas, I., Venetsanopoulos, A.N.: *Nonlinear order statistic filters for image filtering and edge detection*. Signal Processing **10** (1986), pp. 395-413.

6.3 IMAGE SHARPENING AND TEXTURE ENHANCEMENT

The operators of this Section aim at emphasizing those image regions or ordered image point patterns, whose spectrum mostly contains higher spatial frequencies. The aim is to emphasize details, edges, fine textures etc. without emphasizing noise. This task can mostly be solved in a better way by non-linear filters than by linear filters. The effects of the different *non-linear filters* are qualitatively discussed, and in Section 6.3.2 alone a certain systematic description of the operator behavior is given.

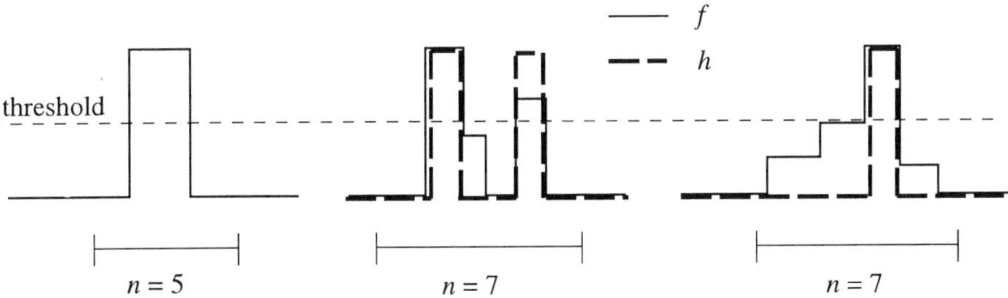

Figure 6.11: Result h of applying the extreme operator to different one-dimensional gray value functions f.

6.3.1 Extreme Value Sharpening

(1) The result of this operator is either the *local gray value minimum* $f_{(1)}$ or the *local gray value maximum* $f_{(a)}$, depending whether the gray value $f(p)$ at the current point p is closer to $f_{(1)}$ or to $f_{(a)}$. This can be achieved by applying a local gray value distribution dependent threshold $(f_{(a)} + f_{(1)}) / 2$ (Variant 1), or by a quantization of all gray values onto the gray values $f_{(1)}$, W (arithmetic average) and $f_{(a)}$ using the thresholds $(W + f_{(1)}) / 2$ and $(f_{(a)} + W) / 2$ (Variant 2).

Image structures, which are rich in details and low in contrast, are emphasized in the resultant image. Unlike with the median operator with this operator the resultant image can contain steeper gray value transitions than the original, without any smoothing effect.

Attributes:

Images:	$M \times N$ gray value images
Operator:	window operator
Kernel:	data driven, logically structured, order dependent

Inputs:

window size n,
Variant $VAR = 1$ or $VAR = 2$.

(2) It is again $a = n^2$. The local values

$$f_{(1)} = \min_{q \in \mathbf{F}(p)} \{f(q)\}, \quad f_{(a)} = \max_{q \in \mathbf{F}(p)} \{f(q)\} \quad \text{and}$$

$$W = \frac{1}{a} \sum_{q \in \mathbf{F}(p)} f(q) = AVERAGE(\mathbf{F}(f, p))$$

are used for defining the operator. For Variant 1, it is

$$h(p) = \begin{cases} f_{(1)} & , \text{if } f(p) - f_{(1)} \leq f_{(a)} - f(p) \\ f_{(a)} & \text{otherwise} \end{cases} .$$

For Variant 2, it is

$$h(p) = \begin{cases} f_{(1)} & , \text{if } \begin{cases} f(p) - f_{(1)} \leq f_{(a)} - f(p) \text{ and} \\ f(p) - f_{(1)} \leq |W - f(p)| \end{cases} \\ W & , \text{if } \begin{cases} |W - f(p)| \leq f_{(a)} - f(p) \text{ and} \\ |W - f(p)| \leq f(p) - f_{(1)} \end{cases} \\ f_{(a)} & \text{otherwise} . \end{cases}$$

(3) The local gray value minimum and maximum act as "attractors" for all gray values between these two extrema, if only these two quantization levels are chosen (Variant 1); the same applies to the case of choosing three quantization levels (Variant 2). The spatial size of the "attractor domain" depends upon the chosen window size. Texture details can remain unchanged, even if they are characterized by high spatial frequencies up to a period of 2 pixels, if they contain local extrema. This represents a difference between this operator and the median operator (cp. Section 6.5.1), or the other rank order operators of the median type (cp. Section 6.5.4, 6.5.5), which generally smooth such texture details.

In general, *spots* with steep edge slope will remain unchanged after filtering. Otherwise shaped impulses are transformed into steep spots, as shown in Figure 6.11. Spots can be deleted if another spot with more than double impulse height occurs in the window $\mathbf{F}(f, p)$.

Figure 6.11 illustrates qualitatively the generation of *steeper edge slopes* for different one-dimensional gray value functions. Gray value shapes with monotone first derivatives, as, e.g., ramps, exponential or power functions, are only shifted by $n / 2$ pixels, cp. Figure 6.12.

Variant 2 offers the possibility of realizing a finer gray value quantization. The spatial shift of such gray value functions with monotone first derivatives is smaller than in case of the Variant 1. Furthermore, in case of Variant 2 there is not such a shift for ramps.

This operator can be used for tasks which demand *region growing* or *gray value agglomeration*, as well as the preservation of image details. It can be used also for *preprocessing before image segmentation* with multilevel thresholding (cp. Section 5.5).

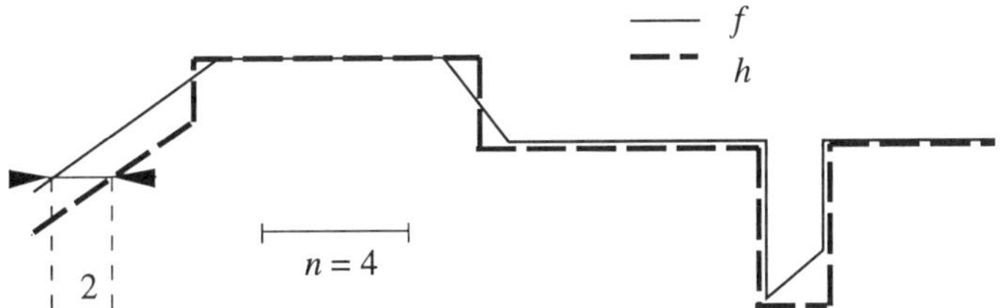

Figure 6.12: Effect of the *extreme value operator* on a monotone gray value function f with resultant gray value function h.

Figure 6.13: Results of the *extreme value operator* for an *aerial view* (top left). Top right: the Variant 1 with a 3×3 window. Below left: the Variant 1 with a 7×7 window. Below right: the Variant 2 with a 11×11 window.

Figure 6.13 shows some results of applying this operator to the *image sharpness enhancement*. The joint computation of maximum and minimum can be speeded up in comparison to the separate computation of maximum and minimum by using procedure *MAXMIN* of Section 3.4.3.

(4) *Control structure*: window operator in Figure 3.6, centered, with odd n.

Inputs and initialization:

```
input of parameter n and VAR;
k := (n – 1) / 2
```

Operator kernel:

```
        {transform F(i, j), with –k ≤ i, j ≤ k, into v = h(x, y), cp. Figure 3.6 }
MAX := 0;     MIN := G – 1 ;        W := 0;
for j := – k to k do
        for i := – k to k do begin
                if (MAX < F(i, j))     then MAX := F(i, j);
                if(MIN > F(i, j))      then MIN := F(i, j);
                if (VAR = 2)  then W := W + F(i, j)
                end {for}
W := W / a ;    v := MIN ;
if    (F(0, 0) – MIN > MAX – F(0, 0))      then v := MAX ;
if (VAR = 2)  then
        if  (| W – F(0, 0) | ≤ |v – F(0, 0) |)    then v := W
```

(5) Lester, J.M., Brenner, J.F., Selles, W.D.: *Local transforms for biomedical image analysis*. Computer Graphics and Image Processing **13** (1980), pp. 17-30.

Zamperoni, P.: *An automatic low-level segmentation procedure for remote sensing images*. Multidimensional Systems and Signal Processing **3** (1992), pp. 29-44.

6.3.2 Unsharp Masking and Space-variant Binarization

(1) This program combines two different operators. For practical reasons such a combination is useful, because both operators share the step of a smoothing operation by means of a *linear low-pass*. These two operators are the following:

(a) binarization of gray value images with an adaptive threshold which is a function of the average local gray value (assumed to be a characteristics of the local background), and

(b) increasing the image sharpness by a linear high-pass filter which results from the image difference between the original image and a low-pass filtered image. This technique is known as *unsharp masking*.

In both cases, the high-pass characteristics of this filter can be varied by a different weighting of the two images at subtraction. An approximate *Gaussian low-pass* is used for low-pass filtering.

Attributes:

Images:	$M \times N$ gray value images
Operator:	window operator
Kernel:	linear (convolution with constant kernel), logically structured in case of Variant 1

Inputs:

window size n,
weighting factor c of the background image, with $0 \leq c < 1$,
Variant $VAR = 1$ for binarization or $VAR = 2$ for image sharpening,
binarization threshold S, with $0 < S \leq G - 1$, in case of Variant 1.

(2) Let $F(1)$, ..., $F(a)$ be the gray values of the given picture window $\mathbf{F}(f, p)$. Let g be the result of the high-pass filtering by subtraction of the original gray value $f(p) = F(a)$ minus the weighted sum of all window gray values $F(z)$. The weights are denoted by $w(z)$, $z = 1, ..., a$:

$$g(p) = \frac{1}{1 - c} \left(F(a) - c \cdot \frac{1}{W} \cdot \sum_{z=1}^{a} w(z) \cdot F(z) \right)$$

with

$$W = \sum_{z=1}^{a} w(z) .$$

The computation of the weights $w(z)$ is explained in point **(3)**. In case of the Variant 1 it is

$$h(p) = \begin{cases} G - 1 & \text{, if } g(p) > S \\ 0 & \text{otherwise.} \end{cases}$$

In case of the Variant 2 it is $h(p) = g(p)$.

(3) The main aim of this operator combination is to increase the image sharpness by means of a linear high-pass filter. This filter is realized by subtracting a low-pass filtered image from the original image. A Gaussian low-pass was chosen for low-pass filtering. In contrast to the box filter, the Gaussian low-pass possesses a limited spectrum of spatial frequencies. Thus the Gaussian low-pass minimizes the aliasing effects. In the Variant 1, the high-pass filtered image is binarized by comparison with a constant gray value threshold. The result of this comparison is independent of the image signal components with low spatial frequencies. Such components correspond often to a slowly varying background. In the Variant 2, i.e. without binarization, one obtains an enhancement of signal components with high spatial frequencies, such as texture, details etc.

The lower *cutoff frequency* of the high-pass filter depends upon the selection of the window size: this cutoff frequency is decreased by increasing the window size.

The selection of the factors $1 / (1 - c)$ and c, as used in the definition of the function $g(p)$, is based on the assumption that the original image and the low-pass filtered image have about the same gray value contrast range. Then, also the resultant image has the same gray value dynamics because the introduced scaling factor is about 1.

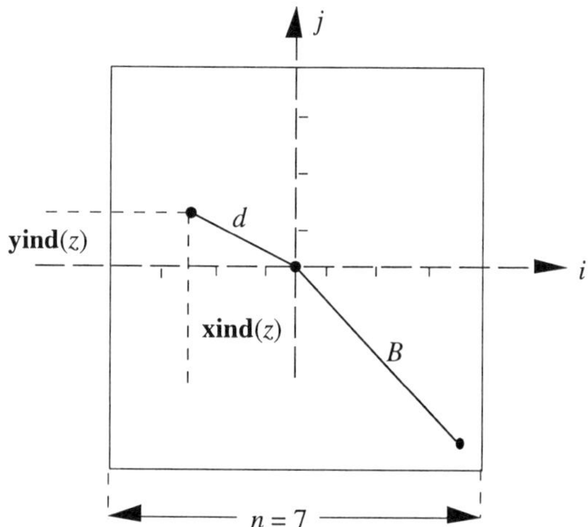

Figure 6.14: Explanation of the computation of the coefficients $w(1)$, ..., $w(a)$ of the Gaussian low-pass filters for unsharp masking.

The coefficients $w(z)$ in the equation defining the function $g(p)$ are computed at the beginning of the program after the input of parameter n. The choice of these coefficients ensures that the Gaussian impulse answer of the low-pass filter drops to $e^{-9/2} \approx 1\%$ of its maximum value at the border of the operator window, i.e. at the distance B from the current point. Figure 6.14 shows an example of such a computation for $n = 7$, $B = 3\sqrt{2}$, $z = 16$, **xind**$(z) = -2$, **yind**$(z) = 1$ and $d = \sqrt{5}$.

The function **exp**$(-s^2 / 2)$ with $s = 3d / B$, and the *Euclidean distance*

$$d = \sqrt{i^2 + j^2}$$

between an image point $q = (x + i, y + j) \in \mathbf{F}(p)$ and the current point $p = (x, y)$, assuming the centered *ij*-coordinate system, are used for the computation of the coefficients. The maximum value of d in $\mathbf{F}(p)$ is

$$B = (n - 1) / \sqrt{2} .$$

Thus the *Gaussian function* drops practically to zero for $d = B$ and $s = 3$. It follows that $\mathbf{exp}(-s^2 / 2) = \mathbf{exp}(-d^2 / B_3)$ with

$$B_3 = \frac{2}{9} B^2 = \frac{(n-1)^2}{9} .$$

The final formula for the computation of the coefficients used in this program is given by

$$w(z) = \mathbf{exp}\left[- \frac{1}{B_3} (\mathbf{xind}(z)^2 + \mathbf{yind}(z)^2) \right], \quad \text{for} \quad z = 1, ..., a .$$

In this equation, $\mathbf{xind}(z)$ and $\mathbf{yind}(z)$ denote the coordinate values i, j as defined by the index z (cp. Figure 3.6, control structure of the centered *ij*-coordinate system).

(4) *Control structure*: window operator in Figure 3.6, centered, with $a = n^2$

Inputs and initialization:

```
input of parameters n, c, and VAR;
if (VAR = 1) then input of threshold S;
                                  {computation of the filter coefficients}
B3 := (n – 1)² / 9 ;      W := 0 ;
for z := 1 to a – 1 do begin
        w(z) := exp[ – ( xind(z) ²+ yind(z)²) / B3 ] ;
        W := W + w(z)
        end {for}
```

Operator kernel:

```
            {transform F(z), with 1 ≤ z ≤ a, into v = h(x, y), cp. Figure 3.6 }
SUM := 0;
for z := 1 to a – 1 do        SUM := SUM + w(z)·F(z);
g := 1/(1–c) · ( F(a) – c · (SUM + F(a)) / (W + 1) );
```

```
call ADJUST(g, w);                                  { ADJUST from Section 2.1 }
if     (VAR = 2)          then   v := w
else     if  (w > S)      then   v := G – 1
                          else   v := 0
```

(5) Ballard, D.H., Brown, C.M.: *Computer Vision*. Prentice-Hall, Englewood Cliffs, 1982.

Wahl, F.M.: *Digital Image Processing*. Artech House, Norwood, 1987.

Zamperoni, P.: *Methoden der digitalen Bildsignalverarbeitung*. Vieweg Verlag, Wiesbaden, 2nd ed., 1991.

6.3.3 Locally Adaptive Scaling for Detail Enhancement

(1) This image processing operator serves for the intensification of low-contrast edges. Gray value scaling factors are computed, which are inversely proportional to the standard deviation inside the operator window.

Attributes:

Images:	$M \times N$ gray value images
Operator:	window operator, in two runs
Kernel:	analytical

Inputs:

window size n,

Variant $VAR = 1$ without smoothing, or $VAR = 2$ with gray value smoothing.

(2) It is $a = n^2$. The average and the standard deviation of the gray values $F(1)$, ..., $F(a)$ in $\mathbf{F}(f, p)$ are denoted by μ and σ (cp. *AVERAGE* and *VARIANCE* in Section 1.3.2), i.e.

$$\mu = \frac{1}{a} \sum_{z=1}^{a} F(z) \quad \text{and} \quad \sigma = \frac{1}{a} \sum_{z=1}^{a} (F(z) - \mu)^2.$$

For Variant 1 it is

$$h(p) = \frac{M_1}{M_2} \cdot \frac{f(p)}{1 + \sigma}$$

and for Variant 2 it is

$$h(p) = \frac{M_1}{M_2} \cdot \frac{\mu}{1 + \sigma} ,$$

where M_1 and M_2 denote, respectively, the average gray values for the full original image f, and for the resultant image before the scaling with the factor M_1 / M_2.

(3) This operator is designed for improving the visibility of low-contrast edges, especially inside the bright image domains. Its basic idea is the observation that the visibility of a pattern depends upon the ratio σ / μ. This suggests of weighting the gray values with a position-dependent scaling factor, proportional to μ / σ. Often, and especially with small size operator windows, this leads also to an enhancement of details with high spatial frequencies. It is practical to chose the operator windows for computing the average and the standard deviation of rather low dimensions to keep the computing time low. The computed values $h(p)$, defined by the equation given above, can lie outside the range $0, ..., G-1$. Therefore it is opportune to perform a global scaling of the resultant image with factor M_1 / M_2, taking account of the limits imposed by the gray value range $0, ..., G-1$.

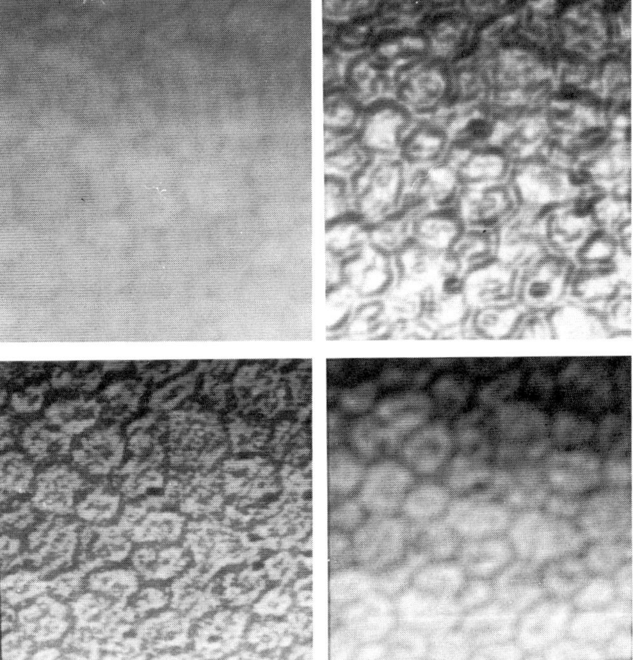

Figure 6.15: Contrast enhancement of an *endothelium microscope image* (top left). Top right: result of a *locally adaptive scaling* (cp. Section 6.3.3) with the Variant 2 and a 7×7 window. Below left: *Adaptive contrast enhancement* (cp. Section 6.3.4) of the original image (top left) with the exponent $r = 0.5$ and a 9×9 window. Below right: *contrast reduction* for the image shown below left with the same operator (i.e. Section 6.3.4) but with $r = 3$ and a 11×11 window.

By aid of this scaling, the gray value range of the resultant image is adjusted to the dynamics of the original image. Since the values M_1 and M_2 can be calculated only at the end of a first run through the image, the scaling (note: this is a point operator) needs a second run having the intermediate image g as input.

In case of Variant 2, the *inverse contrast ratio mapping* includes also a smoothing operation. This *smoothing* is chosen for reducing the high spatial frequency enhancement effect. For this aim, in the Variant 2 the gray value of the current image point is replaced by the average gray value computed by a *box filter*.

The upper half of Figure 6.15 shows an example of contrast enhancement for an *endothelium microscope image*. The original image (left) is of very low contrast. The filtered image (right) allows already the visual recognition of the cellular structure. This filtered image can be the starting point for further image analysis steps.

(4) *Control structure*: window operator in Figure 3.6, centered, with $a = n^2$ for the first run, and point operator in Figure 3.9 for the second run.

Inputs and initialization:

```
input of parameters n and VAR;
M1 := 0;        M2 := 0
```

Operator kernel - first run:

```
                {transform F(z), with 1 ≤ z ≤ a, into w = g(x, y), cp. Figure 3.6 }
μ := 0;         S := 0;
for z := 1 to a do begin
        μ := μ + F(z);  S := S + [F(z) ]²
        end {for};
M1 := M1 + F(a);      μ := μ / a;      S := S / a – μ² ;
if    (VAR = 1)       then  g := F(a) / (1 + √S)
                      else  g := μ / (1 + √S);
call ADJUST (g, w);
M2 := M2 + w
```

Operator kernel - second run:

```
                    {transform w = g(x, y) into v = h(x, y), cp. Figure 3.9}
g := (M1 / M2) · w;      call ADJUST(g, v)
```

(5) Jain, A.K.: *Fundamentals of Digital Image Processing*. Prentice-Hall, Englewood Cliffs, 1989.

6.3.4 Adaptive Contrast Enhancement at Edges

(1) In this operator the measure of contrast enhancement is locally adaptively controlled by an opportunely defined edge gray value E. The value E represents a weighted average of all edge pixels inside the given picture window. The *local contrast* C is defined by the relative deviation between E and the gray value $f(p)$ of the current point. This local contrast should be enhanced. After the calculation of a transformed, enhanced contrast value C', this value is inserted in the inverse formula for determining the gray value in dependence of the contrast. As result of this operator, one obtains a transformed gray value $h(p)$ which features a higher contrast at edges in comparison to the original image.

Attributes:

Images:	$M \times N$ gray value images
Operator:	window operator
Kernel:	logically structured, data driven, order independent

Inputs:

exponent r as a positive real number, with $r < 1$ for obtaining a contrast enhancement.

(2) With the non-centered *ij*-coordinate system (cp. Sections 1.2.1 and 3.3.2) and odd n it holds

$$I = J = (n + 1) / 2 = k + 1 .$$

The local edge gray value E is defined as a weighted and normalized average gray value inside the window $\mathbf{F}(f, p)$. The weight d_{ij} for the gray value $f(x - J + i, y - J + j)$ is a function of an edge value computed at position $(x - J + i,\ y - J + j)$,

$$E = \frac{\sum_{ij} d_{ij} f(x - J + i,\ y - J + j)}{\sum_{ij} d_{ij}} \qquad \text{with} \quad 2 \leq i, j \leq n - 1.$$

As specified above, the variation range of the indices i and j is restricted so that a 3×3 window $\mathbf{F}_{ij} \subset \mathbf{F}(p)$ can be placed in each given image point $(x - J + i,\ y - J + j)$ for the computation of the edge values.

For the computation of edge values the *Prewitt operator* has been chosen here because of its noise-reducing effect, obtained by smoothing orthogonally to the edge direction. Instead of the Prewitt operator another operator could also be used, e.g. as described in Section 6.2.2. The Prewitt operator computes the following edge value d_{XY} for point (X, Y):

$$d_{XY} = \mathbf{max}\{\ |f(X-1, Y+1) + f(X, Y+1) + f(X+1, Y+1) - f(X-1, Y-1) - f(X, Y-1) - f(X+1, Y-1)|,\ |f(X-1, Y+1) + f(X-1, Y) + f(X-1, Y-1) - f(X+1, Y-1) - f(X+1, Y) - f(X+1, Y+1)|\ \}\ .$$

The local contrast C between gray values $f(x, y)$ and E is defined by using the *Canberra distance measure*

$$C = \frac{|f(x, y) - E|}{f(x, y) + E}\ , \quad \text{with} \quad 0 \le C \le 1\ .$$

By inversion of this formula, the gray value $f(x, y)$ follows as a function of E and C,

$$f(x, y) = \begin{cases} E\,\dfrac{1 - C}{1 + C} & , \text{if } f(x, y) \le E \\[2ex] E\,\dfrac{1 + C}{1 - C} & , \text{if } f(x, y) > E\ . \end{cases}$$

For obtaining the desired contrast enhancement the transformed value C' is computed from the measured value C, with $C' \ge C$, by means of a concave gradation function. Being $C \le 1$, the function $C' = C^r$, with $r < 1$ and r given by the user, can be used for this aim.

The result $h(x, y)$ is finally obtained by replacing C with C' in the inversion formula yielding $f(x, y)$.

(3) This method produces a contrast enhancement because the gray values close to the edges are decreased if they are inferior to the weighted edge gray value E. On the other hand, the gray values are increased if they are larger than E. The local contrast C as defined above represents a measure of deviation between value E and the gray value $f(x, y)$ which has to be transformed. The functional dependence between the contrast C and $f(x, y)$ is given under point **(2)**. The increase in contrast based on the application of function C^r can be adjusted by the parameter r. Since it is $C \le 1$, values $r < 1$ have to be selected for contrast enhancement. The resultant gray values $h(x, y)$ are defined by the transformed contrast C' which replaces the value C in the above specified inverse function for $f(x, y)$.

Note that the C / C'-characteristics is convex for $r > 1$. In this case the contrast will decrease.

Two application examples of this operator are illustrated in Figure 6.15. The cell borders are nearly invisible in the original *endothelium microscope image* shown at the top left. This image is very poor in contrast. However the cellular structure is visible after applying this operator with $r = 0.5$ and a 9×9 window. The resultant contrast enhancement is quite strong. Also fine details are strongly enhanced, even at a relatively large distance from the cell borders. For demonstration purposes, this effect can be reduced, thereby decreasing the contrast, by means of a repeated application of the same operator with $r = 3$ and a 11×11 window. In this way irrelevant edges with low contrast can be deleted and strong edges between the cells are only reduced in contrast. The result of this repetitive application of this operator is shown below at right in Figure 6.15.

(4) *Control structure*: window operator in Figure 3.8, non-centered for $n = 3$

Inputs and initialization:

```
input of parameter r with r < 1;
```

Operator kernel:

```
    {transform F(i, j), with 1 ≤ i, j ≤ n, into v = h(x, y), cp. Figure 3.8 }
                           { F(i, j) is equal to BUF(x – k – 1 + i, ind(j))}
S := 0;
Q := 0;
for j := 2 to n – 1 do
    for i := 2 to n – 1 do begin
        d_ij := max{ | F(i – 1, j + 1) + F(i, j + 1) + F(i + 1, j + 1)
                     – F(i – 1, j – 1) – F(i, j – 1) – F(i + 1, j – 1)|,
                   | F(i – 1, j + 1) + F(i – 1, j) + F(i – 1, j – 1)
                     – F(i + 1, j + 1) – F(i + 1, j) – F( i + 1, j – 1) | };
        S := S + d_ij · F(i, j);          Q := Q + d_ij
        end {for}
E := S / Q;
C' := [ |F(k + 1, k + 1) – E| / (F(k + 1, k + 1) + E) ]^r;
if (F(k + 1, k + 1) ≤ E) then   w := E(1 – C') / (1 + C')
                         else   w := E(1 + C') / (1 – C');
call ADJUST(w, v)
```

(5) Beghdadi, A., Le Negrate, A.: *Contrast enhancement technique based on local detection of edges*. Computer Vision, Graphics and Image Processing **46** (1989), pp. 162-174.

The Prewitt operator is discussed in

Zamperoni, P.: *Methoden der digitalen Bildsignalverarbeitung*. Vieweg Verlag, Wiesbaden, 2nd ed., 1991.

6.4 REGION GROWING AND IMAGE APPROXIMATION

The Operators of this Section merge pixels into image regions. These regions are labeled with constant gray values. For example, homogenous image structures can be approximated by means of image segments by properly choosing one of such operators.

6.4.1 Agglomerative Region Growing Operator

(1) This operator agglomerates gray values and therefore it can be used for region growing and image segmentation. By means of successive regular splitting of the gray value scale into subranges, that subrange is computed which contains the highest number of the gray values occurring in the local window $\mathbf{F}(p)$. This selected subrange is considered and the resulting gray value for the window $\mathbf{F}(p)$ is determined by taking the mid-range of this subrange.

Attributes:

Images:	$M \times N$ gray value images
Operator:	window operator
Kernel:	logically structured, data driven

Inputs:

window size n.

(2) Opportunely determined gray value thresholds $u_0{}^{(s)}$, $u_1{}^{(s)}$ and $u_2{}^{(s)}$ are considered during successive iterations $s = 1, 2, \ldots$. Define

$$u_0{}^{(1)} = \min_{q \in \mathbf{F}(p)} \{f(q)\} \quad , u_2{}^{(1)} = \max_{q \in \mathbf{F}(p)} \{f(q)\} ,$$

$$u_1{}^{(s)} = \frac{1}{2}(u_0{}^{(s)} + u_2{}^{(s)}) \quad , \text{for} \quad s \geq 1,$$

$$u_0{}^{(s+1)} = u_0{}^{(s)} + \frac{1}{2}\, k^{(s)} \cdot D^{(s)} \quad , \text{and}$$

$$u_2{}^{(s+1)} = u_0{}^{(s)} + \frac{1}{2}\, (1 + k^{(s)}) \cdot D^{(s)} \quad , \text{for} \quad s \geq 1,$$

with $D^{(s)} = u_2{}^{(s)} - u_0{}^{(s)}$ and $k^{(s)} = 0$ or $k^{(s)} = 1$. These thresholds $u^{(s)}$ are computed from one iteration to the next one in such a way that the condition

$$\underset{q \in \mathbf{F}(p)}{\mathbf{card}} \{u_0^{(s+1)} \le f(q) \le u_2^{(s+1)}\} \Rightarrow \underset{k^{(s)}1=0}{\mathbf{max}}$$

is satisfied. Let t be the smallest value $s + 1 = 2, 3, \ldots$, such that also the condition

$$\underset{q \in \mathbf{F}(p)}{\mathbf{card}} \{u_0^{(t)} \le f(q) \le u_1^{(t)}\} = \underset{q \in \mathbf{F}(p)}{\mathbf{card}} \{u_1^{(t)} \le f(q) \le u_2^{(t)}\}$$

is satisfied. Here, **card**$\{*\}$ denotes the number of gray values $f(q)$ which fulfill the condition $*$. Then, $h(p)$ is defined as follows:

$$h(p) = f(r), \quad \text{for } r \in \mathbf{F}(p) \quad \text{with} \quad |f(r) - u_1^{(t)}| = \min_{q \in \mathbf{F}(p)} |f(q) - u_1^{(t)}| .$$

(3) With respect to its definition and to its effects, this operator may be compared with the operators of Sections 6.4.2 and 6.4.3. All these operators belong to the class of *agglomerative operators* because the resultant gray value is identical to one of the gray values occurring inside $\mathbf{F}(p)$. Thus, no computation of new gray values takes place but already existing gray values can form certain regions (*region growing*).

For selecting the proper nested subranges of gray values, at first the gray value range defined by the local gray value extrema, i.e. the minimum $f_{(1)}$ and the maximum $f_{(a)}$, is split into two subranges of equal size. In the sequel, that half of the full range is considered which contains more gray values than the other half.

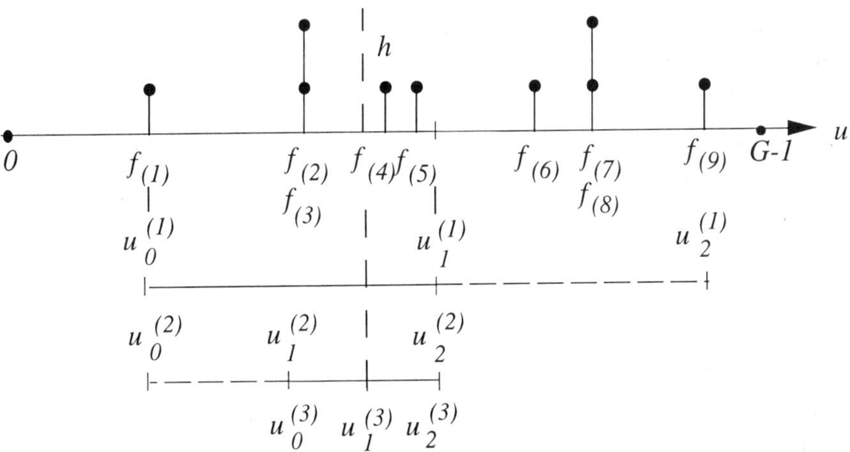

Figure 6.16: Explanation of the agglomeration operator based on a fictitious local gray value distribution.

Figure 6.16 shows an example of an assumed local gray value distribution inside a 3×3 window. It is $a = 9$. The border values of the lower or upper half at the s-th iteration are denoted, respectively, with $u_0^{(s)}$ and $u_1^{(s)}$, and $u_1^{(s)}$ and $u_2^{(s)}$.

Based on the *majority criterion*, one of the two halves is then selected and split again into two halves. Again, the majority criterion is used for these two halves to select one for further processing. This process continues until the number of gray values in both halves is equal. Assume that this happens at the t-th iteration level. In the example of Figure 6.16 it is $t = 3$, and the two halves with the equal number of gray values are from $u_0^{(3)}$ to $u_1^{(3)}$, and from $u_1^{(3)}$ to $u_2^{(3)}$. The mid-range at this iteration level is given by $u_1^{(3)}$. Finally, the resultant gray value $h(p)$ is the input gray value $f(r)$ such that $|f(r) - u_1^{(t)}|$ is a minimum, for image points r inside the given window. In the example in Figure 6.16, the resultant gray value would be $f_{(4)}$.

This operator realizes a region growing process similar to that of the operators of Sections 6.4.2 and 6.4.3. Such a process can be of interest for image segmentation, e.g. in situations where the regions do not consist of textures rich in detail. This can be the case for *homogenous regions* of several natural scenes.

With regard to the computing time there is no essential difference between this operator and the *mode enhancement operator* discussed in Section 6.4.3. The latter requires an ordering process, which is not necessary for the agglomerative region growing described here. However, several counts of gray values comprised between two thresholds, not necessary with the operator of 6.4.3, have to be performed here.

(4) *Control structure*: window operator in Figure 3.6, centered, with $a = n^2$

Inputs and initialization:

```
        input of parameter n
```

Operator kernel (*can also be performed with sequential processing*):

```
                {transform F(z), with 1 ≤ z ≤ a, into v = h(x, y), cp. Figure 3.6 }
MAX := 0;     MIN := G – 1;
for z := 1 to a do begin            {procedure MAXMIN could also be used}
        if (F(z) > MAX) then MAX := F(z);
        if (F(z) < MIN) then MIN := F(z)
        end {for};
BER = MAX – MIN;          u := 0;          w := 2;          flag := 0;
while (flag = 0) do begin
        U0 := MIN + BER· u/w;
        U1 := MIN + BER· (u + 1)/w;
```

```
U2 := MIN + BER· (u + 2)/w;
NUNT := 0 ;            NOBE := 0 ;
for z := 1 to a do begin
        if ((F(z) ≥ U0) and (F(z) ≤ U1)) then NUNT := NUNT + 1;
        if ((F(z) > U1) and (F(z) ≤ U2)) then NOBE := NOBE + 1
        end {for};
if (NUNT = NOBE) then begin
                flag := 1;        MIN := G – 1;
                for z := 1 to a do begin
                        D := |U1 – F(z)|;
                        if (D < MIN) then begin
                                MIN := D;            v := F(z)
                                end {then}
                        end {for}
                end {then}
        else begin
                w := 2w
                if (NUNT > NOBE) then    u := 2u
                                 else    u := 2u + 1
                end {else}
end {while}
```

(5) In a concrete application, *segmentations* of *aerial views* have been realized by means of this operator, cp.

Schöniger, I.: *Die Erkennung von Bodenerosionsschäden in digitalisierten Luftbildern mit Hilfe von Regionenwachstumsverfahren*, Diplomarbeit, Institut for Geography, Technische Universität Braunschweig, 1988.

Zamperoni, P.: *An agglomerative approach to modelling and segmentation of aerial views*. in: Proc. MARI-87, Paris, May 1987, pp. 426-431.

6.4.2 Concavity-filling Operator for Gray Value Images

(1) This operator carries out a filling of *concave regions* in gray value images. Depending upon the number of performed iterations, concavities of different sizes can be filled with this operator. By increasing the number of iterations, a filled region approximates gradually its *convex hull*. Optionally, bright concavities can be filled with points of the dark background, or vice-versa. It is suggested to consider also the sequential processing as an useful alternative for this operator.

Attributes:

Images:	$M \times N$ gray value images
Operator:	window operator
Kernel:	logically structured, data driven, order dependent

Inputs:

number *ITE* of iterations,

Variant $VAR = 1$ for the filling of bright concavities, or $VAR = 2$ for the filling of dark ones,

selection between parallel and sequential processing.

(2) This operator is based on a definition of *local concavities*. This definition is explained under point **(3)**. The notation $f(i, j) = F(z)$ of window gray values corresponds to the Figure 3.5 with $n = 3$ and with the control structure for the centered *ij*-coordinate system.

Here first some bright objects on dark backgrounds are assumed, e.g. each with a constant gray value. Then an image point p is defined to be a *local concavity point* if its gray value $f(p) = F(9)$ satisfies the following condition

$$\exists\ \mathrm{r}: (F(r) > F(9)) \wedge (F(r+1) > F(9)) \wedge (F(r+2) > F(9)) \wedge (F(r+3) > F(9)) ,$$

with $1 \leq r \leq 8$ and $r + s = (r + s)_{mod8}$ for $1 \leq s \leq 3$, and $F(z)$, $z = 1, ..., 8$, for the remaining gray values of its 8-neighbors.

For dark objects on a bright background, the sign > in this condition has to be replaced by the sign <. The corresponding operation is given in brackets in the following equation.

The filling of a concavity at point p means that the value $F(9)$ has to be replaced by the gray value

$$h(p) = \mathbf{min}\ [\mathbf{max}]\ \{F(r), F(r+1), F(r+2), F(r+3)\} .$$

(3) First, a criterion for the recognition of concavity points in a gray value image must be defined. Assume that bright concavities have to be filled. Let $\mathbf{F}'(p)$ be the set of all 8-neighbors (i, j) of the current point p, such that $f(i, j) > f(p)$. From $n = 3$ it follows that $\mathbf{F}'(p) \subseteq \mathbf{F}(p)$. If $\mathbf{F}'(p)$ contains at least four point which are 4-connected, then p is defined to be a *convex image point*. A similar definition applies to dark concavities.

To apply this operator it first has to be specified whether bright gray values are object values and dark gray values are background values, or vice-versa. The aim of the operator is the filling of local concavities. Larger concavities can be filled by increasing the number of iterations. During an image run tests are mode on whether the condition given in **(2)** is satisfied or not. In the positive case, the gray value modification

$$h(p) = \mathbf{min}\ [\mathbf{max}]\ \{F(r), F(r+1), F(r+2), F(r+3)\}$$

is carried out. Otherwise, the original gray value is assigned to the image h. In this way, the image $f(p)$ is altered as little as possible, and as much as necessary. If the condition is satisfied, then the gray values $F(r)$, ..., $F(r+3)$ constitute a *local contour* between object and background, and the value $f(p)$ is transformed (increased or decreased) into an object gray value. Thus, point p becomes a member of the (bright or dark) object.

The filling of large concavities may require many iterations. Choosing the sequential processing can speed-up the filling of large concavities. Of course, sequential processing has the disadvantage that the filling is not isotropic, i.e. it depends upon the scan order.

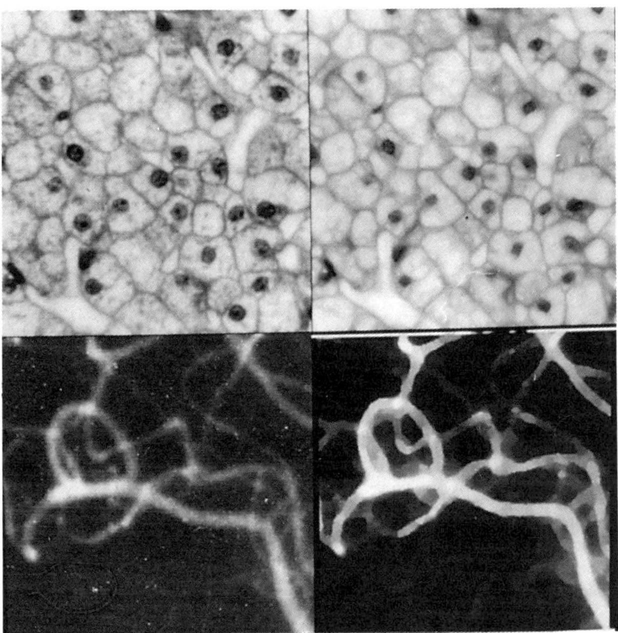

Figure 6.17: Top: original *microscope cell image* (left) and result of the *concavity-filling operator*, filling of bright concavities with 5 iterations (left). Below: *medical contrast image* (left) and result of the *mode enhancement operator* with a 9×9 window.

The sequential operator variant can be improved by using alternately different scan orders in successive iterations, e.g. alternating top-down, left-to-right, and bottom-up, right-to-left. A control structure for applying different scan orders is not included in this book. This control structure can require a more complex programming if the access to image data is row-wise (cp. Section 3.2.3). The direct access to the full image simplifies the implementation of different scan orders. With a row-wise access performing an operation of image mirroring (horizontal and vertical) between two subsequent runs is suggested. This task can be performed by the *mirroring operator* of Section 4.1.1.

As a consequence of the concavity-filling process, the regions are approximated by their minimum convex hulls. With *bilevel images* it is sufficient to represent only the shapes of the generated objects. For gray value images it is necessary to represent also the interior of the approximated convex hulls as a gray value object for further processing. The approximation degree can be controlled by the number of iterations. Such an approximation with convex hulls can be of practical value for certain tasks in image analysis, or in image segmentation. It can also be used for data reduction in *image coding*. In the ideal case of *convex regions*, see **(5)**, the contour can be encoded very compactly by means of a special contour code.

In the upper half of Figure 6.17 a *microscope cell image* is shown (left) together with an image resulting from the concavity-filling operator after 5 iterations (right). The small dark gaps have been filled. The inner areas of the cells have become nearly homogenous. The cell borders can be viewed as "straight dark gaps", which in principle are not or very slightly filled by the operator.

(4) *Control structure*: window operator in Figure 3.6, centered, with $n = 3$, $a = 9$

Inputs:

```
input of parameters ITE and VAR
```

Operator kernel:

```
                {transform F(z), with 1 ≤ z ≤ a, into v = h(x, y), cp. Figure 3.6 }
if (VAR = 2) then begin
        flag := 0;        r := 1;
        while ((r ≤ 8) ∧ (flag = 0)) do begin
                MAX := 0;
                for s := 0 to 3 do begin
                        z := AND (r + s, 7);        { bit-wise AND, cp. 5.5.1 }
                        if (F(9) ≤ F(z))  goto L1;
```

```
                        if (F(z) > MAX) then MAX := F(z)
                        end {for};
                    v := MAX;      flag := 1;
L1                  r := r + 1
                    end {while}
            if (flag = 0) then  v := F(9)
            end {then}
      else
            flag := 0;        r := 1;
            while ((r ≤ 8) ∧ (flag = 0)) do begin
                    MIN := G – 1;
                    for s := 0 to 3 do begin
                        z := AND (r + s, 7);                 { bit-wise AND }
                        if (F(9)  ≥ F(z))  goto L2;
                        if (F(z) < MIN) then MIN := F(z)
                        end {for};
                    v := MIN;      flag := 1;
L2                  r := r + 1
                    end {while};
            if (flag = 0) then v := F(9)
            end {else}
```

(5) The definition of convex image points for gray value images, as used in this Section, is an extension of one of the possible definitions of convexity for bilevel images. For such a "strong" *convexity definition* for bilevel images see, e.g.,

Zamperoni, P.: *Methoden der digitalen Bildsignalverarbeitung*. Vieweg Verlag, Wiesbaden, 2nd ed., 1991.

The definition can be briefly described as follows: A binary object is convex if its contour polygon is a convex polygon of the Euclidean geometry. The contour polygon connects all contour points (or, if the cellular model is used, all center points of contour cells) in the order of a contour scan. As a conclusion of this "strong" convexity definition, *convex binary objects* can have at most eight corner vertices. It is straightforward that such a convex object can be encoded by a monotonous increasing sequence of contour code digits. Such a sequence can be compressed and uniquely defined by eight natural numbers that represent a very compact *run-length-type contour code*.

6.4.3 Mode Enhancement

(1) This operator performs an estimation and an enhancement of modes. A *mode* is defined to be the maximum of a local gray value histogram. Possible applications of this operator are *region growing*, *image segmentation*, or *agglomeration of image points*.

Attributes:

Images:	$M \times N$ gray value images
Operator:	window operator
Kernel:	logically structured, data driven, order statistical

Inputs:

window size n.

This operator is a further example of an *agglomeration operator*, i.e., the resultant gray value is identical with one of the gray values of the observation window $\mathbf{F}(p)$. As a result that gray value is chosen which is nearest to the maximum of the local gray value histogram. For this aim the maximum of the gray value distribution inside $\mathbf{F}(p)$ is estimated on the basis of the observed discrete histogram.

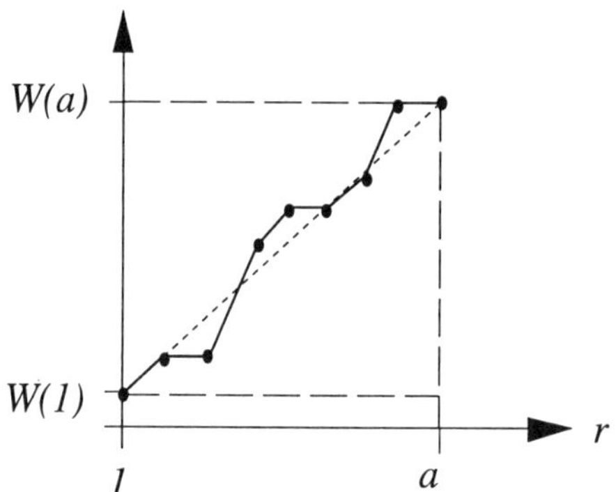

Figure 6.18: An example of a monotonely increasing, ordered gray value function $W(r)$ of the variable r (rank).

This operator initiates a region growing process. The gray value of the local mode successively replaces the remaining gray values which are locally less significant. For implementing this operator, first the gray values $f(i, j) \in \mathbf{F}(p)$ are rank-ordered, and then the ordered sequence is analyzed. For sorting, the procedure *BUCKETSORT* of Section 3.4.7 can be used. This procedure is advantageous for "large values of n"

and can also be used in rank order operators (cp. Section 6.5) demanding a complete ordering of all gray values inside an picture window. For a comparison of the algorithmic complexity of several sorting procedures, see Section 3.1.2, especially for gray value sorting applications.

(2) Let us first define the ordered gray value function $W(r)$, used in the operator definition later on. The discrete rank r, $r = 1, ..., a$, is the variable of this function. The ordered gray values $f(i, j)$ of the given picture window $\mathbf{F}(f, p)$, with

$$W(1) \leq W(2) \leq ... \leq W(r) \leq ... \leq W(a),$$

are the function values. An example of this monotonically increasing function $W(r)$ is given in Figure 6.18.

After the sorting of the gray values $f(i, j)$, for $(i, j) \in \mathbf{F}(p)$, the values of a suitable criterion function $q_L(r)$ are computed, for $r = 1, ..., a$. The resultant gray value $W(\mu)$ of the operator is that gray value which minimizes the function $q_L(r)$. Altogether it is

$$h(p) = W(\mu) \quad , \text{ if } \quad q_L(\mu) = \min_{s = 1, ..., a} q_L(s) \quad ,$$

with

$$q_L(s) = \frac{1}{v-u} \sum_{z=u}^{v} \left| \frac{W(s) - W(z)}{s-z} \right|_{s \neq z} \quad ,$$

for $u = \mathbf{max}\ \{1, s - L\}$ and $v = \mathbf{min}\ \{a, s + L\}$. The function $q_L(s)$ depends upon an integer parameter L (the size of the smoothing window), for which the empirically determined value of $(n + 1)/2$ has been assumed.

(3) This operator is based on the idea that the local probability density function of the gray values can be estimated from the local discrete gray value histogram. The mode, i.e. the absolute maximum, is derived from this estimated density function.

The estimation method rests on the determination of the flattest part of the function graph $W(r)$. If this is observed at position t then this rank t defines the result $W(t)$ of the operator. The rank t is the value which minimizes the steepness of $W(r)$, and the steepness is computed using the differential quotient

$$| (W(s) - W(z)) \ / \ (s - z) |_{s \neq z} \ .$$

For removing of discontinuities due to digitization, the function $W(r)$ is smoothed. This is performed by taking the average of several differential quotients specified by different increments $s - z$. The smoothing range depends upon the parameter L, with

$$L \geq | s - z |_{s \neq z} \ .$$

Thus, the criterion function $q_L(s)$ will change according to the parameter L. As empirical value, it can be suggested to take $L = (n + 1)/2$. Furthermore, it has to be ensured that the summation limits u and v satisfy $u \geq 1$ and $v \leq a$.

In Figure 6.19 three different cases are illustrated. These include the cases with $n = 3$, $a = 9$ and for different criterion functions $q_L(s)$. The rank μ of the resultant gray value $W(\mu) = h(p)$ has to be determined.

The situations arising for different values of the parameter L, $L = 2$ and $L = 3$, are shown in Figure 6.19. For each resultant curve (lower half of figure), the ranks μ_2 and μ_3 of the minimum values of $q_2(r)$ and $q_3(r)$ determine the resultant gray value. In cases (a) and (b) of Figure 6.19, a variation of parameter L has little influence on the result (rank $r = 9$ or $r = 5$). In case (c), the resultant gray value "switches" between the two strong modes, at $\mu = 1$ or $\mu = 8$.

The effect of this operator is illustrated in the lower half of Figure 6.17. On the left, a medical image is shown, and on the right the processed image. In this application this operator has been used for improving the visibility of vessels.

(4) *Control structure*: window operator in Figure 3.6 , centered, with $a = n^2$
Required data array: $W(1 \ldots a)$ for the ordered gray values

Input and initialization:

```
input of parameters n, with odd n;
L := (n + 1) / 2
```

Operator kernel:

```
                {transform F(z), with 1 ≤ z ≤ a, into v = h(x, y), cp. Figure 3.6 }
sorting of a gray values F(1), ..., F(a) in increasing order and storing the
        resultant sequence in array W(r) ,  1 ≤ r ≤ a,;
QMIN := 2^15 – 1;                           { as example of a very large number}
for s := 1 to a do begin
        umin := max (1, s – L);        umax := min (a, s + L);
        Q := 0;
        for z := umin to umax do
                if (s ≠ z)  then         Q := Q + | (W(s) – W(z)) / (s – z) |;
        Q := Q/(umax – umin);
        if (Q < QMIN) then begin  μ := s;   QMIN := Q   end {if}
        end {for};
v := W(μ)
```

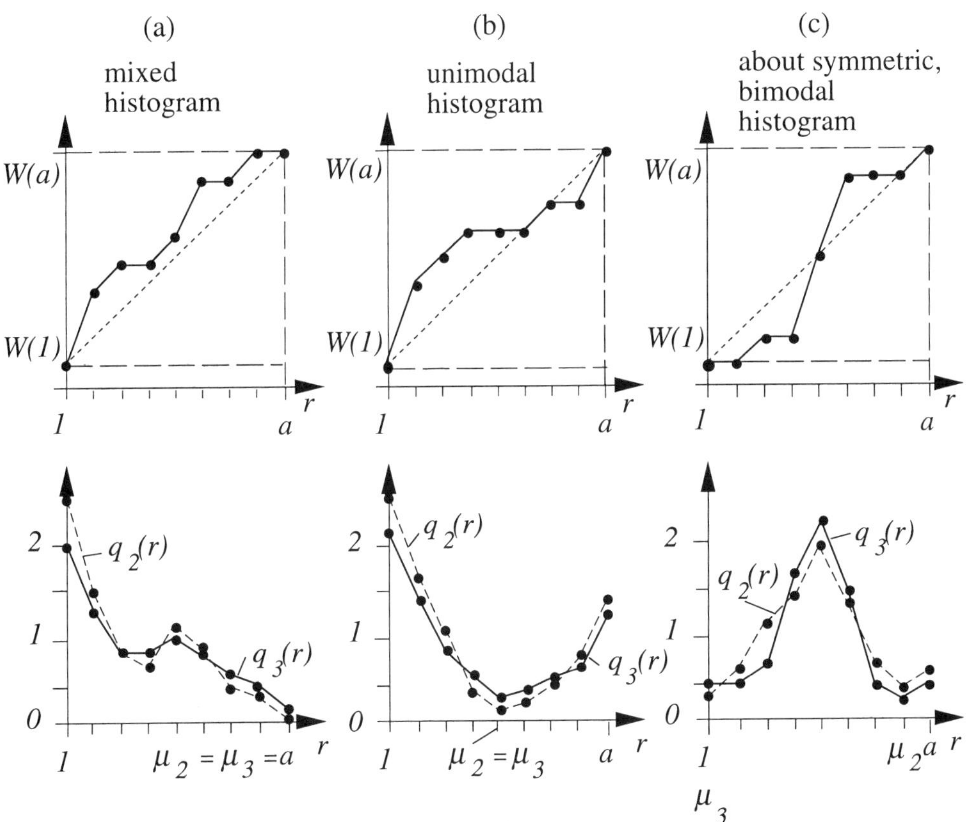

Figure 6.19: Three examples of ordered gray value functions $W(r)$ representing different situations of local gray value distributions. The resultant criterion functions $q_L(r)$ for the mode estimation are shown for $L = 2$ and $L = 3$.

(5) Several systematic probability density function estimation methods are based on discrete random samples. For estimating the maximum of such samples there exist various known measures from statistical pattern recognition, cp. e.g. *Parzen estimation* in

Fukunaga, K.: *Introduction to Statistical Pattern Recognition, 2nd ed.*. Academic Press, New York, 1990 ,

This technique has been used in

Zamperoni, P.: *Feature extraction by rank-order filtering for image segmentation.* International Journal of Pattern Recognition and Artificial Intelligence **2** (1988), pp. 301-319.

for the implementation of region growing operators. For the same aim, a simpler method, which needs lower computing time, was used in **(4)**. For the *sorting problem* of gray values see also

Knuth, D.E.: *The Art of Computer Programming. Vol. 3: Sorting and Searching.* Addison-Wesley, Reading, USA, 1975.

Zamperoni, P.: *Methoden der digitalen Bildsignalverarbeitung.* Vieweg Verlag, Wiesbaden, 2nd ed., 1991.

6.5 RANK-ORDER FILTERING

The operators for *rank-order filtering* are characterized by the property that the resultant gray value is always a function of the rank-ordered gray values of the given picture window $\mathbf{F}(f, p)$. Before computing the final result, a rank ordering of the gray values has to be performed. This can be a full sorting, as described in Sections 3.4.5 to 3.4.7, or just the selection of one or the combination of more rank-ordered gray values, e.g. the minimum, the maximum, or the median (cp. procedure *SELECT* in Section 3.4.4). Rank-order filters can be used for solving a manifold of tasks as *smoothing*, *sharpness enhancement*, or *line extraction*. A systematic treatment of this important class of non-linear filter is given in

Pitas, I., Venetsanopoulos, A.N.: *Nonlinear Digital Filters.* Kluwer Academic Publishers, Boston, 1990.

Examples of *adaptive rank-order operators* for *image improvement* and for *image segmentation* can be found in

Alparone, L., Baronti, S., Carlà, R., Puglisi, C.: *A new adaptive digital filter for SAR images: test of performance for land and crop classification on Montespertoli area.* Proc. IGARSS (International Geographic And Remote Sensing Symposium), Houston, 1992, pp. 899-901.

Bolon, P., Fruttaz, J.L.: *Adaptive order filters: application to edge enhancement of noisy images.* in: Lagunas, M.A. et al. (eds.): Signal Processing V: Theories and Applications, Elsevier, Amsterdam, 1990, pp. 817-820.

Bolon, P., Raji, A., Lambert, P., Mouhoub, M.: *Symmetrical recursive median filters application to noise reduction and edge detection.* in Lagunas, M.A. et al. (eds.), Signal Processing V: Theories and Applications, Elsevier, Amsterdam, 1990, pp. 813-816.

Zamperoni, P.: *An automatic low-level segmentation procedure for remote sensing images*, Multidimensional Systems and Signal Processing **3** (1992), pp. 29-44.

Zamperoni, P.: *Variations on the rank-order filtering theme for grey-tone and binary image enhancement*, in: Proc. ICASSP'89, Glasgow, May 1989, pp. 1401-1404.

Zamperoni, P.: *Adaptive rank order filters for image processing based on local anisotropy measures*, Digital Signal Processing **2** (1992), pp. 174-182.

6.5.1 Median Filtering and Non-linear Sharpening

(1) The Variant 1 is a fast realization of median filtering by means of the *update method.* As Variant 2 has been chosen the *unsharp masking*, i.e. the sharpened image results from the smoothed image weighted by a certain multiplicative factor, and then subtracted from the original image.

Attributes:

Images:	$M \times N$ gray value images
Operator:	window operator
Kernel:	order statistical

Inputs:

Variant $VAR = 1$ or $VAR = 2$ as described above,
window size n,
weighting factor C for the smoothed image for the Variant 2.

(2) The window function *MEDIAN* is defined under point **(5)** of Section 1.3.3. In the Variant 1 it is

$$h(p) = MEDIAN(\mathbf{F}(f, p))$$

for an input image f and a $n \times n$ observation window. In the Variant 2 it is

$$h(p) = C \cdot f(p) \;-\; (C - 1) \cdot MEDIAN(\mathbf{F}(f, p)).$$

Let $R(1)$, ..., $R(a)$ be the gray values of the picture window $\mathbf{F}(f, p)$ sorted in increasing order. Then it holds

$$MEDIAN(\mathbf{F}(f, p)) = R\left(\frac{a+1}{2}\right).$$

(3) The usefulness of the median operator for smoothing is emphasized in many textbooks on digital image processing. The main point is that, typically, median filtering does not blur the step edges. Because of the practical importance of median filtering it was a logical consequence to search for fast implementation algorithms.

The update method, as introduced in Section 3.1.4, example 3.5, is a widespread fast computation technique of general interest, and not restricted only to the median operator. Switching from the current point to the next image point, the placed picture window is shifted of one pixel to the right. For a given window size of $n \times n$, after this shift, there are still $n\ (n - 1)$ image points inside the shifted window which were also elements of the previous window; n image points have been discarded, and n new image points have been included. For exploiting this circumstance, it can be advantageous to utilize some partial results of the previous window position for the computation at the new window position.

In the concrete case of the median operator, the following quantities are used for the realization of the update method: the frequencies *hist*(u) of the gray values u inside $\mathbf{F}(p)$, for $u = 0, 1, ..., G-1$, and the number LM of gray values which are smaller than the previous median (at the old window position). The *frequencies hist*(u) are stored in an array *hist*(1...G). Note that the index is shifted by 1, cp. example 3.5.

For the indices z of the gray values $F(z)$, the spatial ordering of Figure 3.5 is assumed. This is especially relevant for the positions of the n discarded image points, or of the n image points which came into the new window. These indices are stored in a look-up table **ind**(w), for $1 \leq w \leq W$. The size W depends upon the maximum window size n_{max} with $n \leq n_{max}$. The *look-up table* is used for an indirect addressing of the image points. If the maximum window size is n_{max} it is sufficient to use

$$W = 2 \cdot k_{max} \cdot n_{max}, \qquad \text{with} \qquad k_{max} = \frac{n_{max} - 1}{2},$$

look-up table elements. For a given window size only $2n$ table entries are used.

As an example, assume $n_{max} = 7$. Then it follows $W = 42$. Then, the look-up table **ind**(w) contains the following entries

| 4 5 6 0 0 0 0 15 16 17 18 19 0 0 34 35 36 37 38 39 40
1 2 8 0 0 0 0 9 10 11 23 24 0 0 25 26 27 28 46 47 48 |

in its $W = 42$ positions.

In the array *hist*(1...G) the value *hist*(u), for $1 \leq u \leq G$ specifies the gray value frequency of gray value $u - 1$. At the beginning of a line, i.e. for $x = k + 1$, the full local histogram has to be computed. The update method can be used starting with $x = k + 2$ up to $x = M - k$.

The Variant 2 has been combined here with the Variant 1 in order to exploit the fast median algorithm also for another purpose, i.e. for non-linear sharpening. The basic idea is that of unsharp masking, i.e. an unsharp version of the original image is subtracted point-wise from the original image.

Let us consider an unsharp masking where the unsharp image is obtained by means of a *linear low-pass filter*. This corresponds to a linear high-pass with unlimited bandwidth, which would also *enhance the noise*.

Instead, assume that unsharp masking is performed by means of a median-filtered image. This filtered image has stronger higher spatial frequency components in comparison to the linearly low-pass filtered image. These noisy high-frequency components are then subtracted from the original image.

Before *subtraction*, the median-filtered image is weighted by introducing a parameter C. Because the gray value dynamics of the original image and of the resultant image should be nearly the same, the original image is weighted with the constant C and the median-filtered image is weighted with the factor $C - 1$. An increase of C should correspond to an increase of image sharpness.

(4) *Control structure*: window operator in Figure 3.6, centered, with $a = n^2$, but with an opportune modification of the standard control structure: The input of the gray values of $\mathbf{F}(f, p)$ into the array $F(1...a)$ is not performed at the beginning of the operator kernel. The only exception is the beginning of a new line. This modification is due to the update method which is based on operations performed on "old" and "new" gray values. The details are given in the operator kernel.

Required data arrays:

a look-up table **ind**(w), for $1 \leq w \leq W = 2 \cdot k_{max} \cdot n_{max}$, of the indices of those image points which should be discarded after the window shift ($w \leq W/2$), and of the indices of those image points which are newly included into the placed window ($w > W/2$), cp. the example in point **(3)**,

$hist(1...G)$ for the local histogram, cp. function *hist* in Section 1.3.2.

Inputs and initialization:

```
input of the parameter value n, n odd, and input of VAR;
k = (n – 1) / 2;
TH := (a – 1)/2;          K := 1 + [n_max (n – 3)]/2;
if (VAR = 2) then input of parameter C, with 0 ≤ C ≤ 1
```

Operator kernel:

```
{transform F(z), with 1 ≤ z ≤ a, into v = h(x, y), the input of values F(z) is modified
                                                        in comparison to Figure 3.6}
    input of F(a) = f(x, y);
    if (x = k + 1) then begin
        input of gray values F(1), ..., F(a – 1);
        for z := 1 to G do     hist(z) := 0;
        for z := 1 to a do     hist(F(z) + 1) := hist(F(z) + 1) + 1;
        LM := 0;
```

```
                for MED := 1 to G do begin
                        if (LM ≥ TH) then goto L3;
                        LM := LM + hist(MED)
                        end {for}
        else begin
                                        {the window gray values are still the "old" ones}
                for t := K to K + n – 1 do begin
                        hist(1 + F(ind(t))) := hist(1 + F(ind(t))) – 1;
                        if (1 + F(ind(t)) < MED) then LM := LM – 1
                        end {for};
                            {consider the "new" window gray values F(1) ... F(a – 1)}
                for t := K + W/2 to K + n – 1 + W/2 do begin
                        hist(1 + F(ind(t))) := hist(1 + F(ind(t))) + 1;
                        if (1 + F(ind(t)) < MED) then LM := LM + 1
                        end {for};
L1              if (LM ≤ TH) then begin
L2                              if (LM + hist(MED) >TH) then goto L3;
                                LM := LM + hist(MED);
                                MED := MED + 1;
                                goto L2
                                end {then}
                        else begin
                                MED := MED – 1;
                                LM := LM – hist(MED);
                                goto L1
                        end {else}
                end {if}
L3      v := MED – 1;
        if (VAR = 2) then v := C· F(a) – (C – 1) v
```

(5) For the update method in the computation of the window function *MEDIAN*, cp.

Tyan, S.G.: *Median filtering, deterministic properties*, in: Huang, T.S. (ed.), Two-dimensional Digital Signal Processing II. Transforms and Median Filters, Springer, Berlin, 1981,

For the unsharp masking method, cp.

Hall, E.L.: *Computer Image Processing and Recognition.* Academic Press, New York, 1979.

6.5.2 Minimum and Maximum (Erosion and Dilation)

(1) This operator realizes either m_1 iterations of the minimum operator followed by m_2 iterations of the maximum operator, or vice-versa. This also allows to realize special *morphological operations* as *erosion*, *dilation*, *opening* and *closing* for *bilevel images*. A 3×3 window is assumed.

Attributes:

Images: $M \times N$ gray value images or bilevel images
Operator: parallel window operator in several iterations
Kernel: order statistical

Inputs:

numbers m_1 and m_2 of iterations for the operations of maximum and minimum, respectively, i.e. first the number for minimum and then the number for maximum if $VAR = 1$; and vice-versa if $VAR = 2$,

choice of the structuring element centered on the reference point p, with $GEO = 1$ for choosing the 8-neighborhood and with $GEO = 2$ for choosing an "*octagonal neighborhood*" where 8-neighborhood and 4-neighborhood are used in alternations during successive iterations.

(2) Let $\mu = m_1 + m_2$ be the total number of iterations. Let $g^{(r)}(p)$, for $0 \leq r \leq \mu$, be the intermediate result of the r-th iteration. It is $g^{(0)} = f$ and $g^{(\mu)} = h$.

The result of the r-th iteration as well as the resultant image (for $r = \mu$) are defined in dependence upon the input parameters m_1, m_2, VAR and GEO. In the case of $VAR = 1$ it is

$$g^{(r)}(p) = \begin{cases} \min\limits_{q \in \mathbf{F}(p)} \{g^{(r-1)}(q)\} & \text{, if } r = 1, \ldots, m_1 \\ \max\limits_{q \in \mathbf{F}(p)} \{g^{(r-1)}(q)\} & \text{, if } r = m_1 + 1, \ldots, \mu \end{cases} .$$

In the case of $VAR = 2$, the functions **min** and **max** in this definition equation must be interchanged. Furthermore, it is

$$\mathbf{F}(p) = \begin{cases} \mathbf{F}_8(p) & \text{, if } GEO = 1 \\ \mathbf{F}_8(p) & \text{, if } r \text{ odd} \\ \mathbf{F}_4(p) & \text{, if } r \text{ even} \end{cases} \quad \left.\begin{matrix} \\ \\ \end{matrix}\right\} \text{if } GEO = 2$$

where $\mathbf{F}_8 = \{(i, j): |i| \leq 1 \wedge |j| \leq 1\}$ denotes the 8-neighborhood and $\mathbf{F}_4 = \{(i, j): |i| + |j| \leq 1\}$ denotes the 4-neighborhood.

(3) This program allows to perform a wide variety of morphological basic operations for gray value images as well as for bilevel image. This is due to the fact that erosion and dilation applied to bilevel images are particular cases of the minimum and maximum operations applied to gray level images. The morphological operation is chosen by specifying the parameters m_1, m_2 and VAR. For example, $m_1 \geq 1, m_2 = 0, VAR = 1$ means erosion, $m_2 \geq 1, m_1 = 0, VAR = 2$ means dilation, $m_1 = m_2 \geq 1, VAR = 1$ means opening, and $m_1 = m_2 \geq 1, VAR = 2$ means closing. Note that different numbers of erosions and dilations ($m_1 \neq m_2$) can also be realized.

The size of the area of influence at the reference point p is increasing with the iteration number for operations of the same type, cp. Section 3.1.3. This area of influence is always point-symmetric with respect to point p. For example, in the case of $GEO = 1$ it is a square of odd side length (successive expansions of an 8-neighborhood). Note that a t-times repeated use of a structuring element of 3×3 image points will not lead to the same result as using once a structuring element of $(2t+1) \times (2t+1)$ image points, cp. Section 3.1.3. However algorithmically a t-times repetition of a 3×3 window operation if faster than a single $(2t + 1) \times (2t + 1)$ window operation. This is due to the fact that with iterations the computing time increases linearly with $9t$ instead of quadratically with $2t + 1$. In practical experiments it has been observed that the resultant images are nearly the same with t iterations of a 3×3 window, and with the non-iterative solution.

The disadvantage of the 8-neighborhood with a large number of iterations is obvious: Due to the used *maximum metric* the area of influence is a square and not a circle, cp. Section 1.1.1. It can be desirable that this area of influence should be as similar as possible to "an Euclidean circle". Although an ideal circle cannot be obtained, a better approximation than with the maximum metric can be attained by selecting the parameter setting $GEO = 2$. For this aim, the so-called *octagon metric* is chosen. The 8-neighborhood and the 4-neighborhood are used in alternation as structuring elements during successive iteration steps. The alternation of the two metrics related with these two neighborhoods realizes the octagon metric.

To implement the basic morphological operations of opening and closing, during the first m_1 iterations and during the second m_2 iterations the same structuring element has to be used ($m_2 = m_1$). If the octagon metric is chosen and m_1 was selected to be odd, then the seconds m_2 iterations have to start with the 8-neighborhood. The variables $N8$ and $NGEO$ are used in the program to control the correct switching between the 8- and the 4-neighborhood.

A further feature of the program is that a variant of the update method (cp. Section 3.1.4) is used for reducing the computing time; this applies only if the 8-neighborhood is used. In this case, the 3×3 window $\mathbf{F}(p)$ is subdivided into three vertical 1×3 columns $\mathbf{F_1}(p)$, $\mathbf{F_2}(p)$ and $\mathbf{F_3}(p)$, from left to right. Let MIN_1, MIN_2,

MIN_3 and MAX_1, MAX_2, MAX_3 be the minima and maxima of these columns, respectively. The minimum on $\mathbf{F}(p)$ is given as **min** $\{MIN_1, MIN_2, MIN_3\}$, and the maximum in an analogous way. For the next image point, the previous values MIN_2 and MIN_3 are the new values MIN_1 and MIN_2, respectively, and only the new value MIN_3 has to be computed. As an exception, all the values MIN_1, MIN_2, MIN_3 have to be computed if the window is placed at the beginning of a line, i.e. $x = 2$.

(4) *Control structure*: window operator in Figure 3.6, centered, with $n = 3$, used for μ runs

The use of an image memory is necessary if several iterations must be performed. It is practical that each iteration follows the same scheme, no matter whether μ is even or odd. Therefore, in each iteration the intermediate image g is transformed into the array of the resultant image h. It is assumed that the original image has to be saved. At program start, the input image is copied into the intermediate image g. At the end of the r-th iteration, for $r = 1 \ldots \mu$, the resultant image $h^{(r)}$ is copied back from h to g. Finally, after μ iterations the resultant image is stored in the array h.

The window gray values $F(1)$, ..., $F(9)$ have the same meaning as in the default control structure, with the only difference that they are related to image points of the intermediate image g.

Inputs, initialization and start of μ iterations:

```
    input of the parameters m1, m2, VAR and GEO;
    μ := m1 + m2;
    copy the original image f into the array g;
    NGEO := 1;
for r := 1 to μ do begin                      {start of a loop with μ iterations}
    N8 := 1;
    if ((GEO = 2) ∧ (NGEO even)) then N8 := 0
```

Operator kernel of the r-th iteration of the loop:

```
    {transform F(z), with 1 ≤ z ≤ a, from array g into v = h(x, y), cp. Figure 3.6 }
    if (x = 2) then
        if (VAR = 2) then
            if (N8 = 1) then begin
                    M1 := max {F(4), F(5), F(6)};
                    M2 := max {F(3), F(7), F(9)};
                    M3 := max {F(1), F(2), F(8)};
```

```
                        v := max {M1, M2, M3}
                        end {then}
                else
                        v := max {F(1), F(3), F(5), F(7), F(9)};
        else
            if (N8 = 1) then begin
                        M1 := min {F(4), F(5), F(6)};
                        M2 := min {F(3), F(7), F(9)};
                        M3 := min {F(1), F(2), F(8)};
                        v := min {M1, M2, M3}
                        end {then}
                else
                        v := min {F(1), F(3), F(5), F(7), F(9 )}
    else
        if (N8 = 1) then begin
            M1 := M2;     M2 := M3;
            if (VAR = 2) then begin
                        M3 := max {F(1), F(2), F(8)};
                        v := max {M1, M2, M3}
                        end {then}
                else begin
                        M3 := min {F(1), F(2), F(8 )};
                        v := min {M1, M2, M3}
                        end {else}
            end {then}
        else
            if (VAR = 2) then
                        v := max {F(1), F(3), F(5), F(7), F(9)}
                else
                        v := min {F(1), F(3), F(5), F(7), F(9)}
```

Preprocessing for the next iteration of the loop:

```
    NGEO := NGEO + 1;
    if ((r = m1) ∧ (VAR = 1)) then VAR := 2;
    if ((r = m1) ∧ (VAR = 2)) then VAR := 1;
    if ((r = m1) ∧ (GEO = 2) ∧ (m1 odd)) then NGEO := NGEO + 1;
```

if $(r \neq \mu)$ **then** copy the intermediate image of the r-th iteration from array h into array g
end {*for*} {end of the loop}

(5) Rosenfeld, A., Kak, A.C.: *Digital Picture Processing*, Vol. 1 and 2. Academic Press, New York, 1982.

Zamperoni, P.: *Methoden der digitalen Bildsignalverarbeitung*. Vieweg Verlag, Wiesbaden, 2nd ed., 1991.

Pitas, I., Venetsanopoulos, A.N.: *Nonlinear Digital Filters*. Kluwer Academic Publishers, Boston, 1990.

Haralick, R.M., Sternberg, S.R., Zhuang, X.: *Image analysis using mathematical morphology*. IEEE Trans. PAMI-**9** (1987), pp. 532-550.

Rosenfeld, A., Pfaltz, J.L.: *Distance functions on digital pictures*. Pattern Recognition **1** (1968), pp. 33-61.

6.5.3 Rank Selection Filter

(1) Assume that all $a = n^2$ gray values of a placed window $\mathbf{F}(p)$ are sorted in increasing order. The resultant gray value is that in the r-th ranked position, with $1 \leq r \leq a$.

Attributes:

Images:	$M \times N$ gray value images
Operator:	window operator, parallel or sequential
Kernel:	order statistical

Inputs:

window size n,
rank r, with $1 \leq r \leq a$,
choice between parallel or sequential processing

(2) A value $h(p) = f(q)$ has to be computed for the placed window $\mathbf{F}(p)$ such that $q \in \mathbf{F}(p)$, and $f(q)$ is the smallest gray value in $\mathbf{F}(p)$ satisfying

$$\mathbf{card}\{(i,j): (i,j) \in \mathbf{F}(p) \quad \text{and} \quad f(i,j) \leq f(q)\ \} \geq r\,,$$

cp. Section 3.4.4.

(3) A rank selection filter with constant rank is mostly used for *image enhancement* by means of *non-linear smoothing*. The selection of rank r allows a variation range from a *minimum filter* ($r = 1$) to a *maximum filter* ($r = a$). The *median filter* corresponds to the parameter setting $r = (a + 1) / 2$.

Consider an image with *impulse noise*. If the maximum operator is applied to such an image, then the positive noise is enhanced. If it is known that the impulse spike affects at most two image points inside a window, then $r = a - 2$ can be chosen for implementing a *quasi-maximum filter* for *noise suppression*.

This example should illustrate that the value r can be specified with respect to an a-priori known image-and *noise model*, or this value should be optimized with respect to noise suppression.

The sequential processing mode makes sense only for the rank $r = (a + 1) / 2$ (i.e. median filter) because other ranks r are "unstable", i.e. they give rise practically always to streaks in the resultant image. However in some applications sequential median filters did prove to be more efficient than the corresponding parallel operators for *edge-preserving smoothing*.

The procedure *SELECT*, cp. Section 3.4.4, is used for the computation of the gray value of rank r. This is appropriate, because a full sorting is not required.

(4) *Control structure*: window operator in Figure 3.6, centered, with $a = n^2$

Inputs and initialization:

```
input of parameters n, n should be odd, and of r, with 1 ≤ r ≤ a;
selection of parallel or sequential processing
```

Operator kernel:

```
                 {transform F(z), with 1 ≤ z ≤ a, into v = h(x, y), cp. Figure 3.6 }
call SELECT( F(1 ... a), n, r );
v := F(r)
```

(5) Pitas, I., Venetsanopoulos, A.N.: *Nonlinear Digital Filters*. Kluwer Academic Publishers, Boston, 1990.

Bolon, P., Raji, A., Lambert, P., Mouhoub, M.: *Symmetrical recursive median filters application to noise reduction and edge detection*. in Lagunas, M.A. et al. (eds.), Signal Processing V: Theories and Applications, Elsevier, Amsterdam, 1990, pp. 813-816.

Raji, A., Bolon, P.: *Streaking effects of recursive median filters and symmetrical recursive median filters*. in: Cappellini, V. (ed.), Proc. Intern. Conf. on Digital Signal Processing, Firenze, Sept. 1991.

van den Boomgaard, R.: *Threshold logic and mathematical morphology*, in: Cantoni, V. et al. (eds.): Progress in image analysis and processing, World Scientific Publishing, Singapore, 1990, pp. 111-118.
Zamperoni, P.: *Variations on the rank-order filtering theme for grey-tone and binary image enhancement.* in: Proc. ICASSP'89, Glasgow, May 1989, pp. 1401-1404.

6.5.4 Max/min-median Filter for Image Enhancement

(1) This operator *suppresses noise*, *smoothes the image* and *preserves the edge sharpness*. In contrast to the median operator (cp. Section 6.5.3), one-dimensional image structures as *thin lines* (e.g. with a width of a single image point) remain unaltered. Such thin curves would be deleted if a median operator is applied.

Attributes:

Images:	$M \times N$ gray value images
Operator:	window operator
Kernel:	order statistical, order dependent, logically structured

Inputs:

window size n .

(2) The following "one-dimensional" median gray values m_1, ..., m_4 are computed inside an picture window $\mathbf{F}(f, p)$

$$m_1 = MEDIAN\ \{f(k+1, 1), \dots f(k+1, k+1), \dots f(k+1, n)\},$$
$$m_2 = MEDIAN\ \{f(1, k+1), \dots f(k+1, k+1), \dots f(n, k+1)\},$$
$$m_3 = MEDIAN\ \{f(1, 1), \dots f(k+1, k+1), \dots f(n, n)\},$$
$$m_4 = MEDIAN\ \{f(1, n), \dots f(k+1, k+1), \dots f(n, 1)\}.$$

Furthermore, let be $m_0 = Median\ \{f(q) :\ q \in \mathbf{F}(p)\}$. The extreme values

$$MA = \mathbf{max}\ \{m_1, m_2, m_3, m_4\} \quad \text{and} \quad MI = \mathbf{min}\ \{m_1, m_2, m_3, m_4\}$$

are also computed for this placed window. The operator result is

$$h(p) = \begin{cases} MA & \text{, if } |MA - m_0| \geq |m_0 - MI| \\ MI & \text{, if } |MA - m_0| < |m_0 - MI| \ . \end{cases}$$

(3) The conventional two-dimensional *median operator* with a quadratic window does not discriminate between *impulse noise* and thin, one-dimensional patterns as lines or curves. This median operator will delete both, noise and patterns, at the same time, if the one-dimensional patterns do not contain enough image points for enforcing the median. There is however the possibility of avoiding the deletion of one-dimensional image structures by means of a median operator, by using one-dimensional and/or locally adaptive windows. This is the basic idea also of the rank-order operator of Section 6.5.7, which is characterized by an adaptive window size.

The max/min-median operator was also designed to avoide the deletion of one-dimensional thin structures. It uses one-dimensional windows without local size adaptation. This operator performs a median filtering in the four main directions within one-dimensional windows. Thus, small-sized noise patterns are deleted inside these directional windows. The four partial results are combined to yield the final value $h(p)$. This is done in such a way that one-dimensional patterns "get a chance to survive" which would be not the case for the conventional two-dimensional median filter.

(4) *Control structure*: window operator in Figure 3.8, i.e. non-centered.

Based on the control-structure for the non-centered ij-coordinate system, with $1 \leq i, j \leq n$, the gray values of the window are inserted into an one-dimensional array $PF(z)$, with $1 \leq z \leq n^2$. The index z is running bottom-up and left-to-right within the placed window $\mathbf{F}(p)$.

The median has to be computed for different gray values sets of different size at each window position. Thus it is recommended to realize this process by a special routine, say *MEDIAN*(R, Z, *MED*). Here R denotes an additional array for storing Z selected elements from the array *PF*(). Then, the median is computed for these Z elements. A procedure *MEDIAN* can be realized as a special case of the procedure *SELECT* which was given in Section 3.4.4. Therefore this procedure *MEDIAN* need not be illustrated here.

Required data arrays: $PF(1...a)$ for $a = n^2$, and $R(1...n)$.

Inputs and initialization:

```
        input of odd window size n;   k := (n – 1) / 2
```

Operator kernel:

```
            {transform F(i, j), with 1 ≤ i, j ≤ n, into v = h(x, y), cp. Figure 3.8 }
    for j := 1 to n do
            for i := 1 to n do
                    PF(i + n(j – 1)) := F(i, j);
    call MEDIAN(PF(1...a), a, m0);
```

```
for z := 0 to n – 1 do          R(z + 1) := PF(k + 1 + z· n);
call MEDIAN(R(1...n), n, m1);
for z := 1 to n do              R(z) := PF(k· n + z);
call MEDIAN(R(1...n), n, m2);
for z := 0 to n – 1 do          R(z + 1) := PF(z· (n + 1) + 1);
call MEDIAN(R(1...n), n, m3);
for z := 0 to n – 1 do          R(z + 1) := PF(n + z (n – 1));
call MEDIAN(R(1...n), n, m4);
MA := max {m1, m2, m3, m4};         MI := min {m1, m2, m3, m4};
v := MI;
if (|MA – m0| ≥ |MI – m0|) then v := MA
```

(5) The described max/min-median operator was proposed in

Wang, X., Wang, D.: *On the max/median filter*. IEEE Trans. ASSP-**38** (1990), pp. 1473-1475.

This operator is an improvement of the max-median operator, cp.

Arce, G.R., McLoughlin, M.P.: *Theoretical analysis of the max/median filter*. IEEE Trans. ASSP-**35** (1987), pp. 60-69,

which was designed to overcome a disadvantage of the conventional median operator, cp. point **(3)**.

6.5.5 Some Adaptive Variants of the Median Operator

(1) This program combines two *variants of the median operator*, the *k-nearest neighbor median filter* (*KNNM*) and the *median of absolute differences trimmed mean filter* (*MADTM*).

Attributes:

Images:	$M \times N$ gray value images
Operator:	window operator
Kernel:	logically structured, order dependent

Inputs:

window size n,
Variant $VAR = 1$ for the KNNM, and $VAR = 2$ for the MADTM,
parameter Q , odd and with $1 \le Q \le n^2$, for $VAR = 1$.

(2) It is practical for both variants, to define a function

$$R = \Phi\,[r, Z, (F(1)...F(Z))]$$

as follows: assume that the Z elements of the list $(F(1)...F(Z))$ are ranked in increasing order, and R is the r-th gray value in this ranked sequence (i.e., of rank r). The computation of R can be carried out by means of the subroutine *SELECT*, cp. Section 3.4.4.

Furthermore, let $F_{(r)}$ be the gray value of rank r of all of the $a = n^2$ gray values $f(i, j)$ inside the placed window $\mathbf{F}(p)$. For brevity, for $p = (x, y)$ we write $f(i, j)$ instead of $f(x + i, y + j)$. Let ρ be the rank of $f(p)$.

The result of the KNNM filter is

$$h(p) = \Phi\,[q, Q, (F_{(s)}...F_{(s+Q-1)})]$$

with

$$q = \mathbf{integer}\left(\frac{Q+1}{2}\right) \quad \text{and} \quad s = \mathbf{min}\,(n^2 - Q + 1, \mathbf{max}\,(1, \rho - q + 1))\,.$$

Let $\mu_1 = F_{(w)}$ be the 2-dimensional median, i.e. with $w = (a + 1) / 2$. Furthermore, let

$$\mu_2 = \Phi\,[w, a, (|f(-k, -k) - \mu_1| \;...\; |f(+k, +k) - \mu_1|)]$$

be the median of all the absolute deviations of gray values $f(i, j)$ from μ_1. Then, the result of the MADTM filter is the arithmetic average of all gray values $f(q)$ inside the placed window $\mathbf{F}(p)$, which differ from μ_1 by less than μ_2. Formally,

$$h(p) \;=\; \frac{1}{c}\sum f(q) \text{ with summation over all points } q \in \mathbf{F}(p) \text{ with } |f(q) - \mu_1| \le \mu_2,$$

where $c = \mathbf{card}\,\{q\colon\ q \in \mathbf{F}(p) \wedge |f(q) - \mu_1| \le \mu_2\}$.

(3) The result $h(p)_{KNNM}$ of the KNNM filter is the median of those Q gray values inside the placed window $\mathbf{F}(p)$ with ranks closest to rank ρ of the current gray value $f(p)$. For obtaining an univocal median it is practical that Q is chosen to be odd. The value $h(p)_{KNNM}$ can be computed in a simple way by a so-called "weighted median filter".

The *weighted median filter* output is determined starting from a list of gray values, whose median is computed. This modified list is derived from the original list of gray values $f(i, j)$ inside the placed window $\mathbf{F}(p)$, by duplicating each gray value $f(i, j)$ a number $\alpha(i, j)$ of times, and by including it into the list with this multiplicity $\alpha(i, j)$.

A given KNNM filter can be implemented as a weighted median filter if $\alpha(i, j) = 1$, for $(i, j) \neq p$, and $\alpha(p) = a + 1 - Q$ are chosen as weights. This special type of weighted median filter is also called the *center weighted median.*

The MADTM filter can be considered as a *locally-adaptive alpha-trimmed mean filter*, cp. Section 6.5.6. As a special case of *alpha-trimmed mean filter*, the result of the MADTM filter is the arithmetic average of the gray values with ranks in a certain range around the median μ_1. Here only the gray values within the range $\mu_1 \pm \mu_2$ are averaged. The threshold μ_2 is the median of all absolute differences between μ_1 and the gray values inside the placed window $\mathbf{F}(p)$. In comparison to the fixed alpha-trimmed mean filter, this adaptation results into an improved *preservation of the edge sharpness.* This is due to the fact that the averaging operation is restricted to gray values with small mutual numerical (but not necessarily rank) differences. The *suppression of impulse noise* is efficient also because isolated outliers are not included into the averaging operation.

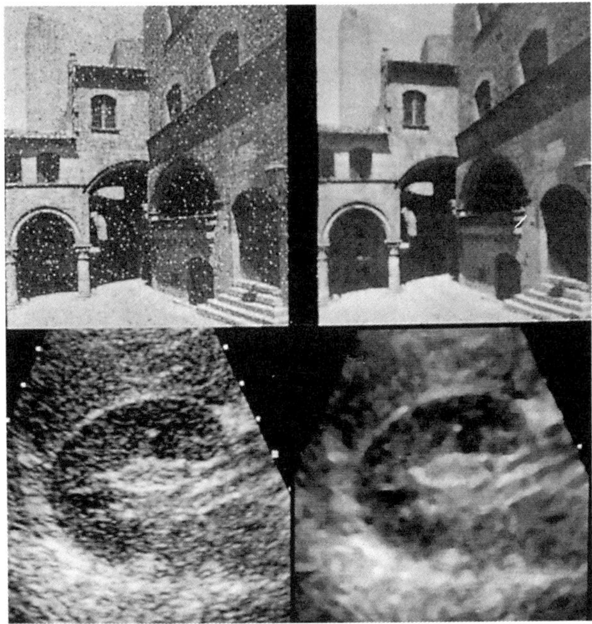

Figure 6.20: Examples of applications of two different variants of the median operator. Top left : original image with *impulse noise*, top right : noise suppression by means of the *k-nearest neighbor median filter* with $n = 5$ and $Q = 15$. Below left: original *ultrasonic image*, below right: suppression of the spurious texture by means of the *median of absolute differences trimmed mean filter*, with $n = 7$.

Figure 6.20 shows two examples of applications of the KNNM and of the MADTM filter. The aims consists in the suppression of impulse noise or of spurious textures which are typical for *ultrasonic images*. In both cases it is desirable at the same time to smooth details and to preserve sharp transitions between homogenous regions.

(4) *Control structure*: window operator in Figure 3.6, centered, with $a = n^2$

Required data arrays:

the array $F(1...a)$ for the gray values inside $\mathbf{F}(p)$, cp. Figure 3.6, has to be extended to $F(1...2\,a)$ for the KNNM filter,

array $D(1...a)$ as temporary memory for the deviations between gray values and the median, for the MADTM filter.

Inputs and initialization:

```
input of parameters n and VAR; a := n²;
w := (a + 1) / 2;
if (VAR = 1) then begin
        input of Q;      a₁ := 2 a – Q;          r := (a₁ + 1) / 2
        end {if}
```

Operator kernel:

```
                {transform F(z), with 1 ≤ z ≤ a, into v = h(x, y), cp. Figure 3.6 }
if (VAR = 1) then begin
        for z := a + 1 to a₁ do          F(z) := F(a);
        v := Φ [r, a₁, (F(1)...F(a₁))]
        end {then}
else begin
        μ₁ := Φ [w, a, (F(1)...F(a))];
        for z := 1 to a do      D(z) := |μ₁ – F(z)|;
        μ₂ := Φ [w, a, (D(1)...D(a))];
        c := 0;           R := 0;
        for z := 1 to a do
                if (|F(z) – μ₁| ≤ μ₂) then begin
                        c := c + 1;
                        R := R + F(z)
                        end {if}
        v := R / c
        end {else}
```

(5) The KNNM filter has been proposed by

Itoh, K., Ichioka, Y., Minami, T.: *Nearest-neighbor median filter.* Applied Optics **27** (1988), pp. 3445-3450.

The MADTM filter has been illustrated in

Cai, Y.L., Chen, C.S.: *An edge preserving smoothing filter based on the robust estimate.* Proc. 8th ICPR, Paris, 1986, pp. 206-208.

For the alpha-Trimmed mean filter, cp. e.g.

Lee, Y.H., Kassam, S.A.: *Generalized median filtering and related nonlinear filtering techniques*, IEEE Trans. ASSP-**33** (1985), pp. 672-683.

There exist many variants of the median operator. See

Pitas, I., Venetsanopoulos, A.N.: *Nonlinear Digital Filters.* Kluwer Academic Publishers, Boston, 1990.

for an overview. The two typical variants described in this Section have been chosen on the basis of results reported in

Campbell, T.G., du Buf, J.M.: *A quantitative comparison of median-based filters*, in: Kunt, M. (ed.), Proc. SPIE Conf. on Visual Communications and Image Processing, Lausanne, 1990, pp. 176-187.

i.e. that these two variants perform especially well with respect to edge preserving smoothing.

6.5.6 General L-Filter in a 3×3 Window

(1) The output of a general-type non-adaptive L-filter is a linear combination of the rank-ordered gray values of the observation window, with user-defined weights. This operator carries out such a L-filter in a 3×3 window.

Attributes:

Images:	M x N gray value images
Operator:	window operator
Kernel:	order statistical

Inputs:

weights $w(1), ..., w(9)$.

(2) Let $F(z)$, for $1 \leq z \leq a = 9$, denote the gray values of a 3 x 3 picture window $\mathbf{F}(f, p)$. Let $F_{(z)}$ be the rank-ordered sequence of these gray values, with $F_{(z)} \leq F_{(z+1)}$ for $z = 1, ..., a - 1$. The result of the operator is

$$h(p) = \frac{1}{S} \sum_{z=1}^{a} w(z) \cdot F_{(z)} \qquad \text{with} \qquad S = \sum_{z=1}^{a} |w(z)| \neq 0 .$$

(3) The statistical and deterministic properties of *one-dimensional L-filters* have been extensively investigated in the nonlinear digital filtering literature as robust estimators in presence of different types of noise. The well-known minimum, maximum or median operator are special cases of the general *L*-filter structure. A further special case is given by the *alpha-trimmed mean filter*. Here it is

$$w(1) = \ldots = w(s) = w(a - s + 1) = \ldots = w(a) = 0$$

and

$$w(s + 1) = \ldots = w(a - s) = 1, \text{ for } 1 \leq s \leq (a - 1) / 2.$$

An alpha-trimmed mean filter builds the arithmetic average of all gray values with the exclusion of the s smallest and s largest gray values. For certain noise distributions it is possible to estimate optimum values of the parameter s with respect to a given criterion.

The alpha-trimmed mean filters used in one-dimensional estimation have been extended to the two-dimensional case of image processing. More general filter structures have been proposed by considering also non-linear sample combinations.

A value of the parameter s specifies the alpha-trimmed mean filter structure. For example, $s = 0$ leads to a *box filter* (i.e. arithmetic average: bad in *edge preserving*, good in suppression of *Gaussian noise*, and bad in the suppression of *impulse noise*), and $s = (a - 1) / 2$ leads to the median filter (good in edge preserving, bad in the suppression of Gaussian noise, and good in the suppression of impulse noise). Besides the alpha-trimmed mean filters there are further interesting filter functions of useful operators, which can be carried out by this program by choosing other sets of weights, as e.g.

- $w(1) = w(a) = 1$, and $w(z) = 0$ otherwise; corresponds to the *mid-range-filter* (optimum *maximum-likelihood estimate* for uniformly distributed noise),
- $w(1) = -1$, $w(a) = 1$, and $w(z) = 0$) otherwise, for obtaining a so-called *range edge detector.*

Because of the small value $a = 9$ it is convenient here to use the procedure *BUBBLESORT* (cp. Section 3.4.6) also for gray value sorting.

(4) *Control structure*: window operator in Figure 3.6, centered, with $n = 3$ and $a = 9$

Required data array: w(1...9) for the weighting coefficients.

Inputs and initialization:

```
S := 0;
for z := 1 to 9 do begin
        input of the weighting coefficients w(z);
        S := S + |w(z)|
        end {for}
```

Operator kernel:

```
                    {transform F(z), with 1 ≤ z ≤ a, into v = h(x, y), cp. Figure 3.6}
                                  {sorting of gray values in place, in array F()}
call BUBBLESORT(F(1 ... 9), 9);
Q := 0;
for z := 1 to 9 do      Q := Q + w(z)· F(z);
call ADJUST(Q / S, v)                                          {ADJUST in 2.1}
```

(5) The *L*-filter and its variants are extensively treated in

Pitas, I., Venetsanopoulos, A.N.: *Nonlinear Digital Filters*. Kluwer Academic Publishers, Boston, 1990.

Bednar, J.B., Watt, T.L.: *Alpha-trimmed means and their relationship to median filters*. IEEE Trans. ASSP-**32** (1984), pp. 145-153.

Lee, Y.H., Kassam, S.A.: *Generalized median filtering and related nonlinear filtering techniques*. IEEE Trans. ASSP-**33** (1985), pp. 672-683.

For the extension of the alpha-trimmed mean filter to two-dimensional signals, see

Restrepo, A., Bovik, A.C.: *Adaptive trimmed mean filters for image restoration*, IEEE Trans. ASSP-**36** (1988), pp. 1326-1337.

6.5.7 Rank Selection Filter with Adaptive Window

(1) This operator produces as result a gray value of a specified rank, which can be arbitrarily chosen between minimum and maximum, in the placed window. However the arguments of this operator are not all the gray values inside the $\mathbf{F}(f, p)$, but only those of image points located in one of four subwindows $\mathbf{F}_u(p)$, for $u = 1, ..., 4$, of the full observation window $\mathbf{F}(p)$. These four subwindows, cp. Figure 6.21, are disjoint and cover completely the full window $\mathbf{F}(p)$ with the only exception of point p. They are characterized by a main direction (horizontal, diagonal from top right to below left, vertical, diagonal from top left to below right), and by a wedge-like shape, pointing from the window border to the point p.

The average absolute gray value deviation from the respective average is computed for each subwindow. Then, the *adaptive rank-order filter* function is applied to the subwindow with minimum (or, optional, maximum) average deviation.

Attributes:

Images: M x N gray value images
Operator: window operator
Kernel: order statistical, order dependent, logically structured

Inputs:

window size n,

rank R of the resultant gray value $h(p)$, with $1 \leq R \leq N(u)$,

Variant $VAR = 1$ for the minimum, or $VAR = 2$ for the maximum gray value deviation as a criterion for the subwindow selection.

(**2**) The size of subwindow $\mathbf{F}_u(p)$ is equal to $N(u)$. For a given window size n, the sizes $N(u)$ of both diagonal subwindows are equal, and the sizes of the horizontal and vertical subwindows are equal. However a diagonal and a horizontal subwindow can have different sizes $N(u)$. Thus it is practical to replace the parameter R, which represents the rank to be selected, with a *percentual rank* r, with $0 \leq r \leq 1$, where $r = 0$ designates the minimum operation and $r = 1$ the maximum operation. The rank R can be computed from r according to the formula

$$R = 1 + \textbf{integer}\ [r \cdot (N(u) - 1)].$$

For fast access to the gray values $F(i, j)$, with $1 \leq i, j \leq n$, in the program, the operator kernel starts with a sorting of these values into a two-dimensional array $g(u, s)$, with $1 \leq u \leq 4$ and $1 \leq s \leq S_M$.

After this sorting, $g(u, s)$ contains the gray values of the subwindow $\mathbf{F}_u(p)$. The value of S_M has to be sufficiently large for the maximum possible window size n_{max}. Roughly, it should hold

$$S_M \approx n_{max}^2 / 4 .$$

The shape of the four subwindows $\mathbf{F}_u$ is shown in Figure 6.21. These four subwindows have symmetry axes $\alpha_1, .., \alpha_4$. Whether an image point (i, j) belongs to such a subwindow $\mathbf{F}_u$ or not, depends upon its minimum (Euclidean) distance $d((i, j), \alpha_u)$ to the symmetry axis α_u:

$$(i, j) \in \mathbf{F}_v \quad \text{if} \quad d((i, j), \alpha_v) = \min_{u=1...4} \{d((i, j), \alpha_u)\} \quad \text{for} \quad (i, j) \in \mathbf{F}(p) .$$

Let μ_u be the average gray value of the subwindow $\mathbf{F}_u$, i.e.

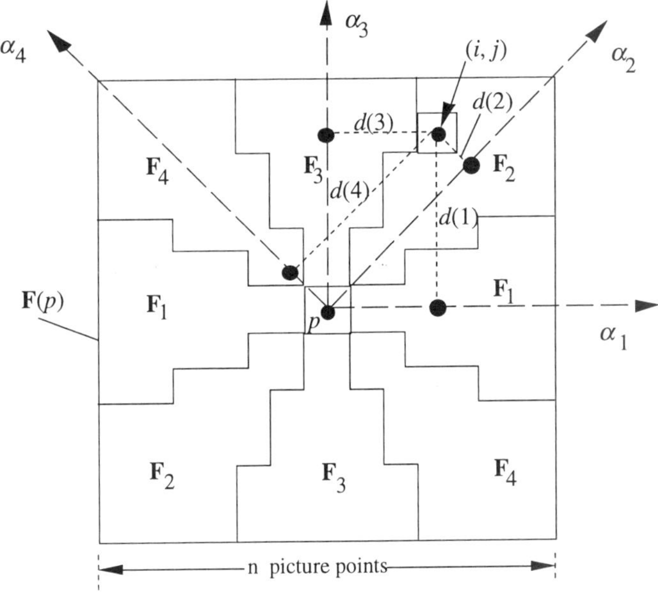

Figure 6.21: Subwindows $F_u\ (p)$, with u = 1,...,4, of the placed window F(p) and their symmetry axes α_u. The distances $d(u)$ to these four symmetry axes are shown for a generic image point (i, j).

$$\mu_u = \frac{1}{N(u)} \sum_{s=1}^{N(u)} g(u, s) \ .$$

The average absolute gray value deviation δ_u of the gray values inside $F_u\ (p)$ from the arithmetical average μ_u, is

$$\delta_u = \frac{1}{N(u)} \sum_{s=1}^{N(u)} |g(u, s) - \mu_u| \ .$$

A full sorting of all $N(u)$ gray values is not required for computing the value with rank R inside the subwindow $F_u\ (p)$. This can be done faster with the procedure *SELECT*, cp. Section 3.4.4. Let

$$Select_R\ (x_1, x_2, ..., x_L), \text{ for } R \leq L,$$

be the value with rank R in the array $(x_1,..., x_L)$. The operator's result is

$$h(p) = Select_R\ (g(v, 1), g(v, 2),..., g(v, N(v))) \quad \text{for} \quad \delta_v = \min_{u=1...4} \{\delta_u\}$$

for the case of $VAR = 1$. In the case of $VAR = 2$, the function **min** has to be replaced by the function **max**.

(3) The rank-order operators are widely used as *edge-preserving smoothing* operators. This can be relevant for image improvement or for early processing in view of *image segmentation*. Besides the *median operator* (input of $r = 0.5$), also the *minimum* and the *maximum operators* (input of $r = 0$ or $r = 1$) have a special significance as basic morphological operators, i.e. as *erosion* and *dilation*. However further adaptive window rank-operators with fixed proportional rank r ($0 < r < 1$ and $r \neq 0.5$) can often be of practical use for image improvement.

The rank-order operators with square windows have the disadvantage that they can also delete large patterns if the smallest pattern dimension (e.g. "thickness") is not larger than the window width. To overcome this disadvantage, a local adaptation is introduced. The selected operator subwindow is oriented in the direction of maximum homogeneity of the gray value function (parameter $VAR = 1$). This orientation adaptation allows the use of large windows (good for smoothing), and yet reduces the deletion of thin or elongated patterns. On the other hand, the Variant $VAR = 2$ allows a stronger smoothing than with a median filter.

The algorithm presented here uses nearly one-dimensional, wedge-shaped windows, as illustrated in Figure 6.21. Strictly line-shaped windows would be more adequate for graphical images (digitizations of line drawings etc.) than to relatively unsharp natural images. Furthermore, the algorithm has been extended from the median filter to arbitrary rank-order filters.

The program starts with the input of the window size n and with the initialization of some data which are dependent upon n. For example, the $n \times n$ data array $DMIN(i, j)$ is considered to be a geometric image of the placed window $\mathbf{F}(p)$ for $p = (x, y)$. For each image point $q \in \mathbf{F}(p)$, $q = (i, j)$ in the non-centered ij-coordinate system, the gray value u is stored into the corresponding position in $DMIN(i, j)$, for which the Euclidean distance $d((i, j), \alpha_u)$ is minimum, for $u = 1,...,4$. Using this information all gray values $f(q)$ are entered into the data array $g(u, s)$ if they belong to the subwindow $\mathbf{F}_u(p)$, for $u = 1,...,4$. Then μ_u and δ_u can be computed for $u = 1,...,4$. After determining the subwindow v with minimum or maximum δ_u-value, the rank-order filtering is performed on this subwindow v. To this aim the procedure $Select_R$ see **(2)**, is applied to the gray values stored in the array $g(v, s)$.

Figure 6.22 gives insight into the effects of some variants of this operator used here for *straight line detection* in aerial views. To reach this aim, on the one hand the background with many small details has to be smoothed, but on the other hand the direction of the smoothing window has to be fitted to the local line structure for avoiding deletions of line patterns. At top-right in Figure 6.22 the results of a simple median filtering are shown. Figure 6.22 at below left shows the results of a line extraction by means of a point-to-point image subtraction. At first an image is generated where dark line-like patterns are deleted by means of a maximum operator (Variant 2). Then, the original image is subtracted from this image. Finally Figure 6.22 at below right shows the result of an *anisotropic smoothing* by means of the median filter, followed by an *anisotropic minimum operator* (Variant 1 in both cases). The dark line-like patterns are enhanced.

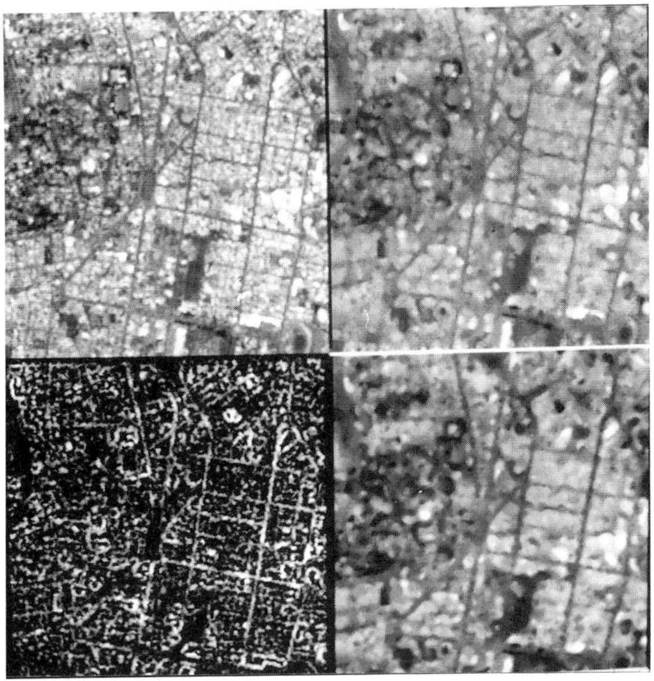

Figure 6.22: Detection of *straight-line patterns* in an *aerial view* by means of some variants of a *rank selection filter with adaptive window*. Top left : original image. Top right : Variant $VAR = 1$ for $n = 5$ and $r = 0.5$. Below left: point-to-point image difference between the Variant 2, with $n = 7$ and $r = 0.7$, and the original image. Below right: two iterations of Variant 1, first with $n = 5$ and $r = 0.5$, and then with $n = 3$ and $r = 0$.

(4) *Control structure*: window operator in Figure 3.8, i.e. non-centered

Required data arrays (as discussed already in **(3)**):

$g(u, s)$ for gray values sorted into the subwindows $u=1,...,4$, for $1 \leq s \leq S_M$,

$DMIN(i, j)$ with $1 \leq i, j \leq n_{max}$ and $1 \leq DMIN \leq 4$ for the assignment of image points (i, j) into the subwindows $\mathbf{F}_u$,

$\mu(u)$ for the average gray values in the subwindows $\mathbf{F}_u$,
$\delta(u)$ for the average absolute deviations in the subwindows $\mathbf{F}_u$,
$D(u)$ as temporary memory for the distances between an image point and the four symmetry axes α_u.

Inputs and initialization:

```
input of parameters n, R and VAR;
K := (n + 1) / 2;
for j := 1 to n do
        for i := 1 to n do begin
                D(1) := |j – K|;          D(2) := 0.7071 · |i – j|;
                D(3) := |i – K|;          D(4) := 0.7071 · |i + j – n – 1|;
                MIND := n_max;
                for z := 1 to 4 do
                        if (D(z) < MIND) then begin
                                MIND := D(z);          Z := z
                                end {if};
                DMIN(i, j) := Z
                end {for}
```

Operator kernel:

```
        {transform F(i, j), with 1 ≤ i, j ≤ n, into v = h(x, y), cp. Figure 3.8}
                        {sorting of gray values into the proper subwindows, and
                                              computation of average gray values}
for z := 1 to 4 do begin
        g(z, 1) := F(K, K);
        N(z) := 2;        {initialization of the indices of the elements in F_z(p)}
        μ(z) := 0
        end {for};
for j := 1 to n do
        for i := 1 to n do
                if ((i ≠ K) ∨ (j ≠ K)) then begin
                        g(DMIN(i, j), N(DMIN(i, j))) := F(i, j) ;
                        μ(DMIN(i, j)) := μ(DMIN(i, j)) + F(i, j);
                        N(DMIN(i, j)) := N(DMIN(i, j)) + 1
                        end {if};
```

```
                {computation of the average deviations in the four subwindows}
for z := 1 to 4 do begin
        N(z) := N(z) – 1;        μ(z) := μ(z) / N(z);    δ(z) := 0
        end {for};
for u := 1 to 4 do begin
        for s := 1 to N(u) do       δ(u) := δ(u) + |g(u, s) – μ(u)|;
        δ(u) := δ(u) / N(u)
        end {for};
if (VAR = 1) then begin
        MIN := largest representable number;
        for u := 1 to 4 do
                if (δ(u) < MIN) then begin
                        MIN := δ(u);        w := u
                        end {if}
        end {then}
else begin
        MAX := 0;
        for u := 1 to 4 do
                if (δ(u) > MAX) then begin
                        MAX := δ(u);        w := u
                        end {if}
        end {else}
R = 1 + integer [r (N(w) – 1)];
v := Select_R (g(w, 1), g(w, 2), ..., g(w, N(w)))
```

(5) Rank-order operators for image improvement and for pre-processing for image segmentation are described e.g., in

Wahl, F.M.: *Digital Image Processing*. Artech House, Norwood, 1987.

Zamperoni, P.: *Methoden der digitalen Bildsignalverarbeitung*. Vieweg Verlag, Wiesbaden, 2nd ed., 1991.

Data-driven rank-order operators with constant rank are discussed in

van den Boomgaard, R.: *Threshold logic and mathematical morphology*. in: Cantoni, V. et al. (eds.): Progress in image analysis and processing, World Scientific Publishing, Singapore, 1990, pp. 111-118.

Zamperoni, P.: *Variations on the rank-order filtering theme for grey-tone and binary image enhancement*. in: Proc. ICASSP'89, Glasgow, May 1989, pp. 1401-1404.

An *anisotropic median filter* with (straight) one-dimensional windows has been proposed by

Nieminen, A., Heinonen, P., Neuvo, Y.: *A new class of detail-preserving filters for image processing*, IEEE Trans. PAMI-**9** (1987), pp. 74-90.

The principle of the window direction adaptation has also been used in

Presetnik F.F., Filipovič, M.: *Adaptive median filtering of images degraded by speckle noise*, in Lacoume, J.L. et al. (eds.), Signal Processing IV: Theories and Applications, Elsevier, Amsterdam, 1988, pp. 651-654.

within the context of a specific application. The adaptation was controlled by the local image data.

6.5.8 Rank-order Transformation (Contrast Stretching)

(1) This operator causes a stretching of the local contrast. The strength of this stretching increases with decreasing the size of the operator windows $\mathbf{F}(p)$. The contrast stretching is realized by means of a local gray value equalization inside the window $\mathbf{F}(p)$, i.e. the gray values are uniformly mapped onto the full gray value scale 0, ..., $G-1$. Texture elements or details can be enhanced with this operation up to a certain lowest spatial frequency, determined by the window size.

Attributes:

Images:	M x N gray value images
Operator:	window operator
Kernel:	data driven, order statistical

Inputs:

window size n.

(2) The resultant image h is defined as

$$h(p) = \frac{r-1}{a-1} \cdot (G-1)\,, \qquad \text{with} \quad r = u + \textbf{integer}\left(\frac{w}{2}\right),$$

where u and w denote the number of gray values $f(i, j)$, with $(i, j) \in \mathbf{F}(p)$, satisfying the conditions $f(i, j) < f(p)$ or $f(i, j) = f(p)$ respectively.

(3) Let us consider the continuous gray value scale $[0, G-1]$. This scale is uniformly subdivided into a "nominal gray values"

$$g_z = \frac{G-1}{a-1}\, z \quad \text{with} \quad z = 0, ..., a-1.$$

In practice, these values g_z are rounded to the next integers.

Assume that $f(p)$ has the rank r, with $0 \leq r \leq a-1$, among the a gray values of $\mathbf{F}(p)$. Then this operator maps $f(p)$ onto the gray value g_r, having in the discrete nominal gray value scale the same rank as $f(p)$ in the set of the gray values of $\mathbf{F}(p)$. Thus the local contrast in $\mathbf{F}(p)$ is stretched to the full gray value scale. A local minimum results into $g_0 = 0$, and a local maximum results into $g_{a-1} = G - 1$.

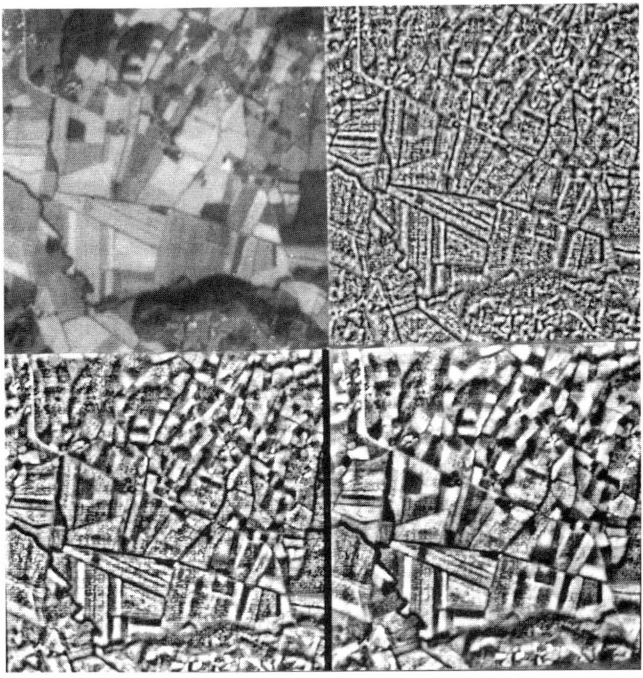

Figure 6.23: Example of an application of the rank order transformation for an *aerial view*. Top left : original image, top right : 7×7 window, below left: 13×13 window, below right: 21×21 window.

This operator realizes a local gray value equalization. This can be qualitatively illustrated taking the example of homogenous regions as follows: For such region in each picture window $\mathbf{F}(f, p)$ there exists exactly one smallest, one second smallest, ..., one largest gray value. After this rank-order transformation, the gray values of the originally homogenous region are uniformly distributed over the full gray value scale.

The global *gray value equalization*, as discussed in Section 5.1.4 can be seen as an extreme case of this rank-order transformation, i.e. if n is taken equal to N.

There is a certain similarity between the equations for $h(p)$ in Section 5.1.4, with $r = 1$, and for $h(p)$ here in point **(2)**. However the sum frequencies in 5.1.4 are here replaced by the exact numbers $r-1$ or $a-1$ of gray values in $\mathbf{F}(f, p)$ which are smaller than $f(p)$ or smaller than the local maximum respectively.

The determination of the rank r of $f(p)$ is not unique if gray value $f(p)$ occurs several times in $\mathbf{F}(f, p)$. In this case, half of the values $f(p)$ are considered to be less than $f(p)$, and half of these values are considered to be greater than $f(p)$. While defining $h(p)$ in **(2)** the normal case is also considered i.e. the value $f(p)$ appears just once in $\mathbf{F}(f, p)$.

The example in Figure 6.23 illustrates that the strength of detail enhancement increases if smaller windows are chosen.

(4) *Control structure*: window operator in Figure 3.6, centered, with $a = n^2$

Inputs:

```
input of odd parameter n
```

Operator kernel:

```
        {transform F(z), with 1 ≤ z ≤ a, into v = h(x, y), cp. Figure 3.6 }
u := 0; w := 0,
for z := 1 to a do begin
        if (F(z) < F(a)) then u := u + 1;
        if (F(z) = F(a)) then w := w + 1
        end {for};
r := u + w / 2;
v := [(r – 1) (G – 1)] / (a – 1)
```

(5) Zamperoni, P.: *Methoden der digitalen Bildsignalverarbeitung*. Vieweg Verlag, Wiesbaden, 2nd ed., 1991.

Fahnestock, J.D., Schowengerdt, R.A.: *Spatially variant contrast enhancement using local range modification*. Optical Engineering **22** (1983), pp. 378-381.

Kim, V., Yaroslavskii, L.: *Rank algorithms for picture processing*. Computer Vision, Graphics, and Image Processing **35** (1986), pp. 234-258.

6.5.9 Anisotropy-controlled Adaptive Rank-order Filters

(1) This program combines two variants of an *adaptive rank-order filter*. They can be used for *image enhancement* or for preprocessing before *image segmentation*, especially in the case of complex natural images. Adaptation in dependence of the local image data allows certain requirements to be met which would otherwise be incompatible:

- smoothing of *stochastic noise* (for image improvement) or of *texture patterns* (for image segmentation),
- preservation or enhancement of *edge sharpness* or of *edge contrast*,
- preservation of structured details of one pixel width, e.g. of *thin lines*.

In both variants, the filter adaptation is controlled by means of the *local anisotropy measure* as previously defined in Section 6.6.1.

The Variant 1 can be considered as a gradual *extreme sharpening operator* (cp. Section 6.3.1). The resultant gray value is an adaptively determined element of the set of the rank-ordered gray values of $\mathbf{F}(f, p)$, depending on a local anisotropy measure. Unlike with the conventional extreme sharpening, the output gray value is not restricted to be only either the minimum or the maximum.

As Variant 2 is presented an *adaptive k-nearest-neighbor filter*. The resultant gray value is the average of those K gray values inside $\mathbf{F}(f, p)$ with closest value to $f(p)$. The number K is anisotropy dependent.

Attributes:

Images:	M x N gray value images
Operator:	window operator
Kernel:	order statistical, order dependent, logically structured (in case of Variant 1)

Inputs:

window size n,
Variant $VAR = 1$ or $VAR = 2$.

(2) There are $T = 2(n - 1)$ different *digital straight line segments* which contain the point p and connect two border points of the $n \times n$ window $\mathbf{F}(p)$. A digital straight line segment has a width of one pixel. Figure 6.24 shows two of such segments for a 7×7 window with $T = 12$. Furthermore, let M_t, with $t = 1, ..., T$, be the average gray value of the t-th segment multiplied by n. For example, using $k = (n - 1) / 2$ it holds

$$M_1 = \sum_{t=-k}^{k} f(x + t, y - t) \quad \text{and} \quad M_{k+1} = \sum_{t=-k}^{k} f(x + t, y).$$

The control structure for addressing all the image points of each digital straight line segment is explained in **(3)** and **(4)** in detail. Furthermore, let

$$MI = \min_{t=1\ldots T} \{M_t\} \quad \text{and} \quad MA = \max_{t=1\ldots T} \{M_t\}.$$

As in Section 6.6.1, a measure α of local anisotropy is defined as follows,

$$\alpha = \begin{cases} \dfrac{MA - MI}{MA + MI} & \text{, if } MA \neq 0 \\ 0 & \text{otherwise.} \end{cases}$$

It follows $0 \leq \alpha \leq 1$. In the case of the Variant 1, the operator result $h(p)$ is given by

$$h(p) = F_{(R)},$$

where $1 \leq R \leq a$ and $F_{(R)}$ is the R-th rank-ordered gray value, with

$$R = \begin{cases} \langle c + \alpha(c-1) \rangle & \text{, if } f(p) > (F_{(a)} + F_{(1)}) / 2 \\ \langle c - \alpha(c-1) \rangle & \text{, if } f(p) \leq (F_{(a)} + F_{(1)}) / 2 \end{cases}$$

for

$$c = \frac{a+1}{2}.$$

The sign $\langle\ \rangle$ stands for the rounding operation, i.e. "next integer of". In Variant 2 it is

$$h(p) = \frac{1}{K} \sum_{z=1}^{K} F_{(R_z)},$$

where $F(R_1)$, ..., $F(R_K)$ are the K gray values with value closest to $f(p)$ or with rank closest to R_z. The parameter K is defined by

$$K = \langle n + (1 - \alpha)(a - n) \rangle .$$

(3) In the literature there are many examples of adaptive rank-order filters, especially of *L-filters* (cp. Section 6.5.6) with adaptive coefficients. Several of the proposed methods aim at an optimum estimate of an uniform gray value in the presence of differently distributed noise. An optimality criterion has to be defined for this approach as well as a noise model. Typically, the filter parameters are adaptively determined, cp. references to *Bolon*, *Lee/Kassam* and *Restrepo/Bovik* in **(5)**.

The preservation of the edge sharpness and the suppression of *impulse noise* is also included into the optimization process in most of the cases, cp. references to *Bolon*, *Lee/Tantaratana*, *Petit* and *Pitas/Venetsanopoulos* in **(5)**.

Many of these methods are focused on the quality of the estimation, i.e. they are directed on a good smoothing of stochastic noise. The disadvantages of such methods are algorithmic complexity and the loss of structured details, e.g. of thin lines. High algorithmic complexity often restricts their practical use.

In comparison with such methods, the operators described in this Section are characterized in general by a lower performance in *noise smoothing* and by an about equally good *edge preservation*. They are superior in the preservation of details and in the algorithmic simplicity. The latter property is typical for the rank-selection filter (*L*-filter with only one coefficient not equal to zero), i.e. the Variant 1 of the program. A rank-selection filter is necessary for the preservation of details because a linear combination of several gray values leads in general to a deletion or to a strong blurring of thin lines. On the other hand, a rank-selection filter does not allow the estimation of a uniform gray value affected by noise as efficiently as with a *L*-filter with a coefficients not equal to zero.

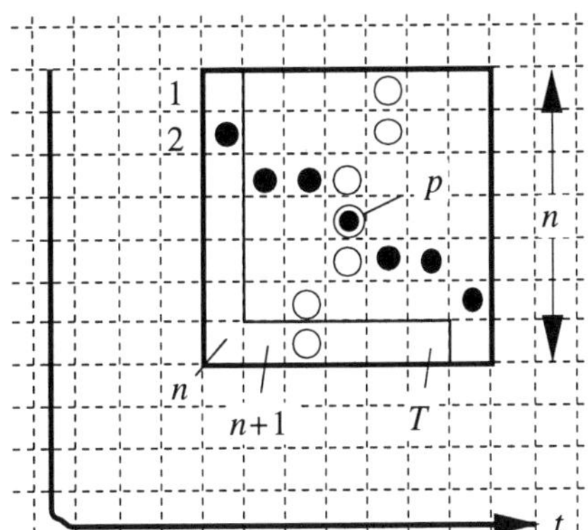

Figure 6.24: Operator window $\mathbf{F}(p)$ with $n = 7$ and $k = 3$. There are $T = 12$ possible *digital straight line segments* through point p, for corresponding values of the index t. Two segments are shown, for $t = 2$ and for $t = n + 2 = 9$:

• $t = 2$ with $\Delta Y = 1 + k - t = +2$,

∘ $t = n + 2 = 9$ with $\Delta X = t - n - k = -1$.

The detail-preserving properties of the *extremum sharpening operator* (cp. Section 6.3.1) are exploited in the Variant 1 in *high anisotropy situations* ($\alpha \approx 1$, e.g. for thin lines). In this case the formula in **(2)** yields the result $h(p) \approx F_{(1)}$ or $h(p) \approx F_{(a)}$, depending upon the value of $f(p)$. This is exactly what is needed to preserve a line pattern at location p. At a point p with *low anisotropy* ($a \approx 0$, e.g. in regions of nearly constant gray value) the operator performs like a median filter ($R = c$). A median filter is efficient in smoothing impulse noise and broadly distributed noise, but not concentrated or normally distributed noise, cp. the references to *Bolon*, *Pitas/Venetsanopoulos*, *Melsa/Cohn* and *Restrepo/Bovik* in **(5)**.

The operator of Variant 2 is characterized by a better smoothing and by a less good preservation of details in comparison to Variant 1. This operator is suitable for an image model where the smallest structures, which have to preserved, are segments of $n \times 1$ image points inside a $n \times n$ window. For such patterns it is $a \approx 1$; the formula for parameter K in **(2)** leads to $K \approx n$. The resultant value $h(p)$ represents then the average inside a line segment. The number of gray values involved in the averaging operation increases if α decreases. For $\alpha = 0$, all gray values in $\mathbf{F}(f, p)$ are arguments of the averaging operation, i.e. this is true for a region with about a constant gray value. In such a region the operator behaves like a box filter.

The *anisotropy measure* α can be defined axiomatically. Let be $0 \leq \alpha \leq 1$. Let us assume $\alpha = 0$ for a constant gray value inside the picture window $\mathbf{F}(f, p)$, and $\alpha = 1$ for a thin white line on a black background (or vice-versa). In **(5)** a paper by *Zamperoni* is cited where different alternatives for an analytical definition of α are discussed. The definition as used here reflects a dependency upon the average local gray value, but also has other advantages. In correspondence to visual perception a pattern of constant contrast will obtain a higher α-value if it is embedded in a dark image region.

The correct addressing of gray values inside the T digital straight line segments of $\mathbf{F}(p)$ is now explained. See Figure 6.24 for an example. In the program, averages M_t, for $t = 1, ..., T$, have to be computed for these T segments.

An index $t = 1, ..., T$ is used for designating these segments. The non-centered ij-coordinate system is used for points in the placed window $\mathbf{F}(p)$. The coordinates $DX(z, t)$ and $DY(z, t)$ of points on the t-th segment are calculated for $z = -k, ..., +k$, and for $t = 1, ..., T$, and stored in the proper data array. As illustrated in Figure 6.24, there are two different situations for t , one for $t \leq n$ (the segment begins at the leftmost column), and one for $t > n$ (the segment begins at the lowest row). This computation can be split into two subprocesses, one for $t = 1, ..., n$ and one for $t = n + 1, ..., T$. By means of the relative pixel addresses stored in the data arrays $DX(z, t)$ and $DY(z, t)$, the gray values of each segment can be read, then the averages $M_1, ..., M_T$ are computed, and finally the extreme values MI and MA are calculated.

Only one rank selection is necessary for Variant 1. However a full ranking is necessary for Variant 2. With respect to algorithmic complexity, the procedure *SELECT* (cp. Section 3.4.4) is suggested for Variant 1. However for simplicity the procedure *BUCKETSORT* (cp. Section 3.4.7) has been chosen for both variants.

Figure 6.25 shows above an application example of Variant 1, and below of Variant 2. At the left two original *aerial views* are displayed. The lower image is a *SAR image*. The processed image above at right has been obtained with a 9×9 window. A 5×5 window has been used for the image below right. In both cases, regions with texture (above: regions without buildings, below: *speckle noise*) are smoothed. The edge sharpness and the details (settlements, streets, channels) have been essentially preserved.

(4) *Control structure*: window operator in Figure 3.6, centered, with $a = n^2$ and $k = (n-1)/2$

Required data arrays:

$DX(-k...k, 1...T_{max})$ and $DY(-k...k, 1...T_{max})$ for the relative coordinates of pixels inside the segments $1...T$, with $T_{max} = 2 \cdot (n_{max} - 1)$ and where n_{max} is the largest allowed window size,

$F(1...a_{max})$, with $a_{max} = n_{max}^2$ for the rank-ordered gray values of $\mathbf{F}(f, p)$,

$Q(1...G)$ for the local gray value histogram *hist*:.

Inputs and initialization:

```
input of odd parameter n, and VAR;   a = n²;
MED := (a + 1) / 2;
for u := 1 to G do    Q(u) := 0;
T := 2(n – 1);
for t := 1 to n do begin
        ΔY := 1 + k – t;
        for z := –k to k do begin
                DX(z, t) := z;            DY(z, t) := ⟨–(z Δ Y) / k⟩
                end {for}
        end {for};
for t := n + 1 to T do begin
        ΔX := t – n – k;
        for z := –k to k do begin
                DY(z, t) := ⟨–(z Δ X) / k⟩;     DY(z, t) = z
                end {for};
        end {for}
```

Remark: The sign ⟨ ⟩ denotes the "nearest integer of".

Operator kernel:

```
        {transform F(i, j), with –k ≤ i, j ≤ k, into v = h(x, y), cp. Figure 3.8}
    MA := 0;        MI := n(G – 1);
    for t := 1 to T do begin
        SUM := 0;
        for z := –k to k do
            SUM := SUM + F(DX(z, t), DY(z, t));
        if (SUM < MI) then MI := SUM;
        if (SUM > MA) then MA := SUM
        end {for};
    α := (MA – MI) / (MA + MI);
    for j := –k to k do
        for i := –k to k do
            Q(1 + F(i, j)) := Q(1 + F(i, j)) + 1;
    Z := 1;
    for u := 1 to G do
        if (Q(u) ≠ 0) then begin
L1          F(Z) := u –1;        Z := Z + 1;     Q(u) := Q(u) – 1;
            if (Q(u) > 0) then goto L1
            end {if};
    if (VAR = 1) then begin
        if ((F(a) – F(0, 0)) < (F(0, 0) – F(1)))
            then  R := ⟨ α (MED – 1) ⟩ + MED
            else  R := MED – ⟨ α (MED – 1) ⟩;
        v := F(R)                                        {value of h(p)}
        end {then}
    else  begin
        K := n + ⟨ (1 – α) (a – n) ⟩;
        for z := 1 to a do
            if (F(0, 0) = F(z)) then begin
                RK := z;      goto L2
                end {if};
L2      SUM := F(RK);      RO := RK + 1;       RU := RK –1;
        for z := 2 to K do
            if (RU < 1) then begin
                          SUM := SUM + F(RO);      RO := RO + 1
```

```
                    end {then}
               else if (RO > a) then begin
                    SUM := SUM + F(RU);        RU := RU -1
                    end {then}
               else if ((F(RO) - F(RK)) < (F(RK) - F(RU))) then
                    begin
                    SUM := SUM + F(RO);        RO := RO + 1
                    end {then}
               else  begin
                    SUM := SUM + F(RU);        RU := RU -1
                    end {else};
     v := SUM / K                              {value of h(p)}
     end {else}
```

(5) For the operators described here (Variant 1 and Variant 2) see

Bolon, P.: *Filtrage d'ordre, vraisemblance et optimalité des prétraitements d'image*. TS (Traitement du Signal) **9** (1992), pp. 225-250.

Lee, Y.H., Kassam, S.A.: *Generalized median filtering and related nonlinear filtering techniques*. IEEE Trans. ASSP-**33** (1985), pp. 672-683.

Restrepo, A.., Bovik, A.C.: Adaptive trimmed mean filters for image restoration. IEEE Trans. ASSP-**36** (1988), pp. 1326-1337.

The preservation of edge sharpness and the suppression of impulse noise was included into the optimization process in the above cited paper by *Bolon* as well as in

Lee, Y.H., Tantaratana, S.: *Decision-based order statistic filters*. IEEE Trans., ASSP-**38** (1990), pp. 406-420.

Petit, J.L.: *Amélioration d'image par filtrage statistique d'ordre adaptatif*. Dissertation, University of Aix-Marseille, Nov. 1990.

Pitas, I., Venetsanopoulos, A.N.: *Nonlinear Digital Filters*. Kluwer Academic Publishers, Boston, 1990.

The extremum sharpening filter has been used by

Lester, J.M., Brenner, J.F., Selles, W.D.: *Local transforms for biomedical image analysis*. Computer Graphics and Image Processing **13** (1980), pp. 17-30.

for tasks in image processing. The adaptive variant as discussed here has been proposed in

Zamperoni, P.: *Adaptive rank order filters for image processing based on local anisotropy measures*. Digital Signal Processing **2** (1992), pp. 174-182.

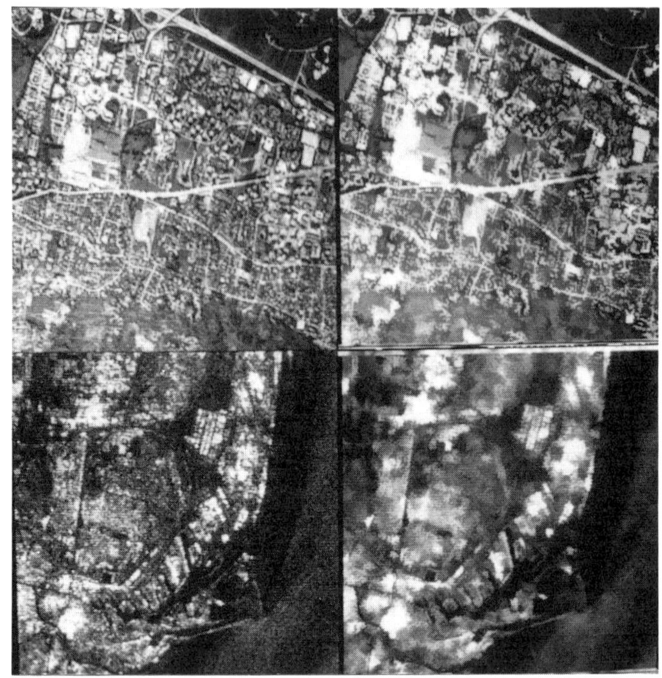

Figure 6.25: *Anisotropy-controlled adaptive rank-order filters*. Top: Variant 1, i.e. gradual *extremum sharpening operator*, left: original *aerial view*, right: result of a processing using a 9×9 window. Below: Variant 2, i.e. averaging of K nearest neighbors, left: original *SAR image*, right: result of a processing using a 5×5 window.

With respect to gray value estimation under different noise distributions, reference can be made to the above cited sources by *Bolon*, *Pitas/Venetsanopoulos*, *Restrepo/Bovik* as well as to

Melsa, J.L., Cohn, D.L.: *Decision and Estimation Theory*. McGraw-Hill, New York, 1978.

6.6 LINE EXTRACTION FILTERS

This Section contains two operators for line-like patterns. A *line* is roughly defined to be an elongated image region with about constant gray value, and the width of this region should be small in comparison to the window width n. The first operator aims at the enhancement, and the second at the deletion of line-like patterns.

6.6.1 Line Extraction

(1) This operator can be used for the selective *enhancement of line-like patterns.*

Attributes:

Images:	$M \times N$ gray value images, no bilevel images
Operator:	window operator
Kernel:	data driven, order dependent

Inputs:

window size n,
Variant $VAR = 1$, or $VAR = 2$.

(2) The centered ij-coordinate system is assumed for the gray values $F(i, j)$, for $-k \le i, j \le k$, of the picture window $\mathbf{F}(f, p)$. Furthermore, consider four straight stripes going through the window's current point p. These stripes have a one-pixel width. There are one horizontal, one vertical, and two diagonal stripes. The average gray values of these four stripes, multiplied by n, are denoted as M_z, for $1 \le z \le 4$. It holds

$$M_1 = \sum_{t=-k}^{k} F(t, 0) \ , \qquad M_2 = \sum_{t=-k}^{k} F(0, t) \ ,$$

$$M_3 = \sum_{t=-k}^{k} F(t, t) \qquad \text{and} \qquad M_4 = \sum_{t=-k}^{k} F(t, -t) \ .$$

Let

$$MA = \max_{z=1...4} \{M_z\} \quad \text{and} \quad MI = \min_{z=1...4} \{M_z\}.$$

A measure α of local anisotropy is defined as follows:

$$\alpha = \begin{cases} 0 & , \text{if } MA = 0 \\ \dfrac{MA - MI}{MA + MI} & \text{otherwise} \end{cases} \ .$$

The operator result is given by

$$h(p) = \begin{cases} \alpha \cdot f(p) & , \text{if } VAR = 1 \\ \alpha \cdot (G - 1) & , \text{if } VAR = 2 \ . \end{cases}$$

(3) The *anisotropy measure*, as defined above, is equal to 0 in regions of constant gray value, and is equal to 1 for a one-pixel thin, n-pixel long straight line through the current point. Line segments of the input image are extracted according to their length, and this length can be adjusted by means of the input parameter n. This anisotropy measure does not discriminate between bright and dark lines. If α is used as factor for the gray value $f(p)$ of the reference point p, then only bright lines are preserved. The line preservation depends upon the degree of correspondence between the window gray value pattern and the ideal line model. For extracting only dark lines, a straightforward solution consists in starting from the inverted image g of the original image, $g(p) = G - 1 - f(p)$, and then proceeding as described for Variant 1.

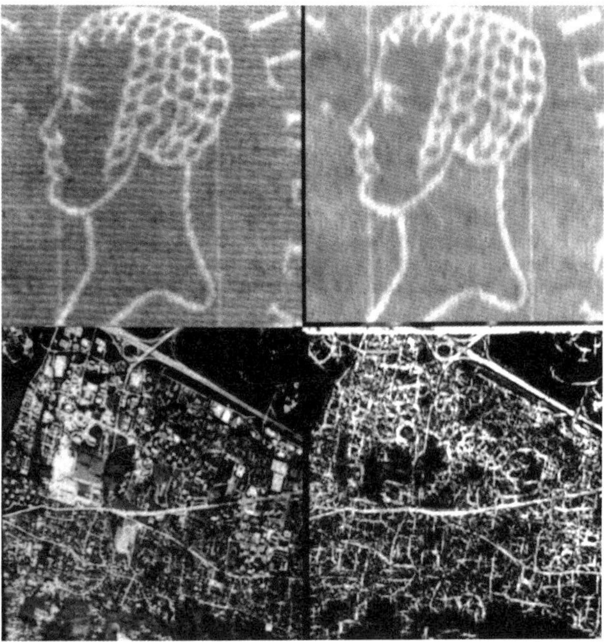

Figure 6.26: Examples of operators for *filtering line-like structures*. Above: suppression of horizontal line-like structures with the operator of Section 6.6.2, left: original image, right: processing by means of a 5×7 window. Below: *line extraction* with the operator of Section 6.6.1, Variant 1, left: original image, right: processing by means of a 9×9 window.

Variant 2 can be used if bright as well as dark lines have to be enhanced at the same time. In this case, α is used as factor of the symbolic gray value $G - 1$ (white) instead of the gray value $f(p)$. This second variant can also be used advantageously if lines of only one type are present, i.e. either bright or dark ones. If bright lines have a gray value close to $G - 1$, and if dark lines have a gray value close to 0, then the resultant image (applying Variant 2) has a strong contrast, and further contrast enhancement is mostly not necessary. Figure 6.26, below, shows an example of the application of this operator to the extraction of roads in an aerial view.

(4) *Control structure*: window operator in Figure 3.6, centered, with $k = (n - 1) / 2$

Inputs and initialization:

```
input of the odd parameter n, and of VAR;
g := G – 1
```

Operator kernel:

```
        {transform F(i, j), with –k ≤ i, j ≤ k, into v = h(x, y), cp. Figure 3.6 }
for z := 1 to 4 do      M_z := 0;
for z := – k to k do begin
        M1 := M1 + F(z, 0);             M2 := M2 + F(0, z);
        M3 := M3 + F(z, z);             M4 := M4 + F(z, – z)
        end {for};
MA := max {M1, M2, M3, M4};
if (MA = 0) then        v := 0
else begin
        MI := min {M1, M2, M3, M4};
        if (VAR = 1) then g := F(0, 0);
        v := g · [(MA – MI) / (MA + MI)]
        end {else}
```

(5) Zamperoni, P.: *Adaptive rank order filters for image processing based on local anisotropy measures*. Digital Signal Processing **2** (1992), pp. 174-182.

6.6.2 Suppression of Line Patterns

(1) Assume that horizontal or vertical line-like superposed patterns of one pixel width have to be considered as *interferences*. The function of this operator is of subtracting such noise patterns from the original image.

Attributes:

Images:	M x N gray value images
Operator:	window operator
Kernel:	linear, order dependent

Inputs:

window width m and window height n, both as odd numbers,

Variant $VAR = 1$ for eliminating horizontal line-like patterns, or $VAR = 2$ for vertical ones

(2) Assume the non-centered ij-coordinate system for the gray values $\mathbf{F}(f, p)$, with $1 \leq i \leq m$ and $1 \leq j \leq n$, of the picture window $\mathbf{F}(f, p)$. For the resultant image h it is

$$h(p) = \begin{cases} f(p) + \frac{1}{m\,n} \cdot \sum\limits_{q \in F(p)} f(q) - \frac{1}{m} \cdot \sum\limits_{i=1}^{m} F\left(i, \frac{n+1}{2}\right) & \text{, if } VAR = 1 \\ f(p) + \frac{1}{m\,n} \cdot \sum\limits_{q \in F(p)} f(q) - \frac{1}{n} \cdot \sum\limits_{j=1}^{n} F\left(\frac{m+1}{2}, j\right) & \text{, if } VAR = 2\,. \end{cases}$$

(3) In some applications it can be desirable to eliminate line-like interferences arising during the process of image acquisition. Systematic interference noise of remote sensing is a typical example of such a situation, cp. references in **(5)**. Such noise can be suppressed on the basis of a simple image model. The model assumes the alternatives of horizontal or vertical line-like noise, where the line thickness should be much smaller than the window size.

In general, non-square windows are used in this program for ensuring a flexible match to the image content. According to the image model used here, a deviation between the average gray value μ of $\mathbf{F}(f, p)$ and the average gray value μ_s of a (horizontal or vertical) stripe through p of one pixel width, is attributed to the noise which has to be eliminated. Therefore, the value $f(p)$ is corrected by means of the difference $\mu - \mu_s$.

The window parameters m and n have to be chosen according to the noise model. In case of horizontal noise, e.g., the window height n should be a multiple of the line thickness. So it can be ensured that the value of μ is practically not influenced by the presence of noise, and it can be suggested to choose $m < n$, e.g. if the noise lines are fragmentary (with gaps), as the window should be homogenous with respect to noise.

Figure 6.26, above, shows an example of the application of this operator to the suppression of horizontal line-like noise. The line-like noise has an irregular structure.

(4) *Control structure*: window operator in Figure 3.6, i.e. non-centered

Inputs and initialization:

```
input of the odd parameters m and n;
input of VAR;
b := (m + 1) / 2;        c := (n + 1) / 2
```

Operator kernel:

```
{transform F(i, j), with 1 ≤ i ≤ m and 1 ≤ j ≤ n, into v = h(x, y), cp. Figure 3.6}

μ := 0;           μs := 0;
for j := 1 to n do begin
      μ1 := 0;
      for i := 1 to m do begin
            μ1 := μ1 + F(i, j);
            if ((VAR = 2) ∧ (i = b)) then   μs = μs + F(i, j)
            end {for}
      μ := μ + μ1;
      if ((VAR = 1) ∧ (j = c)) then   μs := μ1
      end {for};
μ := μ / (m· n);
if (VAR = 1) then μs := μs / m
else   μs := μs / n;
v := F(0, 0) + μ – μs
```

(5) Chavez, P.: *Simple high-speed digital image processing to remove quasi-coherent noise patterns*. U.S. Geological Survey Computer Center Division, internal report.

McDonnell, M.J.: *Box-filtering techniques*. Computer Graphics and Image Processing **17** (1981), pp. 65- 70.

7 GLOBAL OPERATORS

Global operators have been introduced in Section 1.4.4. In general, their implementation requires a higher computational complexity. However some basic operators of the global type should be included in a library of image processing procedures. A global operator has no specific window **F** such that output data $h(p)$ depend only from pixel values of input image f inside the placed window $\mathbf{F}(p)$. For global operators it is impossible to define an operator kernel for repeated applications at different positions in the input image f. If the window **F** is selected to be very large in comparison to M and N, then the number of border points increases. Border points would require special treatment. The whole input image should be available in the working memory for direct access in the case of global operators.

In general, for a given point p of the resultant image h there is no restriction to the size of the subimage of f which is sufficient for the computation of $h(p)$. However some global operators can be realized with several iterations of local operators on the input image f, or on adequate intermediate images. For example, a few iterations of sequential local operators can be sufficient for carrying out some topological operators. Information "propagates" over the whole image as an effect of utilizing the sequential (*IIR*) processing mode.

7.1 TOPOLOGICAL OPERATORS

Component labeling and thinning are the operators considered in this Section. Connectedness properties characterize the *topology of image segments*. For example, an image segment may have several holes or cavities, and these holes or cavities may contain further image segments. Such a segment is considered to be topologically complex. A connected region without holes or cavities is topologically simply-connected. The number of image segments or of connected components is a (simple) topological feature of an image. The process of connected component labeling provides an iconic representation of this feature. Each image segment is assigned exactly one (symbolic) gray value.

Topological features are invariant with respect to topological mappings. Deformations of image segments are topological mappings, in general. The generation of a skeleton by thinning is a special topological mapping.

7.1.1 Connected Component Labeling

(1) Let us consider a *bilevel image*. This operators aim consists in labeling of *8-components* (cp. Section 1.1.1). Each 8-component is assigned a unique label, and this label is the pixel value of all points of this 8-component. Labels can be, e.g., pseudo colors (cp. Section 5.6.3) or symbolic gray values. This process allows image segments to be emphasized visually even in the case of very complex and interleaved shapes.

Attributes:

Images:	bilevel images
Operator:	global, topological, realization in two iterations of a sequential local operator
Kernel:	logically structured

Inputs:

Variant $VAR = 1$ with screen display of the list of equivalent labels, $VAR = 2$ otherwise.

(2) The method is better described by the algorithm itself instead of giving a formal mathematical definition of the realized transformation. First the operations which have to be realized for each pixel in the first iteration through the input image are explained. The first iteration is sequential, from bottom-left to top-right.

The pixels of the first iterations input image must have been classified into background and object points. Background points are ignored. Background pixels have the gray value u_1 and object pixels have the gray value u_0. Assume that $u_1 = G-1$ and $u_0 = 0$. The labeling is only performed for object pixels. These object pixels belong to different 8-components.

Let **F** be a 3×3 window. The centered *ij*-coordinate system is assumed for image windows $\mathbf{F}(h, p)$. Image h is the sequentially produced resultant image. The gray values inside $\mathbf{F}(h, p)$ are described by $H(i, j)$, for $-1 \leq i, j \leq 1$, where $H(0, 0) = h(p)$.

Assume $f(p) = u_0$. Besides the gray value $f(p)$ at the current point p the algorithm uses also the gray values $H(-1, 0)$, $H(-1, -1)$, $H(0, -1)$, $H(1, -1)$ of the previously processed 8-neighbors p_1, p_2, p_3 and p_4 of the current point p. These gray values are not original gray values anymore. Because of the sequential processing mode they are already symbolic gray values, i.e. labels of image segments.

The *symbolic gray values* z, with $1 \le z \le G-2$, are the labels. Thus, a confusion with the gray values $u_0 = 0$ and $u_1 = G-1$ of the original bilevel image is not possible.

The algorithm is described as follows:

For the current point p, it is tested whether a label had been previously assigned to at least one of its 8-neighbors p_c, for $c = 1,..., 4$:

(a) If this is not true, then p is marked with the smallest label which had not yet been used so far (initialization of a "new" segment).

(b) If one or several image points p_c had already been labeled with labels z_r, for $r = 1, ..., R \le 4$, then p is assigned the label z with

$$z = \min_{r=1...R} \{z_r\} .$$

At the same moment, it is entered in a special equivalence table that the labels z and z_r are labeling the same image segment.

(3) The procedure is designed for an image processing system which is restricted to $log_2\, G$ bits used for gray value representation. Thus, there are only $G-2$ labels available, i.e. at most $G-2$ image segments can be labeled without equivocation. If more than $G-2$ image segments are contained in an image, then multiple label assignments must be taken account of.

The labels are assigned in the first iteration as described under point **(2)**. This labeling process has been the object of several studies, see references at point **(5)**. For example, the first iteration can assign more than one label to some image segments featuring contour concavities which are open downwards or to the left. However, an *equivalence table* is also generated during the first iteration. This table allows to trace equivalent labels at the end of the first iteration. Then, a unique label is selected for each image segment, e.g. that with the smallest value. In principle, the process of connected component labeling is already achieved after the first iteration, since the labeled resultant image and the equivalence table together allow a labeling without equivocation. However, a second iteration is mostly used for obtaining a clearer iconic representation of the result. In this second iteration, each segment gets a unique label based on the equivalence table.

Tab. 7.1 illustrates the structure of an equivalence table, which can be implemented, for example, with two arrays $g1$ and $g2$, where $g1(w)$ is a label of a neighbor of the current point p, and $g2(w)$ is the label assigned to p and taken from a neighbor on the basis of the rules. Let G_m be the number of used labels, i.e. $G_m \le G-2$. It is $w = 1, 2, ..., G_m$. The inequality $g2(w) < g1(w)$ is ensured by the labeling process.

index w	label $g1(w)$	label $g2(w)$
1	...	...
2	...	...
3	m	r
4	n	$m \to r$
...	...	...
...	b	$m \to r$
...	c	$n \to m \to r$
...	...	...
G_m		

Table 7.1: Example of some entries in an equivalence table.

Table 7.1 schematically shows a situation where several pairs of equivalent labels $g1(w)$ and $g2(w)$ are first assigned to the same image segment, which, after label-redistribution, is characterized by the smallest label r.

After the first iteration, the two arrays $g1$ and $g2$ are subject to a transformation as a preprocessing before the second iteration. By effect of this transformation, the smallest label of each segment is determined as the new value of $g2(w)$. This transformation reduces the number of different labels from G_m to Z.

The second iteration uses the results of this table processing. The labels $g1(w)$ are replaced by the labels $g2(w)$ of the equivalence table. Thus, the result is a one-to-one mapping between labels and segments.

Assume that the input image f contains only few 8-connected image segments. Then the symbolic gray values $z = 1, 2, ..., Z$ represent only a small fraction of the whole gray level scale. A visual display of the labeled image on a screen would result in a bad visual discrimination between the different image segments. A contrast stretching of the resultant image, from the gray value scale $0, ..., Z$ to the full gray value scale $0, ..., G - 1$, is advantageous in this case. The Sections 5.1.2, 5.1.3 and 5.1.4 contain programs for contrast stretching.

For obtaining a good visual discrimination without a separate contrast stretching, the algorithm assigns final labels such that they are about uniformly distributed over the full gray value scale. The program uses the symbolic gray values $1, 1 + N_z, 1 + 2N_z, ..., 1 + (Z - 1)N_z$ as labels with

$$N_z = \textbf{integer}\ [0.5 + (G - 2) / Z].$$

(4) Assume a binary input image with background gray value $G - 1$ and with object gray value 0. The program labels the object regions. We will further assume that all border points of the input image belong to the background, otherwise transform the rows and columns at the image border into background pixels before processing. The *BRESENHAM* procedure can be used for such a preprocessing, cp. Section 3.4.10.

Control structure: window operator in Figure 3.6, i.e. centered, with sequential processing and $n = 3$, for the first iteration, and point operator for the second iteration

Required data arrays: two arrays $g1(1...G_{max})$, $g2(1...G_{max})$ for the equivalence table with sufficiently large G_{max}, e.g. $G_{max} = 254$.

Inputs and initializations:

```
ensure that the image border of f belongs to the background
        {e.g. call procedure BRESENHAM for the two border rows and
                                        the two border columns}
input of parameter VAR;
w := 1;        {initialization of the index of the equivalence table}
z := 0      {symbolic gray value of the first available label minus one}
```

Operator kernel of the first iteration:

```
{compute v = h(x, y) from F(i, j), with -1 ≤ i, j ≤ +1, and from H(i, j), with
                                                        -1 ≤ i, j ≤ +1}
if (f(p) ≠ G - 1) then begin                               {object pixel}
    if (h(p1) = h(p2) = h(p3) = h(p4) = G - 1) then begin
        z := z + 1;
        {output a message in case of z = G - 2, and stop the program}
        v := z
        end {then}
    else begin
        v := min {h(p1), h(p2), h(p3), h(p4)};
        if not (h(p1) = h(p2) = h(p3) = h(p4)) then
            for c := 1 to 4 do begin
                if ((v ≠ h(pc)) and (h(pc) ≠ G - 1)) then begin
                    for i := 1 to w do
                        if ((v = g2(i)) ∧ (h(pc) = g1(i)))
                                                then goto L1;
                    w := w + 1;
        {output a message in case of w = Gmax, and stop the program}
```

```
                                                g2(w) := v;     g1(w) := f(p_c)
                                                end {if}
L1                                        end {for}
                           end {else}
                    end {then}
```

Preprocessing before the second iteration:

```
        Z := z;             G_m := w;
        if (VAR = 1) then output of equivalence table, i.e. of g1 and g2, on the
                                                                         screen;
                                                {processing of equivalence table}
L2      flag := 0;
        for r := 1 to G_m do
                for s := 1 to G_m do
                        if ((g1(r) = g2(s)) and (s > 1)) then begin
                                g2(s) := g2(r);         Z := Z – 1;
                                flag := 1
                                end {then};
        if (flag = 1) then goto L2;
        N_z := ⟨ [0.5 +(G – 2) / Z] ⟩
```

Operator kernel of the second iteration (new assignment of labels):

```
                {transform v = h(x, y), into a new value v = h(x, y), cp. Figure 3.9}
        for r := 1 to G_m do
                if (v = g1(r)) then     v := g2(r);
        v := N_z · v
```

(5) The described connected component labeling method is equivalent to the "classical" approach in

Rosenfeld, A., Kak, A.C.: *Digital Picture Processing*, Vols. 1 and 2. Academic Press, New York, 1982.

This approach is also described, e.g., in

Wahl, F.M.: *Digital Image Processing*. Artech House, Norwood, 1987.

Zamperoni, P.: *Methoden der digitalen Bildsignalverarbeitung*. Vieweg Verlag, Wiesbaden, 2nd ed., 1991.

For a possibly faster program see

Dillencourt, M.B., Samet, H.: *Connected-component labeling of binary images.* Technical Report CS-TR-2303, University of Maryland, 1989.

However the latter method is more complex.

7.1.2 Thinning of Binary Images

(1) This program combines two operators for the computation of *skeletons* of *bilevel images*. The first method is slower in computing speed but more accurate. The second method needs less iterations but the computed skeleton may satisfy the requirements (cp. points (i) ... (v), below) worse than the first method for certain input images.

Attributes::

Images:	bilevel images
Operator:	global, topological, realization in several iterations
Kernel:	order dependent, logically structured

Inputs:

Variant $VAR = 1$ for the first method, or $VAR = 2$ for the second.

(2) A qualitative description of the aims of thinning is clearer and more opportune than a mathematical definition, which would be rather problematic at this place. These aims can be described as requirements to be met by a skeleton.

In general, a thinning operator extracts a network of thin curves from a bilevel image. The thin curves should be placed "about in the middle" of the objects and should describe the "skeleton" of the objects. This skeleton characterizes the shape of the object regions. General strong criteria about the quality of the resultant images do not exist because of the broad variety of object shapes and specific aims of image processing. However the following requirements specify some properties which are widely accepted to be typical for skeletons:

(i) The skeleton should consist of curves with a width of a single image point (curve thickness property).
(ii) The topological connectedness relations of the skeleton should be identical to those of the original image, i.e. the number of 8-components of object segments should remain the same (topology preserving property).
(iii) The skeleton curves should lie in the middle of the objects (medial axis property).
(iv) In the case of thick or rugged objects there should be not too many irrelevant "skeleton branches" (noise robustness property).
(v) The skeletonization process should converge to a stable skeleton after a certain number of s iterations (convergence property).

A skeletonization procedure deletes step-by-step all "superfluous" object points in an image for achieving these aims. A point p can be deleted if the number of 8-connected segments inside a 3×3 window $\mathbf{F}(p)$ remains constant after deletion, and if p is no end point of a curve.

(3) The *medial axis* property (iii) leads to the conclusion that only border points of objects should be deleted during the repeated iterations of a skeletonization procedure. Each iteration may consist of four successive subcycles, and each subcycle considers only points either at the Eastern, Southern, Western or Northern border of the objects.

The *connectivity number* of 8-connected segments, or the crossing number of 4-connected segments inside a placed 3×3 window is a value of a window function which can be used as deletion criterion. If this value is equal to 1, then the current point p can be deleted, provided that p is no curve end point, cp. *Yokoi/Toriwaki/Fukumura* in **(5)**.

Several papers as *Arcelli/Cordella/Levialdi*, *Bourbakis*, *Chin/Wan/Stover/Iverson*, *Eckhardt/Maderlechner* and *Suzuki/Abe*, cp. **(5)**, follow a different approach. Here, the deletion criterion is based on a comparison of $\mathbf{F}(p)$ with a set of templates. Our program also follows this approach.

The "classical" skeletonization algorithms use four subcycles in each iteration, and a 3×3 window, cp. *Arcelli /Cordella /Levialdi* and *Rosenfeld /Kak* in **(5)**. Faster algorithms as proposed by *Bourbakis* and *Chin /Wan /Stover /Iverson*, cp. **(5)**, use only two subcycles, or even just one subcycle but 3×4 or 4×3 windows. Several further complex algorithms exist in the literature which aim at an efficient search of deletable points: *Pavlidis*, *Arcelli /Sanniti di Baja* and *Xia*, cp. **(5)**, describe methods where image segments are repeatedly "peeled off" taking into consideration the requirements (i) to (v) in **(2)**.

Piper, cp. **(5)**, describes a skeletonization algorithm based on run-length encoded image segments.

These algorithms do not perform iconic operations, i.e. image-to-image transformations which are the specific subject of this book. *Tamura* and *Jaisimha/Haralick/Dori*, cp. **(5)**, give comparative evaluations of several skeletonization algorithms.

In choosing the two skeletonization algorithms of this Section the following criteria have been relevant. The first method by *Arcelli/Cordella/Levialdi*, cp. **(5)**, using four subcycles, is slower, and it represents a basic algorithm, which satisfies the requirements (i) to (v) of **(2)**. Furthermore, a simple program realizes this method. The stable resultant image is always a *minimum skeleton*, i.e. it contains no points which could still be deleted. The minimum skeleton property is especially valuable in those practical applications where the resultant skeleton image has to be encoded by a curve tracing algorithm. Such an encoding allows a compact vector representation instead of the original iconic image representation.

The second algorithm by *Chin /Wan /Stover /Iverson*, cp. **(5)**, has no subcycles, therefore it is faster, but sometimes the requirements (iii) and (iv) are not satisfied. Then the resultant skeleton is asymmetric and has short spurious branches. Therefore, the original paper proposes some postprocessing of the resultant image. This algorithm takes account of all four directions at the same time, i.e. in each iteration, and thus it has no subcycles.

Figure 7.1 and Figure 7.2 give a qualitative impression of which skeletonization results can be obtained with both variants for the same original image.

A bilevel image f is assumed as input image for the program in **(4)**. The gray value 0 characterizes the background, and a value $u > 0$ characterizes the object segments. Before starting with the skeletonization, the program maps the gray value u onto the gray value 1. Therefore, the following discussion is based on a binary image f with gray values 0 and 1.

The Variant 1 performs successive subcycles 1 to 4 until the resultant image converges. The value of the parameter W_4 signalizes the occurrence of this situation. This value is equal to 1 if no object point was deleted during the last four subcycles. The program assigns value 1 to a variable W if at least one object point was deleted during a subcycle. In this case the variable W shows that the resultant image is not yet converging. At the beginning of each subcycle, the variable W is reset to 0. Note that if no deletion occurs in four consecutive subcycles, the convergence has been attained.

The program compares full 8-neighborhoods with templates. There are two templates for each subcycle, with respect either to an Eastern, Southern, Western or Northern border point. The program deletes an object point if its full 8-neighborhood is identical to one of these eight templates. Table 7.2 specifies these templates. The symbol X denotes an image value which can be either 0 or 1.

The Variant 2 has no subcycles but iterations. All iterations are identical. Table 7.3 specifies the *templates* used in these iterations. The program deletes an object point if its full 8-neighborhood is identical to one of the templates A to H, and if it is not embedded into the 4×3 template I or 3×4 template J.

Variant 2 has a simpler convergence condition than Variant 1, since all subcycles are identical. If variable W has value 0 at the end of an iteration then the resultant image is stable.

The comparison between the full 8-neighborhood of the current point and the relevant templates takes place in the same way for both variants. The comparisons are carried out between 8-bit binary numbers encoding the template patterns as well as the current 8-neighborhood. The don't-care value X is also taken account of by the binary templates, as these are constructed in such a way as to outmask neighbors which are irrelevant for the comparison. The spatial ordering of indices $z = 1,\ldots, 8$ follows Figure 3.5 for the binary values in the 8-neighborhood $F(1),\ldots, F(8)$ of a point p, i.e.

4	3	2
5	p	1
6	7	8

Thus, these eight binary values can be encoded by the binary number

$$Z = \sum_{z=1}^{8} F(z) \cdot 2^{z-1}$$

North		South		West		East
0 0 0 X 1 X X 1 1		1 1 X X 1 X 0 0 0		0 X 1 0 1 1 0 X X		X X 0 1 1 0 1 X 0
X 0 0 1 1 0 X 1 X		X 1 X 0 1 1 0 0 X		0 0 X 0 1 1 X 1 X		X 1 X 1 1 0 X 0 0

Table 7.2: Table of templates for Variant 1.

0 0 0 1 1 1 X 1 X	0 1 X 0 1 1 0 1 X	X 1 X 1 1 1 0 0 0	X 1 0 1 1 0 X 1 0	X 0 0 1 1 0 X 1 X	0 0 X 0 1 1 X 1 X
A	B	C	D	E	F

	X 1 X 0 1 1 0 0 X	X 1 X 1 1 0 0 0 X	X X X X 0 p 1 0 X X X X	X 0 X X p X X 1 X X 0 X	
	G	H	I	J	

Table 7.3: Table of templates for Variant 2.

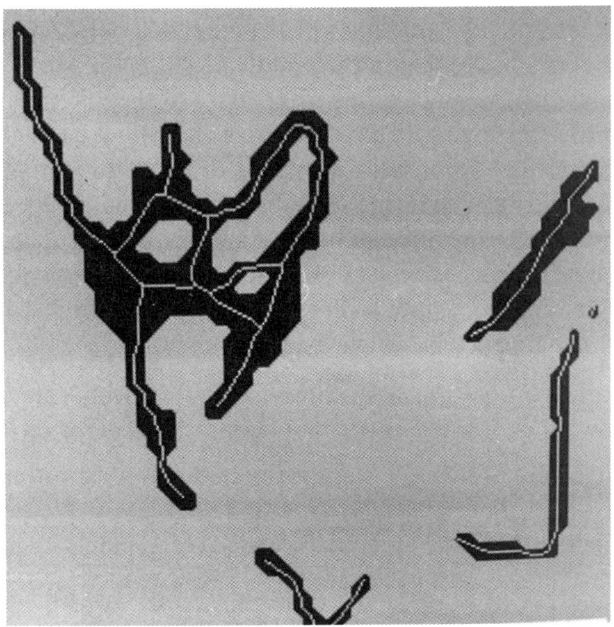

Figure 7.1: Skeletonization result obtained with Variant 1.

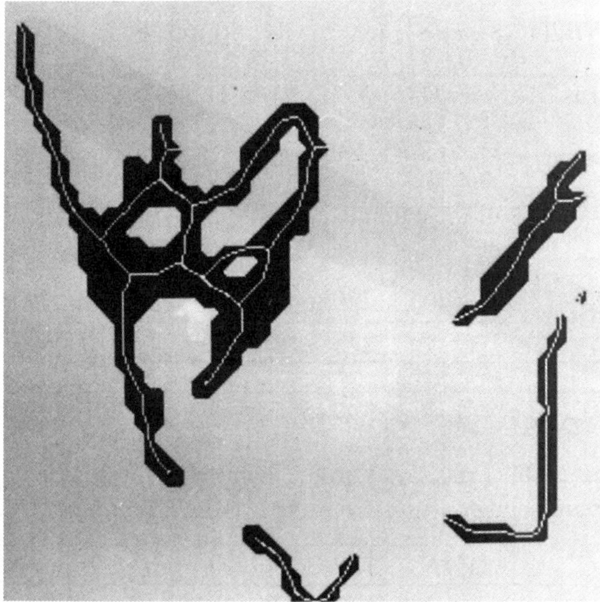

Figure 7.2: Skeletonization result obtained with Variant 2.

The Variant 2 needs also the consideration of the extended neighborhoods I and J. The current point $p = (x, y)$ is located in the middle of the 3×3 kernels of these extended neighborhoods. The line $y - 2$ of the original, or of the intermediate image, is stored in an array $Q(1...M)$ being needed for comparisons with the template pattern J.

Both variants need several iterations and/or subcycles through the image. The necessity of storing an intermediate image requires an additional image array g. At the beginning the program copies the original image f into the intermediate image g, and at the end of each subcycle or iteration it copies back the intermediate resultant image h into the intermediate image g. Image g plays the part of the original image in each subcycle or iteration, and image h is always the resultant image.

(4) We use the centered ij-coordinate system with the control structure explicitly given here. The following data arrays and line store arrays are defined:

$Q(1...M)$ for the image values of line $y - 2$ which is needed for the comparison with the extended pattern J of Variant 2,

the binary encoded pattern templates (number Z in **(3)**) of Variants 1 and 2, i.e.

$$NB1(1...8) = (192, 80, 12, 5, 3, 65, 48, 20),$$

for the directions (North, North, South, South, West, West, East, East), cp. Table 7.2,

$$NB2(1...8) = (81, 69, 21, 84, 80, 65, 5, 20),$$

for the patterns (A, B, C, D, E, F, G, H) in Table 7.3,

$$NB3(1, 2) = (1, 64)$$

for the 3×3 kernels of the extended neighborhood templates I and J in Table 7.3,

$$MASK1(1...8) = (206, 87, 236, 117, 59, 93, 179, 213)$$

and

$$MASK2(1...8) = (95, 125, 245, 215, 87, 93, 117, 213)$$

for the don't-care bit masks of Variant 1 and Variant 2, respectively, associated with the encoded configurations $NB1$ and $NB2$, and finally

$$MASK3(1, 2) = (17, 68)$$

for the don't-care bit masks for the extended Variant 2 neighborhoods I and J.

An index c controls the proper choice of encoded templates in $NB1$ and $NB2$ as well as that of the encoded don't-care masks in $MASK1$ and $MASK2$, in dependence of the current subcycle.

```
                                                   {Input and initialization}
      copy image f into the intermediate image g;
      PASS := 0;                                   {number of realized subcycles}
      c := 1;                                         {initialization of index}
      W4 := 0;
L1    W := 0,
      for y := 3 to N – 1 do begin
            for x := 2 to M – 2 do begin
                                                       {start of operator kernel}
                  if ((x = 2) ∧ (VAR = 2)) then begin
                        read line y – 2 of the image g into array Q(1...M);
                        for r := 1 to M do      {generation of binary image}
                              if (Q(r) ≠ 0) then Q(r) := 1
                        end {then};
                  if (f(x, y) = 0) then          h(x, y) := 0
                  else  begin
                        read 8-neighbors of p into F(z), z = 1...8, in the order
                              of Figure 3.5;
                        for z := 1 to 8 do        {generation of binary image}
                              if (F(z) ≠ 0) then F(z) := 1;
                        Z := 0;
                        for z := 1 to 8 do     Z := Z + 2^(z–1) F(z);
                        if (VAR = 1) then
                              for z := c to c + 1 do
                                    if (AND(Z, MASK1(z)) = NB1(z))
                                    then  begin
                                                       {bitwise AND, cp. 5.5.1}
                                          W := 1;          h(x, y) := 0;
                                          goto L2
                                          end {then}
                        else  begin
                              if
                  (( AND(Z,MASK3(1)) = NB3(1) )  ∧  ( f(x+2, y) = 0 ))
                                    then goto L2;
```

```
                          if
                    (( AND(Z,MASK3(2)) = NB3(2) )  ∧  ( Q(x) = 0 ))
                                  then goto L2;
                          for z := 1 to 8 do
                                  if ( AND(Z, MASK2(z)) = NB2(z) )
                                        then begin
                                        W := 1;        h(x, y) := 0;
                                        goto L2
                                  end {then}
                          end {else}
                     end {else}
L2                end {for};                    {end of operator kernel}
                                    {preprocessing for the next iteration}
            if ((VAR = 2) ∧ (W = 0)) then goto L3;
            if (W = 0)    then         W4 := W4 + 1
                          else         W4 := 0;
            if (W4 = 4) then goto L3;
            PASS := PASS + 1;
            c := 1 + 2 · [remainder of division of PASS by 4];
            copy intermediate resultant image h into the intermediate image g ;
            goto L1
L3          end {for}
```

(5) The fundamental skeletonization approach consists in successive deletions of object points in consecutive subcycles. It is described in

Rosenfeld, A., Kak, A.C.: *Digital Picture Processing*, Vols. 1 and 2. Academic Press, New York, 1982.

Zamperoni, P.: *Methoden der digitalen Bildsignalverarbeitung*. Vieweg Verlag, Wiesbaden, 2nd ed., 1991.

Arcelli, C., Cordella, L., Levialdi, S.: *Parallel thinning of binary pictures*. Electronic Letters **11** (1975), pp. 148-149.

Arcelli, C., Cordella, L., Levialdi, S.: *More about a thinning algorithm*. Electronic Letters **16** (1980), pp. 51-53.

Other skeletonization approaches use the connectivity number or the crossing number as deletion criterion, cp.

Yokoi S., Toriwaki, J., Fukumura, T.: *An analysis of topological properties of digitized binary pictures using local features*. Computer Graphics and Image Processing **4** (1975), pp. 63-73.

Arcelli et al. (cited above) and

Bourbakis, N.G.: *A parallel-symmetric thinning algorithm*, Pattern Recognition **22** (1989), pp. 387-396.
Chin, R.T., Wan, H.K., Stover, D.L., Iverson, R.D.: *A one-pass thinning algorithm and its parallel implementation.* Computer Vision, Graphics, and Image Processing **40** (1987), pp. 30-40.
Eckhardt, U., Maderlechner, G.: *Parallel reduction of digital sets.* Siemens Forschungs- und Entwicklungs-Berichte, Bd. **17** (1988), N. 4, pp. 184-189.
Suzuki, S., Abe, K.: *Binary picture thinning by an iterative parallel two-subcycle operation.* Pattern Recognition **20** (1987), pp. 297-307.

propose comparisons with templates as a deletion criterion. *Bourbakis*, *Chin/Wan/Stover/Iverson* and *Suzuki/Abe* (cited above) describe faster skeletonization algorithms with two or one subcycle instead of four subcycles for iteration. Skeletonization by repeated "peeling" is described by

Pavlidis, T.: *A thinning algorithm for discrete binary images.* Computer Graphics and Image Processing **13** (1980), pp. 142-157.
Arcelli, C., Sanniti di Baja, G.: *A width-independent fast thinning algorithm.* IEEE Trans. PAMI-**7** (1985), pp. 463-474.
Xia, Y.: *Skeletonization via the realization of the fire front's propagation and extinction in digital binary shapes.* IEEE Trans. PAMI-**11** (1989), pp. 1076-1086.

For skeletonization based on run-length encoded image segments see

Piper, J.: *Efficient implementation of skeletonisation using interval coding.* Pattern Recognition Letters **3** (1985), pp. 389-397.

Comparative evaluations of several skeletonization algorithms are given in

Tamura, H.: *A comparison of line thinning algorithms from digital geometry viewpoint.* Proc. 4th IJCPR, Kyoto, 1978, pp. 715-719.
Jaisimha, M.Y., Haralick, R.M., Dori, D.: *Quantitative performance evaluation of thinning algorithms in the presence of noise.* in: Aspects of Vision From Processing (C. Arcelli, L.P. Cordella, G. Sanniti di Baja - eds.), World Scientific, Singapore 1994, pp. 261-286.

7.1.3 Thinning of Gray Value Images

(1) This procedure transforms a gray value image into an image consisting of a network of thin lines. The resultant lines are placed about in the middle of the object (i.e. bright on dark background) regions. In the sense of Section 1.1.2, a gray value image can be regarded also as a topographical terrain map of a relief with mountains, valleys etc. Following this intuitive model, the resultant lines are at the location of mountain ridges or of watersheds (*watershed transformation*), and the gray value intensities of these curves still correspond to the original heights in the relief.

The gray value skeleton is a schematic representation of the image content focusing on "mountain ridges". This representation may be of benefit for image analysis. The gray value skeleton does not allow a unique reconstruction of the input image.

Attributes::

Images:	gray value images
Operator:	global, topological, realization in several iterations by means of local operators
Kernel:	order statistical

(2) An exact and concise formal definition of a *gray value skeleton* would be very problematic. For better clarity, we give here a qualitative description based on the properties (i) to (v) of point **(1)** in Section 7.1.2. However some modifications are necessary. The curve thickness property (i) is mostly required to hold only in a weak sense. Since the resultant image values are not bilevel anymore, the topology preserving property (ii) requires exact definitions of connectedness for gray value images. Altogether, the requirements (i) to (v) outline the operators "coarse aims", and problems can arise if these aims are understood as hard requirements set to the resultant gray value image.

Some skeleton algorithms for gray value images are actually based on exact definitions, cp. e.g. *Dyer/Rosenfeld* and *Wang/Abe* in **(5)**. However for practical purposes it seems that the high algorithmic complexity of such methods is not justified, since the features of the results strongly depend upon the input image. The method of **(4)** follows *Peleg/Rosenfeld*, cp. **(5)**, and it is algorithmically simple. This algorithm is based on an iterated application of the operations of erosion and dilation for gray scale images (i.e. minimum and maximum), cp. Section 6.5.2, which are well-known basic operations of mathematical morphology.

The computations of local minima or maxima realize the operations of *erosion* and *dilation* as already discussed in Section 6.5.2. The *structuring element* is the full 8-neighborhood centered on the current point. Let $f^{(r)}$ be the resultant image of r dilation iterations of an input image f, and let $f^{(-r)}$ be the resultant image of r erosion iterations of an input image f. It holds

$$f^{(-r+1)} \geq (f^{(-r)})^{(1)}$$

where $\geq$ means that this relation is true for all point-wise comparisons between these two images. The difference image

$$d_r(x, y) = f^{(-r+1)} - (f^{(-r)})^{(1)},$$

has only non-negative gray values. Such a difference image leads to the *top-hat transformation*, cp. Section 2.7. The bright image points of this difference image are exactly those points which have been eroded in the r-th erosion.

The program starts with $r = 1$. Successive applications of the top-hat transformation, for $r = 1...R$ extract the local maxima labeled with their original gray values. The resultant image h is the union of these local maxima. An addition or, as in the program, the logical **OR** combines the partial results into the resultant image h, i.e.

$$h(x, y) \;=\; \max_{r=1...R} \{d_r(x, y)\}.$$

Figure 7.3: Two examples of skeletons of gray value images. Left: original *aerial view* and image of a rose. Right: gray value skeletons after ten iterations of the program.

(3) This program requires two temporary image arrays g_1 and g_2. It copies the original image into g_1 at program start, and it copies the current input image $f^{(-r+1)}$ of the r-th iteration into g_1 at the beginning of this r-th iteration. The result of the r-th iteration is the r-th erosion, it is computed in g_2 during the r-th iteration, and then it is copied into g_1. The program assumes a very simple image data structure, in which the images g_1 and g_2 can only be read line-by-line (cp. Section 2.3). If a direct access of these arrays is possible, the program can easily be simplified.

Erosion as well as subsequent dilation, both in a 3×3 window, are performed during each iteration through the image. Each iteration computes the local minima $MIN(i, j)$ in nine placed 3×3 windows with reference points $(x + i, y + j)$, for $-1 \le i, j \le +1$, which cover a 5×5 window with reference point $p = (x, y)$, for all non-border points p of the image raster **R**. The maximum **max**$\{MIN(i, j)\}$ is the result, computed at each position of the current point $p = (x, y)$.

Figure 7.3 shows two examples of gray value skeletons.

(4) *Control structure*: special, in several iterations, here given explicitly.
Required data arrays:

line store arrays $BUF(1...M, 1...5)$, $BUFIN(1...M)$, $BUFOUT(1...M)$, assuming a row-wise image input,

array **data ind**(1, 2, 3, 4, 5) with initialization values for the indirectly addressed row indices, cp. Section 3.3.1,

image arrays g_1 and g_2 for intermediate results,

line store arrays $Z1(1...M)$ and $Z2(1...M)$ as data buffers.

```
                                              {input and initialization}
      input of the number R of iterations;
      copy the original image f into the image array g1;
      PA := 0;
      for y := 1 to N do begin
            for x := 1 to M do   Z2(x) := 0;
            write line store array Z2(1...M) into line y of h;
            end {for};
      for y := 1 to 5 do
            read line y of the original image f into the line store array BUF(1...M,
                  ind(y));

                                                  {start of iterations}
L1    for y := 3 to N - 2 do begin
            read line y of the original image f into the line store array Z2(1...M);
            for x := 3 to M - 2 do begin
                                        {beginning of the operator kernel}
                  MAX := 0;
                  for j := -1 to 1 do
                        for i := -1 to 1 do begin
                              MIN := G;
```

```
                    for i1 := –1 to 1 do
                         for j1 := –1 to 1 do
                              if (g1(x + i + i1, y + j + j1) < MIN)
                              then
                              MIN := g1(x + i + i1, y + j + j1);
                    if ((i = 0) ∧ (j = 0)) then      Z1(x) := MIN;
                    if (MIN > MAX)    then    MAX := MIN
                    end {for};
          H := g1(x, y) – MAX;          BUFOUT(x) := max(H, Z2(x));
                                   {end of the operator kernels}
          end {for};
     write line store array BUFOUT(1...M) into line y of h;
     write line store array Z1(1...M) into line y of image g2;
     if (y < N – 2) then begin
          read line y+3 of the image g1 into array BUF(1...M. ind(1));
          rot := ind(1);
          for z := 1 to 4 do     ind(z) := ind(z + 1)
          ind(5) := rot
          end {then}
     end {for};
PA := PA + 1;
if (PA < R) then begin
     copy image g2 into image g1;
     goto L1
     end {then}
```

(5) There exist several rather complex gray scale skeleton computation approaches in the literature. These algorithms have been discussed in

Dyer, C.R., Rosenfeld, A.: *Thinning algorithms for gray-scale pictures*. IEEE Trans. PAMI-**1** (1979), pp. 88-89.

Wang, C., Abe, K.: *A method for gray-scale image thinning: the case without region specification for thinning*. Proc. 11th Intern. Conf. on Pattern Recognition, Den Haag, 1992, pp. 404-407.

The program of **(4)** follows

Peleg, S., Rosenfeld, A.: *A min-max medial axis transformation*. IEEE Trans. PAMI-**3** (1981), pp. 208-210.

which represents a relatively simple solution.

7.2 GEOMETRICAL CONSTRUCTIONS

These constructions assume preprocessed images as input. Certain contours or isolated image points $p_1, p_2, ..., p_n$ have a label, say gray value $G - 1$, in these input images. The geometrical construction process adds some geometrical figures into these images, or analyzes the given contours or sets of points. The results can be visualized as images with certain graphical overlays (cp. Figure 2.9), e.g. also with colored curves or straight lines in the original gray value image.

7.2.1 Contour Following for Binary Images

(1) The program traces contours of objects in a *bilevel image*. It stores the encoded contours (*Freeman code*) into a file. This program could also be extended to an image-to-image transformation, e.g. by computing geometrical features (convex hull, diameter etc.) for these contours and by visualizing these features. These possible extensions are not introduced here, because this would require the development of a composite program taking account of too many different geometrical features.

The program performs a complete *object search* in the input image. If there are several objects (i.e. 8-components), then the program finds all of them, and it computes the contour codes of all objects.

The input image can also be a gray value image. In this case, a global threshold discriminates between object points (gray value higher than the threshold) and background points (gray value lower or equal than the threshold.

The resultant file of encoded contours allows an error-free reconstruction of the bilevel input image.

Attributes::

Images: bilevel images or gray value images
Operator: global, geometrical

Inputs:

binarization threshold S_1 for the recognition of a reliable starting point of the contour,
binarization threshold $S_2 \leq S_1$ for the determination of the following contour point during contour tracing,
coordinates Y_{min}, Y_{max}, X_{min} and X_{max} of a rectangular field if the object search has to be restricted to such a region of interest.

(2) The contour of a binary object consists of those object points which have at least one 4-neighbor in the background region (cp. Section 1.1.1). Any contour point can be chosen as initial point for contour tracing. A *contour chain* goes from an initial point back to this point, it connects all the contour points, and it consists of a sequence of steps, each step into one of eight possible directions. The numbers 0 to 7 encode these eight directions, cp. Figure 7.4.

Figure 7.4: Coding scheme for contour steps in an 8-neighborhood.

Thus, a digit sequence r_1, r_2,..., r_N, with $0 \leq r_i \leq 7$ and $1 \leq i \leq Z$, which should satisfy certain formal rules, describes without equivocation the contour of an object (*8-component*). The additional specification of the absolute coordinates of the chosen initial point allows an exact positioning of the object inside the image raster.

(3) The contour encoding algorithm is a fundamental procedure. It is often used in the context of shape analysis, of geometrical feature extraction of data-reducing image code generation, or of geometrical transformations of bilevel images. Contour codes have the advantage that geometrical features of the object can be computed by means of numerical manipulations of the contour code chain, sometimes just based on simple syntax rules. Algorithms on contour code chains are very time-efficient in general, and they do not need access to the image data. A contour code chain of geometrically, or morphologically transformed objects can directly be computed starting from the original code sequence. The contour code chain allows an exact reconstruction of the object, see below.

The whole process of contour code determination consists of two phases, of object finding and of contour tracing. The first phase searches the image line-by-line until a reliable initial point of the next object is found. The second phase traces the contour of this object, and stores the resultant code sequence into a file. Both phases alternate during the whole process until all objects are encoded.

We explain the contour tracing algorithm given in point **(4)**. At first, the main tasks are briefly described and possible solutions are sketched:

1. Assume that several objects are contained in the image. Subsequent contours have to be stored as number sequences into one file in sequential order. It has to be ensured that start and end of each contour code chain can be detected at decoding time.

2. Previously detected objects may also intersect forthcoming scan lines in an object search phase. Thus, detected objects have to be labeled as "invisible" for the further object search phases.
3. Assume bright objects on a dark background in a gray value image. Some minor variations of background or object gray values (e.g. shading, sloped transitions) can be expected. Therefore, the detection of a new initial point of the next object uses for safety a higher threshold S_1, and contour tracing uses a lower threshold S_2 for the determination of the next contour point. The hysteresis range $S_2 \ldots S_1$ ensures a certain robustness against the mentioned variations.
4. *Holes* inside objects should be recognized as holes, and should be labeled as such. The contour of a hole is an *interior object contour*, cp. Figure 7.6. Such an interior object contour connects object points. The program determines the (outer) object contour at first, it stores the code of this object contour, and then it processes the holes of the object. This order corresponds to the order in which the contours are detected in the first phase of object finding. The program uses a special assumption for the sake of reducing the program complexity: The initial point of an interior contour is not allowed to be at the same time an (outer) object contour point.

The contour tracing phase determines the digit sequence of the contour code and stores it in an array $E(1 \ldots L_{max})$, where L_{max} denotes the maximum allowed contour length of an object. For example, $L_{max} = 2000$ can be used as default for an image size of 512×512 image points. The contour code, as generated in $E(1 \ldots L_{max})$, is added into the resultant file after the completion of the contour tracing phase for an object, or a hole.

A special data format is used for the array E, and thus for the generated contour chains plus additional data:

$E(1)$	total number of contour steps
$E(2)$	x-coordinate of initial point
$E(3)$	y-coordinate of initial point
$E(4)$	label: object or hole?
$E(5)$	direction r_1 of the first contour step
$E(6)$	direction r_2 of the second contour step
...	...
$E(Z + 4)$	direction r_Z of the last contour step

The value $E(1)$ allows the parsing of the resultant file into contour data of the single objects and holes. In the resultant file, these data are stored consecutively.

In general, the direct access to the image data simplifies the program, since contours of arbitrary shape have to be traced. We assume this direct access in **(4)**.

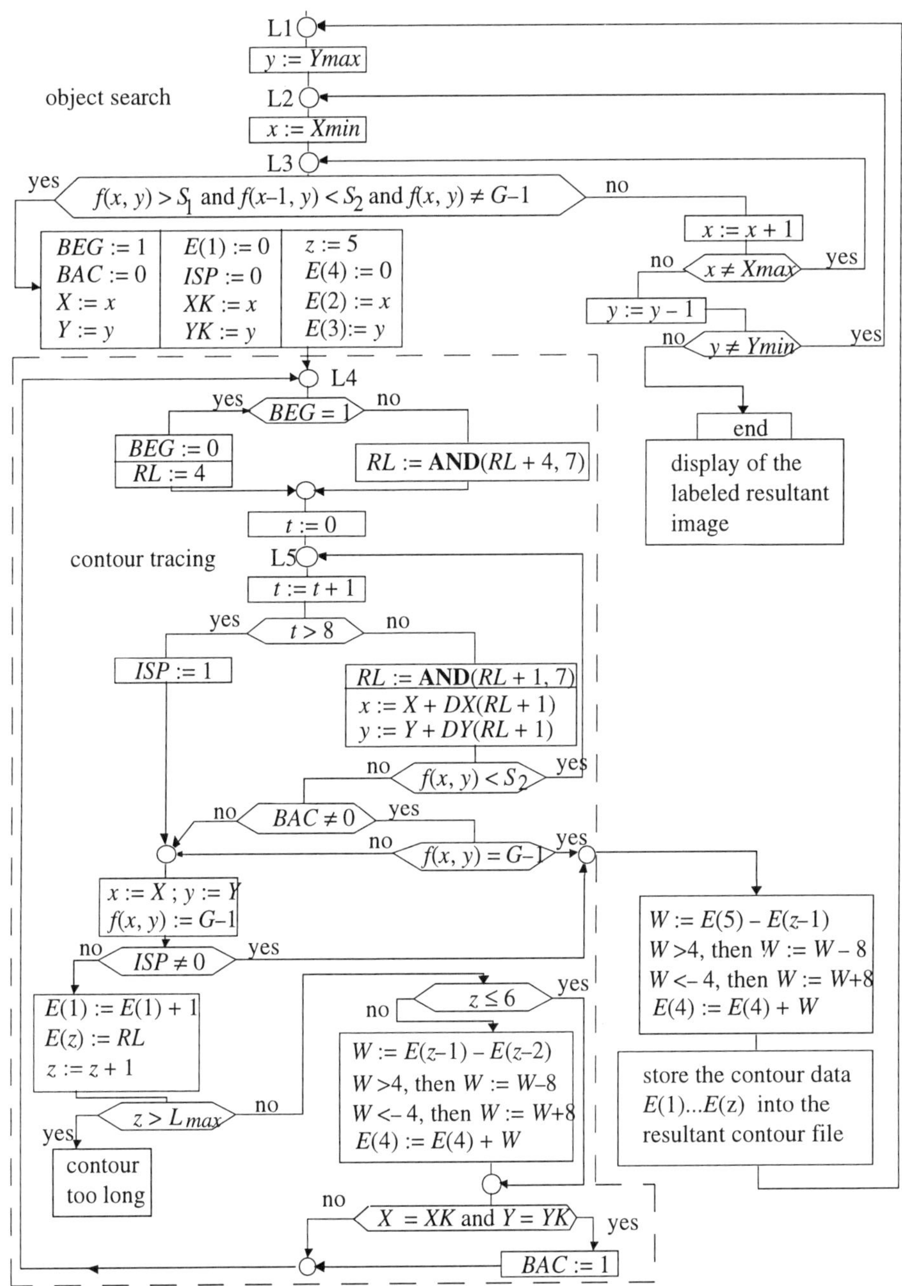

Figure 7.5: Flow chart of the contour tracing program. **AND** denotes the bitwise logical operation, cp. Section 5.5.1.

Thus, the gray values $f(x, y)$ are stored in a two-dimensional array. Other control structures with row-wise data processing are also possible in principle, but need a very frequent row-wise data exchange. - The flow chart in Figure 7.5 illustrates the complete contour tracing process of point **(4)**.

The object search on *scan lines* takes place from top to bottom, i.e. from $y = Y_{max}$ to $y = Y_{min}$, and from left to right, i.e. form $x = X_{min}$ to $x = X_{max}$. The symbolic gray value $G - 1$ labels the contours of previously treated objects. Therefore, the gray values of the original image must be previously limited to a maximum of $G - 2$. These labeled object contours are not detected again in subsequent object searching phases. A transition from a gray value lower than S_2 to a gray value higher than S_1 points to a new object. Then, the phase of contour tracing starts again.

Some variables are initialized at the beginning of the contour tracing phase. They are updated during this phase. Their meaning is as follows:

- *X*, *Y* are the coordinates of the current contour point,
- *XK*, *YK* are the coordinates of the initial point of the contour,
- *BEG* is equal to 1 if the current contour point is also the initial point,
- *ISP* is equal to 1 if the (degenerated) object is just an isolated point,
- *z* is the index of the next free location of the array *E*,
- *BAC* is equal to 1 if the current contour point did reach the initial point again (This situation does not imply that the whole contour is already traced. For example, if the initial point (*XK*, *YK*) is on a "thin neck" between two "object parts" then this point will be visited twice during contour tracing. However the used row-wise scanning for object search normally avoids such situations since the uppermost, leftmost object point is the expected initial point. This can only be not true if the gray value of the uppermost, leftmost object point is in the range between S_1 and S_2).
- *E*(1) ... *E*(4) are initialized or updated with the corresponding data of the current contour.

The contour tracing phase uses further variables in performing the single steps. Their meaning is as follows:

- *RL*, with $0 \le RL \le 7$, denotes the direction of the previous contour step. The current contour point (*X*, *Y*) is reached via this contour step. All the 8-neighbors of point (*X*, *Y*) are tested consecutively for checking whether their gray value exceeds the threshold S_2. Seen from point (*X*, *Y*), the first tested direction is $RL + 5$ (modulo 8). This corresponds to the direction of *RL*, rotated by $180° + 45°$. A default value $RL = 4$ is used at the beginning of contour tracing. The variable *BEG* is set to 0, i.e. the next contour point is assumed not to be the initial point. (This would be the case only if the object is an isolated point.)

- *DX*, *DY* denote the *x*- and *y*- increments for going from point (*X*, *Y*) in the selected direction *RL* to the next contour point. Two look-up tables *DX*(1 ... 8) and *DY*(1 ... 8) specify the values of these two increments. The value of *RL* lies between 0 and 7, and *RL* + 1 is used as table index.
- The variable *t* counts the number of neighbors of (*X*, *Y*) tested so far, without finding the next contour point. If the value of *t* is equal to 8 then the object is an isolated point.
- *W* denotes the angle between the direction of the previous contour step and the direction of the current contour step. This angle is an integer multiple of 45°, and it has a positive or negative sign. The sum of these *W*-values is stored in *E*(4). The *W*-value of the last step to (*XK*,*YK*) is added at the end of the contour tracing. The resultant value of *E*(4) is an integer multiple of 8 (note $8 \cdot 45° = 360°$), and its sign specifies whether the contour encloses an object ($E(4) \geq 0$) or a hole ($E(4) < 0$).
- The variable *z* counts the number of positions occupied by the current contour in the array *E*.

The content *E*(1) ... *E*(*Z*+4) of the array *E* is added to the resultant file of contour codes at the end of the contour tracing phase. Then, the search phase for the next object or hole is started. The problem is, at which position in the image should the search start? Objects and holes can occur in any topological complexity, e.g. an object has a hole, this hole contains an object, this has a hole again etc. The program in **(4)** assumes the worst case, and it always starts the object search phase at (X_{min} , Y_{max}).

Finally, the last object search phase leads to the point (X_{max}, Y_{min}) without detecting a new contour. The contour points are labeled with $G - 1$, i.e. they appear as bright lines in the resultant image. This labeling is carried out during the contour tracing process.

With some gray value images a constant threshold with hysteresis may fail to discriminate between objects and background. In this case, more sophisticated thresholding methods, based on more complex and/or specific image models, should be tried. Examples of such approaches are cited in **(5)**.

Computer graphics offers methods for reconstructing images from contour codes. These methods are not described in this book. A general approach can be as follows: The contours can be drawn into an image. Each contour starts at its point (*XK*, *YK*). In this way, only the silhouettes of the objects and of the holes are drawn. Well-known methods for the filling of closed curves, cp. references in **(5)**, can be used to label the interior of objects with a given gray value label. Holes of an object follow this object in the computed contour list. This ordering is the same as that to be used in this filling process, i.e. at first the object is filled, then the holes are filled with a different gray value label.

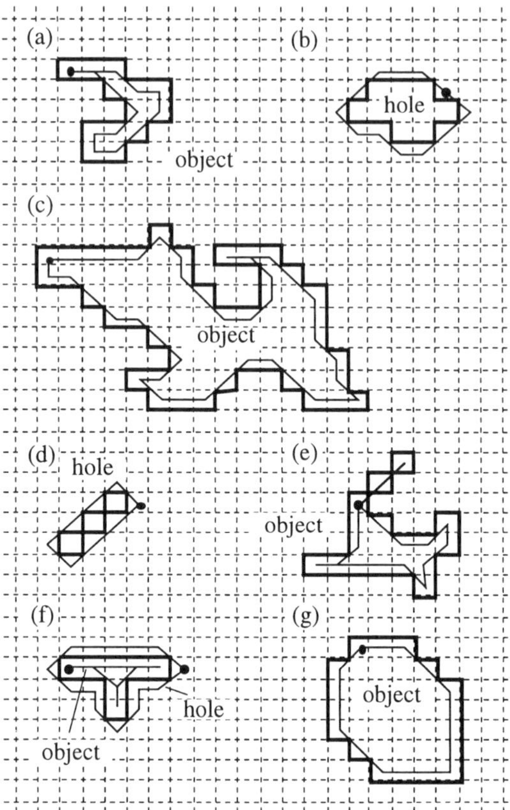

Figure 7.6: Examples of contour paths for objects and holes, where • denotes the initial point. Example (f) shows both possibilities for the same region, i.e. the object contour as well as the hole contour. Examples (c) and (e) assume that the uppermost, leftmost object point was not found as initial point in the object search phase, due to gray value variations. The sum of all angles, calculated in $E(4)$, is equal to 0 in the cases (a), (c), (e) and (f), it is equal to 8 in case of (g), and it is equal to -8 in the cases (b) and (d).

Figure 7.6 shows some examples of objects and holes, together with their contour paths. Note that a connected set of image points has different contours depending upon whether it is considered to be an object or a hole.

(4) *Control structure*: special, explicitly given here

Required data arrays:

$f(1 \ldots M, 1 \ldots N)$ for the gray values of the input image,

$E(1 \ldots L_{max})$ for the current contour code, with the data format as described under point **(3)**,

DX (1 ... 8) = (1, 1, 0, – 1, – 1, – 1, 0, 1) and *DY* (1 ... 8) = (0, 1, 1, 1, 0, – 1, – 1, – 1) as look-up table for the increments of *x*- and *y*-coordinates of the 8-neighbors of a contour point (*X*, *Y*), respectively.

```
                                                {start of image scanning}
L1   y := Y_max;
L2   x := X_min;
L3   if (f(x, y) > S_1 and f(x – 1, y) < S_2 and f(x, y) ≠ G – 1) then begin
                                                         {contour tracing}
               BEG := 1;      ISP := 0;      BAC := 0;
               E(1) := 0;     z := 5;
               X := x;        XK := x;       E(2) := x;
               Y := y;        YK := y;
               E(3) := y;     E(4) := 0;
L4             if (BEG = 1) then begin
                                   BEG := 0;     RL := 4
                              end {then}
                    else
                         RL := AND(RL + 4 , 7);
                                                  {bitwise AND, cp. 5.5.1}
               t := 0;
L5             t := t + 1;
               if (t > 8) then           ISP := 1
                    else  begin
                         RL := AND(RL + 1, 7);
                         x := X + DX(RL + 1);        y := Y + DY(RL + 1);
                         if (f(x,y ) < S_2) then goto L5;
                         if (BAC ≠ 0) then
                              if (f(x, y) = G – 1) then goto L6
                         end {else};
               x := X;        y := Y;
               f(x, y) := G – 1;
               if (ISP ≠ 0) then goto L6;
               E(1) := E(1) + 1;        E(z) := RL;     z := z + 1;
               if (z > L_max) then
                    output of message "contour too long" and program stop;
```

```
                if (z > 6) then begin
                        W := E(z – 1) – E(z – 2);
                        if (W > 4) then W := W – 8;
                        if (W < – 4) then W := W + 8;
                        E(4) := E(4) + W
                        end {then}
                if (X = XK and Y = YK) then BAC := 1;
                goto L4;
L6              W := E(5) – E(z – 1);
                if (W > 4) then W := W – 8;
                if (W < – 4) then W := W + 8;
                E(4) := E(4) + W;
                store E(1) ... E(z) into resultant file;
                goto L1
                end {then}
            else begin
                                                        {object search}
                x := x + 1;
                if (x ≠ X_max) then goto L3;
                y := y – 1;
                if (y ≠ Y_min) then goto L2;
                                                        {no further object}
                display of the image with labeled contours on the screen
                end {else}
```

(5) Definitions, properties and applications of contour encoding can be found, e.g., in the textbooks

Ballard, D.H., Brown, C.M.: *Computer Vision*. Prentice-Hall, Englewood Cliffs, 1982.
Pavlidis, T.: *Algorithms for Graphics and Image Processing*. Springer, Berlin, 1982.
Zamperoni, P.: *Methoden der digitalen Bildsignalverarbeitung*. Vieweg Verlag, Wiesbaden, 2nd ed., 1991.

See also

Wahl, F.M.: *Digital Image Processing*. Artech House, Norwood, 1987.

and *Zamperoni* (as cited above) for special contour tracing algorithms if the objects can not be isolated with a constant gray value threshold. *Filling algorithms* for closed curves are given in *Pavlidis* (as cited above).

7.2.2 Delaunay Triangulation and Voronoi Diagram

(1) An image with labeled (isolated) points is the input of this program. The program constructs in this image either the Delaunay triangulation ($VAR = 1$), the Delaunay triangulation and the Voronoi diagram ($VAR = 2$), or just the Voronoi diagram ($VAR = 3$). The program generates the bounded Voronoi diagram, i.e. the Voronoi diagram without (finite parts of) the infinite rays. The program has time complexity $\mathbf{O}(n^2)$ for an input of n points.

Attributes::

Images: bilevel images of isolated object points, gray value images with labeled (isolated) points

Operator: global, geometrical

Inputs:

list of isolated image points (e.g., result of a previous image processing, or set of interactively selected points in an image),

Variant $VAR = 1$, $VAR = 2$ or $VAR = 3$.

(2) Let $\mathbf{P} = \{p_1, p_2, ..., p_n\}$ be a finite set of points in the Euclidean plane. The *Voronoi cell* of a point p_i of this set consists of those points of the Euclidean plane which are not closer to any other point of the set $\mathbf{P}$ than to the point p_i. A Voronoi cell is denoted by $\mathbf{V}(p_i)$, and it is topologically a closed set of the Euclidean plane. It is the topological closure of the set of those points q of the Euclidean plane which are closer to p_i than to any other point of the set $\mathbf{P}$. Formally, it is

$$\mathbf{V}(p_i) = \{q : q \text{ is a point of the real plane and } d_2(q, p_i) \leq d_2(q, p_j), \quad \text{for all } j = 1, 2, ..., n\}.$$

The function $d_2(q, p)$ denotes the *Euclidean distance* between points q and p. In principle, any metric in the Euclidean plane could be used instead of d_2. For example, for two points p and q (i.e. extreme case $n = 2$) the bisector of the straight line segment pq separates both Voronoi cells if the Euclidean metric is assumed.

The Voronoi cells are convex regions. The cell $\mathbf{V}(p_i)$ contains point p_i. The border of a Voronoi cell is either a closed polygonal chain or an open polygonal chain with an infinite ray at both ends, i.e. the Voronoi cell is either a bounded (compact) region or an unbounded region. In Figure 7.7, the points 3 and 5 have bounded Voronoi cells, and the points 1, 2, 4, 6, and 7 have unbounded Voronoi cells. These points with unbounded Voronoi cells are exactly the extreme points of the convex hull of this set of points.

The union of all borders of the Voronoi cells $\mathbf{V}(p_1)$, $\mathbf{V}(p_2)$, ..., $\mathbf{V}(p_n)$ constitutes the *Voronoi diagram* of the given set $\{p_1, p_2, ..., p_n\}$ of points. It consists of (infinite) rays starting at one *Voronoi point* (end point of the ray), and of straight line segments limited by two *Voronoi points* (both end points of the segment).

The Voronoi diagram defines a *neighborhood relationship* between the elements of the given set $\mathbf{P} = \{p_1, p_2, ..., p_n\}$ of points in the real plane. Two points p_i and p_k of the set of points are said to be *Voronoi neighbors* if and only if their Voronoi cells $\mathbf{V}(p_i)$ and $\mathbf{V}(p_k)$ have at least two points in common, i.e. if these cells do not have only a single Voronoi point, but a border segment (a straight line segment or an infinite ray) in common.

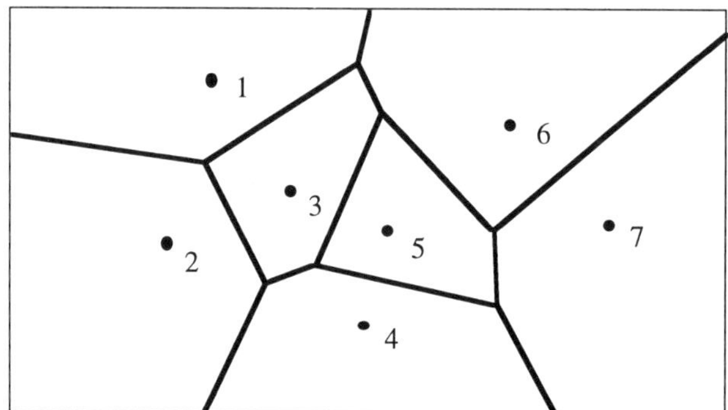

Figure 7.7: Example of a Voronoi diagram.

A set **P** of points is *free of circles* if four points of the set always satisfy the following constraint: These points are either not cocircular (i.e. they do not lie on the same circle), or there exists another point of **P** in the interior of the circle if they are cocircular, cp. Figure 7.8.

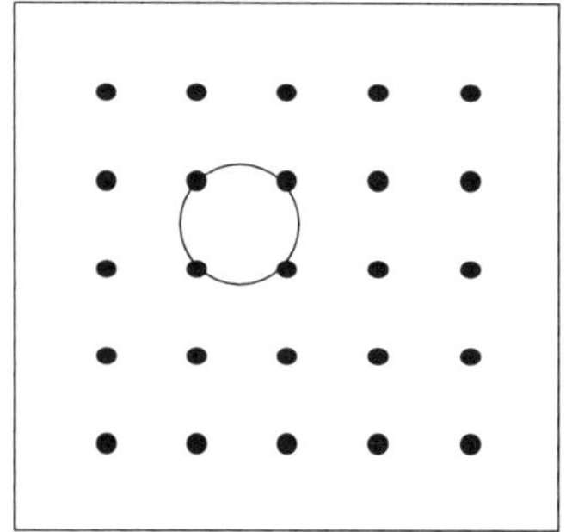

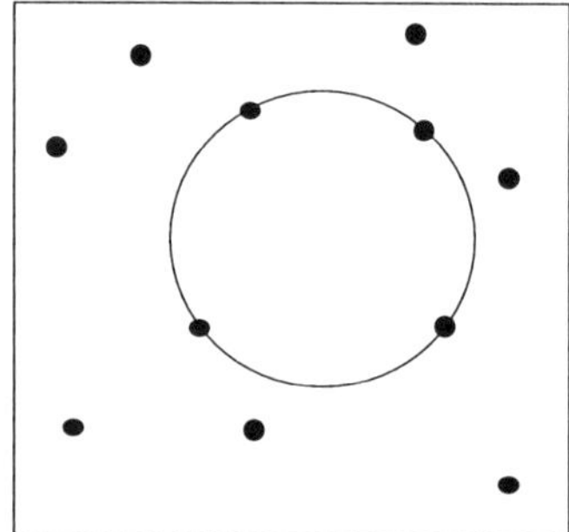

Figure 7.8: Two sets of points which are both not free of circles.

If a set **P** of points is free of circles, then a triangulation of this set results by drawing straight line segments between all pairs of Voronoi neighbors. This is the *Delaunay triangulation* of **P**, cp. Figure 7.9.

The neighborhood graph of the Voronoi neighborhood is called *Voronoi-dual*. If a set of points is free of circles then it holds that exactly three Voronoi cells are incident in each Voronoi point. In this case the Voronoi dual is a Delaunay triangulation of the given set of points. If a given set **P** is not free of circles, then a triangulation of this set **P** can be achieved by inserting further edges into the Voronoi dual. In general, these additional edges can be arbitrarily selected.

(3) The (isolated) points $p_1, p_2, ..., p_n$ can represent very different objects. They can be the result of a certain image processing, e.g. significant points of traced contours (cp. Section 7.2.1). Also (two-dimensional) feature maps can be represented as images. In this case, a point (x, y) would represent a certain object with two computed feature values x and y. In Section 2.8 (cp. Figure 2.9) it is suggested to use Voronoi cells as *shape estimates* of biological cells if only the nucleoli are visible. In general, neighborhood graphs can be used also within the scope of *clustering processes*.

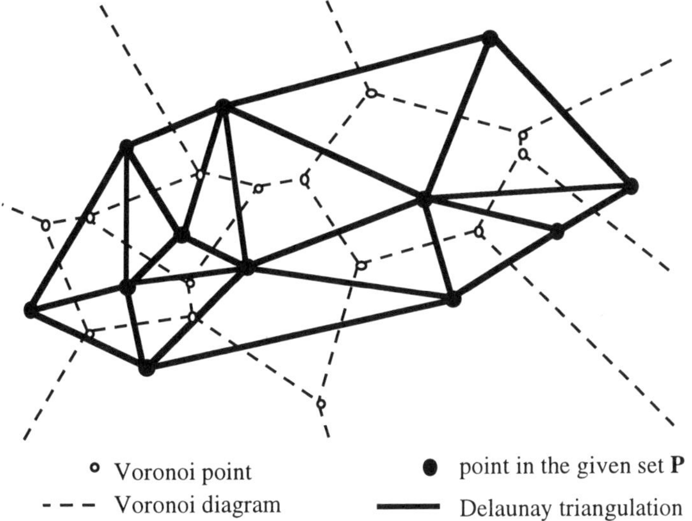

Figure 7.9: Delaunay triangulation and Voronoi diagram.

The rays of a Voronoi diagram can also be represented by straight line segments assuming a *virtual Voronoi point* as second end point. These virtual Voronoi points can be defined as intersection points with a given rectangle, e.g. the border of the region of interest of the image. A graphical representation of these rays is not included in the program at point **(4)**. However the extension of this program to the computation of rays and their virtual Voronoi points is straightforward. A Voronoi point is on the border of at least three Voronoi cells. A virtual Voronoi point is incident with only two Voronoi cells. A Voronoi diagram is characterized without equivocation by straight line segments or by their end point pairs (Voronoi points or virtual Voronoi points). The maximum number of Voronoi points is equal to $2n - 5$ if the set of points has n elements. The maximum number of border lines (straight line segments or rays) is equal to $3n - 6$. Note that the Voronoi diagram and the Voronoi dual are special planar graphs with fixed location in the Euclidean plane, cp. Figure 7.9.

The computation of a Voronoi diagram could be realized as follows: Let $\mathbf{P} = \{p_1, ..., p_n\}$ be the set of n image points. The process consists of several iterations. Around each point p_i are generated regions "growing with constant speed". If two growing regions meet together in one image point, then this image point belongs to the Voronoi diagram. The growing process can be realized via iterated dilations. However this method is very time consuming on sequential processors, and the structuring element used for dilation corresponds to a specific metric, cp. Section 6.5.2. A metric defined by a certain discrete unit circle always differs from the Euclidean metric d_2. Furthermore, the Voronoi diagram is not explicitly computed in such a process.

The program in **(4)** computes the Delaunay triangulation, and the Voronoi diagram follows from this triangulation. In fact, the program computes a triangulation for any point set **P**. If this set is not free of circles, then some edges are added to the Voronoi dual for achieving the triangulation. The program can also be modified in such a way that these additional edges are not generated.

The Delaunay triangulation consists of *Delaunay triangles*. The vertices of these triangles are points of the given set **P**. Each triangle contains no further points of **P** in its interior or on its border. Let **P** be a circle-free set. Three points p_i, p_j, p_k of **P** define a Delaunay triangle if there is no further point of **P** in the interior of that circle which is circumscribed to the triangle p_i, p_j, p_k. Some geometrical fundamentals of the program in **(4)** are given in the following, exploiting this property of the Delaunay triangles.

Three points p_1, p_2, p_3, with $p_i = (x_i, y_i)$, define a triangle $p_1p_2p_3$. The area of this triangle is given by

$$F = \frac{1}{2} \cdot |S(p_1, p_2, p_3)|$$

with

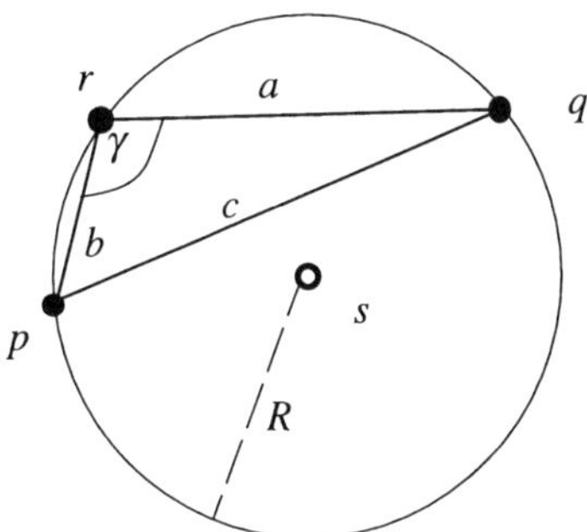

Figure 7.10: A Delaunay edge pq connects two points of the given set of points. If p, q and r define a Delaunay triangle, then it follows that the center point of the circle, circumscribed to p, q and r, is a Voronoi point.

$$S(p_1, p_2, p_3) = x_1(y_2 - y_3) + x_2(y_3 - y_1) + x_3(y_1 - y_2) .$$

A non-zero value $S(p_1, p_2, p_3)$ is either negative or positive, depending upon the orientation of the triangle $p_1p_2p_3$ (i.e. clockwise or counter-clockwise).

Figure 7.10 shows a triangle pqr with side lengths a, b and c. The straight line segment pq has the length c. Assume that pq is a straight line segment of the Delaunay triangulation, i.e. a *Delaunay edge*. The points p and q, and a third (non-collinear) point r define a circle with radius R, center point s and periphery angle γ. The values of R and γ allow a unique selection of a third point r such that pqr is a Delaunay triangle. The selection criterion consists in minimizing $R \cdot \cos\gamma$. This follows from the above cited property of Delaunay triangles. Since

$$\cos\gamma = \frac{a^2 + b^2 - c^2}{2ab} \qquad \text{and} \qquad R = \frac{abc}{4\,F}$$

it follows that this minimization is equivalent to minimizing

$$\frac{c\;(a^2 + b^2 - c^2\;)}{8 \cdot F} .$$

Note that $c/8$ remains constant during the search of the third point r, because p and q do not change. Therefore, the quotient

$$K\;(p, q, r) \quad = \frac{a^2 + b^2 - c^2}{S(p, q, r)}$$

can be used for solving this minimization problem. In the program, S is either selected to be always positive or negative. If it is positive then the quantity $K(p, q, t)$

has to be minimized. If it is negative then this quantity has to be maximized. This optimization criterion is used in the program. For brevity we also use the definition

$$G(p, q, r) := a^2 + b^2 - c^2 ,$$

where a, b, c have the geometrical meaning associated to p, q, r as shown in Figure 7.10. If pqr is a Delaunay triangle then the center point s is a Voronoi point.

The program computes a first Delaunay triangle at the beginning. The function value $S(p, q, r)$ of this triangle is either positive or negative, and this is encoded by $U = +1$ or $U = -1$ in the program. Then, this value (i.e. the first triangle) determines all the optimization decisions of the program, i.e. whether $K(p, q, r)$ has to be minimized or maximized.

A Delaunay edge corresponds either to one (if the Delaunay edge is on the convex hull of the set of points) or to two Voronoi points. A Delaunay edge and a corresponding Voronoi point form a *Delaunay pair*. Assume that the computed Delaunay triangle pqr has the orientation U (i.e. as determined by the first triangle), and this triangle defines the Voronoi point s. Then, the Delaunay pairs of this triangle are stored in a list **L** for further search of Delaunay triangles. The list **L** is called the *Delaunay list* in the program. It can be implemented as a stack in the computer memory. It is very important for the algorithmic processing of this list that the Delaunay pairs are entered into this list with opposite orientation $-U$, i.e. $[qp, s]$, $[pr, s]$ and $[rq, s]$.

Let p_1, p_2, p_3, with $p_i = (x_i, y_i)$ and $y_1 \neq y_2$, be three points which define a circle. Then, the center (x_s, y_s) of this circle has the coordinates

$$x_s = \frac{(y_3^2 - y_1^2 + x_3^2 - x_1^2)(y_2 - y_1) - (y_2^2 - y_1^2 + x_2^2 - x_1^2)(y_3 - y_1)}{2 \cdot [(x_1 - x_2)(y_3 - y_1) - (x_1 - x_3)(y_2 - y_1)]}$$

$$y_s = \frac{x_1 - x_2}{y_2 - y_1} \cdot x_s + \frac{y_2^2 - y_1^2 + x_2^2 - x_1^2}{2(y_2 - y_1)} .$$

If $y_2 = y_1$ then it follows that $y_1 \neq y_3$ (otherwise the points would be collinear, i.e. they would not from a triangle), and it holds

$$y_s = \frac{x_1 - x_3}{y_3 - y_1} \cdot x_s + \frac{y_3^2 - y_1^2 + x_3^2 - x_1^2}{2(y_3 - y_1)} .$$

These formulas follow from the consideration of two intersecting bisectors, say of the straight line segments p_1p_2 and p_1p_3.

(4) The program has time complexity $\mathbf{O}(n^2)$. There exist also more complex, but asymptotically optimum methods of time complexity $\mathbf{O}(n \log n)$. However the program given here can be implemented more easily, and it should be of sufficient speed for practically relevant values of n.

The program computes either only the Delaunay triangulation ($VAR = 1$), this triangulation and the Voronoi diagram ($VAR = 2$), or just the Voronoi diagram ($VAR = 3$). It computes a bounded Voronoi diagram without graphical representations of the rays. The program inputs are the list of points and the variant specification.

It is assumed that there are at least three different points p_1, p_2, and p_3 in the set **P** of input points. If **P** is free of circles the program calculates the Delaunay triangulation. If **P** is not free of circles then the program also computes a certain triangulation by triangulating m-sided polygons of the Voronoi dual, in a certain order. The program can be modified in such a way that always the Voronoi dual is computed, cp. remarks at the end of **(4)**.

The Delaunay list **L** contains Delaunay pairs during the computations. This list is modified by deletions and insertions in the course of the algorithm computation. The chosen list type (e.g. as a stack) influences the order of construction. The list **L** is empty at the beginning, and the program stops as soon as **L** is empty again.

The points p_1, p_2, p_3 are assumed to be represented in an array **ARR**. Let

$$\mathbf{ARR}(i, 1) = x_i \quad \text{and} \quad \mathbf{ARR}(i, 2) = y_i \quad \text{for } p_i = (x_i, y_i).$$

The program works on unsorted array values. However some sorting could be used as preprocessing to speed up the search for the next Delaunay triangle in the program. For example, if the points have been generated by scanning an image, then the points are already arranged in a certain spatial order. Such sophisticated improvements of the computing time are not discussed in the sequel.

The *BRESENHAM* procedure (cp. Section 3.4.10) is suggested for drawing Voronoi or Delaunay edges in the image raster. This procedure can be implemented for generating graphically different lines.

```
procedure DELAUNAY (ARR: point_array);
        var L: point-list;
begin {DELAUNAY}
                {initialization of the Delaunay list L with three Delaunay pairs}

        let p be any point in P, cp. (ARR(1, 1), ARR(1, 2));
        let q be the nearest neighbor of p in P;
                                {in general this needs a complete search through P}
```

```
if a third point r is found by minimization of values K(p, q, i), for all
                                        points i with S(p, q, i) >0
then  U := +1                                      {positive orientation}
else    if a third point r is found by maximization of values
                         K(p, q, i), for all points i with S(p, q, i) < 0
        then  U := −1                              {negative orientation}
        else  stop;                         {all points of P are collinear}

let s be the center point (Voronoi point) of the circle defined by the
        Delaunay triangle pqr;
insert the Delaunay pairs [qp, s], [pr, s] and [rq, s] in L;
                 {inverse orientation of edges with respect to U}
if  (VAR ≤ 2)  then  draw the Delaunay edges qp, pr and rq;
                 {construction of Delaunay and/or Voronoi diagram}

while (list L is not empty) do begin
        take a Delaunay pair [pq,v] out of the list L;
        delete [pq,v] in L;

{the straight line through p and q divides the plane into two half planes;
    a Delaunay triangle was found before in one of these half planes;
                the search considers only the other half plane}
        if ( U = +1) then K := +∞   else K := −∞;      { K := U*∞  }
        k := 0;            {initialization of a pointer to the third point}
        flag := 0;
        for  i : =  1  to  n  do              {pointer i to points in ARR}
                if (i does not point on p or q)  then begin
                        let r be the point (ARR(i,1),ARR(i,2));
                        G := G(p,q,r);           S := S(p,q,r);
                        if (( U = +1 and S > 0 and G/S < K)  or
                        ( U = −1 and S < 0 and G/S > K)) then begin
                          {G/S is minimized for a positive value of U,
                         and it is maximized for a negative value of U}
                                flag := 1;        k := i;
                                K := G/S
                                end {then}
                        end {then};
        if ( flag = 1 ) then begin
                          {processing of the new Delaunay neighbors}
```

```
            r := point with index k;
            s := center point of circle defined by p, q, r;
                                                {Voronoi point s}
            if (VAR ≤ 2) then
                  draw the Delaunay edges pr and rq ;
            if (VAR ≥ 2) then draw the Voronoi edge vs ;
              if (there exists an u such that [pr, u] or [rp, u] is
                                                  already in L)
                  then begin
                    if (VAR ≥ 2) then draw the Voronoi edge su;
                    delete [pr, u] or. [rp, u] in L
                    end {then}
                  else insert the Delaunay pair [pr, s] into L
                            {edge with orientation opposite to U};
              if (there exists an u such that [rq, u] or [qr, u] is
                                                  already in L)
                  then begin
                    if (VAR ≥ 2) then draw the Voronoi edge su;
                    delete [rq, u] or [qr, u] in L
                     end {then}
                  else insert the Delaunay pair [rq, s] into L
                            {edge with orientation opposite to U}
            end{then}
        end {while}
  end {DELAUNAY}
```

In this program it is possible that identical Delaunay edges are drawn several times. This could be avoided by an additional test (edges *pr* or *rq* exist already?). This multiple drawing of edges can practically be neglected.

Sometimes it may be desirable to compute the Voronoi dual. In general, the minimization ($U = +1$) or the maximization ($U = -1$) of values $K(p, q, r)$ can lead to several points r with minimum or maximum value. If there are at least two solutions r_1, r_2 then p, q, r_1, r_2 form a circle without any further point in its interior. Such m-sided polygons, $m \geq 4$, are not triangulated.

The generation of (parts of) the infinite rays of the Voronoi diagram can easily be added to the program by means of some operations in the case of $flag = 0$. Then, the bisector of pq and the border of the active window define a virtual Voronoi point as their intersection point s, where the orientation of the triangle pqs has to be tested. For example, a straight line segment can be drawn for $U = +1$ and $S(p, q, s) > 0$.

(5) This $\mathbf{O}(n^2)$ algorithm was proposed in

Hufnagl, P., Schlosser, A., Voss, K.: *Ein Algorithmus zur Konstruktion von Voronoidiagramm und Delaunaygraph*. Bild und Ton **38** (1985), pp. 241 - 245.

An algorithm with $\mathbf{O}(n \log n)$ time complexity can be found in

Preparata, F.P., Shamos, M.I.: *Computational Geometry*. Springer, New York, 1985.

7.2.3 Hough Transformation for Straight Lines

(1) This program approximates sets of image points by straight line segments.

Attributes::

Images: bilevel images with isolated object points, or gray value images with labeled pixels (e.g. edge pixels, skeleton pixels)

Operator: global, geometrical

Inputs:

a list of all image points which have to be approximated by opportunely constructed straight line segments

(2) Simple geometrical elements (e.g. straight lines, circles, polygons) can be described in the spatial domain by parameter sets. However a parameter set can also be considered as a point in the multidimensional space spanned by the parameters, interpreted as coordinates of this space, and such a parameter domain is called the *Hough space* of the considered geometrical units. A point in the spatial domain can belong to many geometrical objects, e.g. to all straight lines passing through this point. Thus, a point in the spatial domain individuates a manifold of parameter tuples in the Hough space, e.g. the parameters of all straight lines passing through this point. Figure 7.11 illustrates this relation for the *straight line* representation scheme $y = ax + b$, i.e. points (x, y) in the spatial domain are mapped onto sets of parameter tuples (a, b).

Assume that two points in the spatial domain share the same geometrical unit. This unit has parameters in the intersection of the parameter sets associated with the given two points.

In general the task can be described as follows: A set of points (in an image) is given. These points have to be clustered in the parameter space so that each cluster "corresponds" to a unit belonging to a specific class of geometrical units. In general, the solution requires the following steps:

(1) Compute the sets of parameter tuples in the Hough space for all the given points in an image. Discretize the parameters.

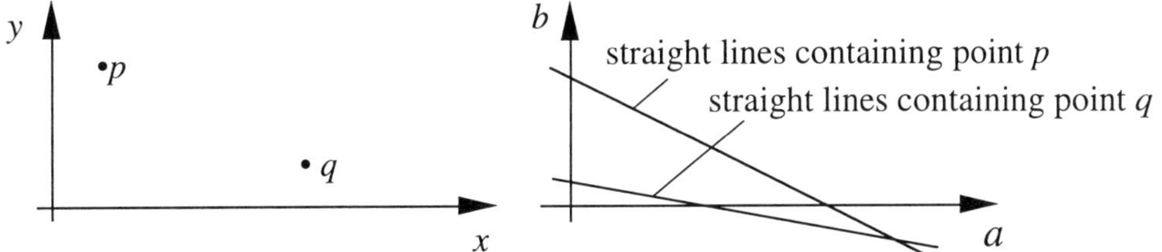

Figure 7.11: Two points in the spatial domain and their associated point sets in the Hough space for the straight line parametrization $y = ax + b$.

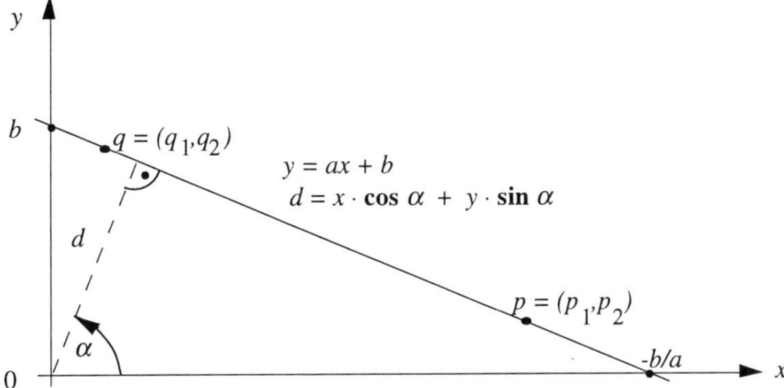

Figure 7.12: A straight line $d = x \cdot \cos\,\alpha + y \cdot \sin\,\alpha$ in the spatial domain containing two image points p and q.

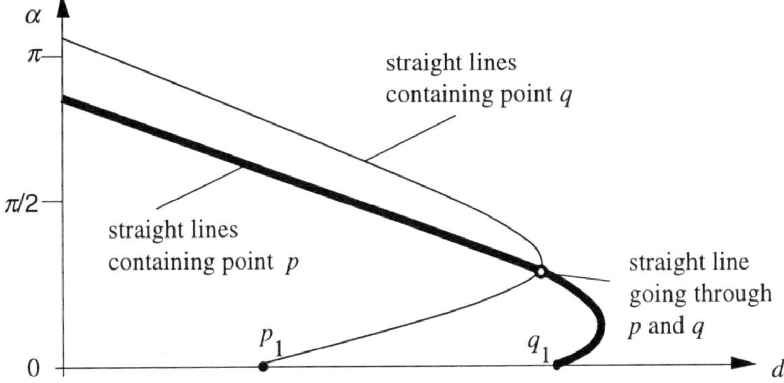

Figure 7.13: The parameter sets are curves in the (d, α)-space. The intersection point of both curves characterizes exactly the straight line passing through both points.

(2) Consider the intersections of parameter tuple sets. Each discrete grid cell in the Hough space is assigned a value counting the number of tuples of different geometrical units in which this point is contained.
(3) Identify all points in the Hough space with a counter value exceeding a given threshold with a "recognized" unit in the spatial domain.

This Section treats the specific task of straight line recognition. For example, assume that isolated edge points have to be approximated through a straight line segment. The class of geometrical units are straight lines. The program uses the (d, α)-parameter space of the *Hessian normal form*

$$d = x \cdot \mathbf{cos}\ \alpha + y \cdot \mathbf{sin}\ \alpha$$

as Hough space, cp. Figure 7.12 and 7.13. The intersection point of both curves

$$d = p_1 \cdot \mathbf{cos}\alpha + p_2 \cdot \mathbf{sin}\alpha \quad \text{and} \quad d = q_1 \cdot \mathbf{cos}\alpha + q_2 \cdot \mathbf{sin}\alpha$$

of the Hough space corresponds to that straight line $y = ax + b$ passing through points p and q in the spatial domain, for points $p = (p_1, p_2)$ and $q = (q_1, q_2)$. In comparison to the (a, b)-space of Figure 7.11, the (d, α)-space has the advantage that the relevant parameter set is bounded. The parameter a can take any value in the (a, b)-space.

The algorithm performs the following steps for recognizing collinear points by means of the (d, α)-space:

(*a*) Discretize the domain of d and α, e.g. α between 0° and 180° in steps of 5°. Estimate the maximum value d_{max} of d. The image diagonal defines this value, e.g. $d_{max} = 724.08$ for 512×512 images. Discretize d between 0 and d_{max}, e.g. between 0 and 723 in steps of 3. As the result of this digitization the Hough space has dimensions $Hx = 1, ..., Hx\_max$ for d and $Hy = 1, ..., Hy\_max$ for α, cp. Figure 7.14. The program maps input point sets into this Hough space, where the "gray value" at point (d, α) represents the score of given points in the spatial domain which lie on the straight line $d = x \cdot \mathbf{cos}\ \alpha + y \cdot \mathbf{sin}\ \alpha$.
(*b*) Select the relevant points in the input image f, e.g. as a result of edge detection or skeletonization.
(*c*) Optional: Select a region of interest in f, e.g. a $m \times m$ window, such that the recognition of collinear points is restricted to this region of interest.
(*d*) Increment the counter values of all points of the Hough space which are on (or close to) a curve

$$d = p_1 \cdot \mathbf{cos}\ \alpha + p_2 \cdot \mathbf{sin}\ \alpha$$

for all the input points (p_1, p_2).

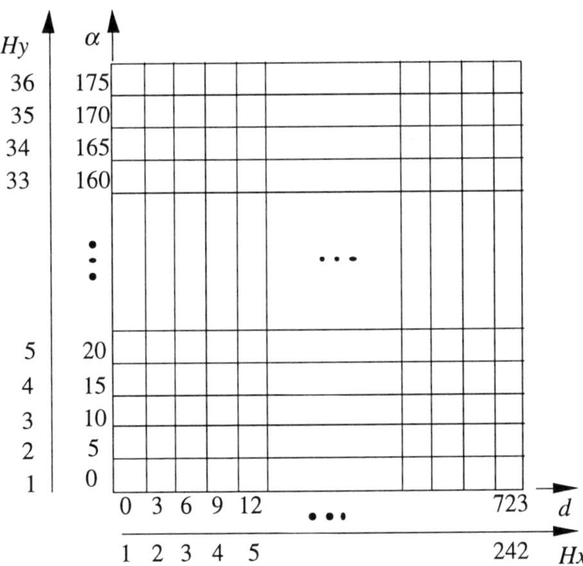

Figure 7.14: Example of a Hough space for images of size 512×512, with Hx_max = 242 and Hy_max = 36.

(*e*) Analyze the Hough space. Search for points with high counter value (high "gray value") or for local clusters of high counter values. Compute a center point for such clusters. As a refinement of this analysis, the computed straight lines can be considered in the spatial domain, and step (*e*) can be eventually repeated with slightly modified parameter sets, e.g. for avoiding to obtain nearly parallel lines if this is a constraint in the specific application.

(*f*) STOP or back to step (*c*).

The Hough space is also called *accumulator field*. The algorithm in **(4)** starts with the initialization of the Hough space in the sense of (*a*). Then steps (*d*) and (*e*) are carried out. The selection of points in the image, i.e. point (*b*), can be the result of an edge detector, cp. Section 6.2, or of other early processing operators. The size of the window in (*c*) should be larger than an a-priori estimate of the relevant length of straight line segments.

(4) *Control structure*: special, here explicitly given
*Required data array: Hough(*1...*Hx_max,* 1...*Hy_max) for integer values*

Inputs and initialization of the Hough space

input of parameters *MAX_NUM* and *S_inkr* (cp. below);

```
α_step := 5;    d_step := 3;                        {examples of step sizes}
Hx_max := d_max / d_step;
Hy_max := 180 / α_step;
            {in some cases 180 should be replaced by 360, e.g. if it is necessary
 to distinguish between "above of a straight line" or "below of a straight line"}
array Hough(1...Hx_max, 1...Hy_max): integer;
for Hx := 1 to Hx_max do
        for Hy := 1 to Hy_max do Hough(Hx, Hy) := 0
```

Mapping of all relevant image points (x, y) into the Hough space (e.g. during edge point detection)

```
MAX := 0;            {counts the current maximum value in the Hough space}
for all relevant points (x, y) in the image do
        for Hy := 1 to Hy_max do begin
{variation of all α-values, i.e. an unique d-value is computed for each α-value}
                α := (Hy – 1) · α_step;                           {α in degrees}
                d  := x · cos α  + y · sin α;
                                        {if necessary, compute α in rad }
                Hx := integer(d / d_step  +  1.5);
                        {d starts at 0 and Hx starts at 1, i.e. 1.5 is added
                        instead of 0.5 for computing the nearest integer}
                if (( Hx ≥ 1 ) and ( Hx ≤ Hx_max )) then begin
                                Hough(Hx, Hy) := Hough(Hx, Hy) + 1;
                        if MAX < Hough(Hx, Hy) then
                                        MAX := Hough(Hx, Hy)
                        end {then}
                        else
        the parameter is not in the initialized range of d-values (increase d_max ?)
                end {for}
```

Detection of straight lines in the Hough space:

```
preprocessing of the Hough space for enhancing local maxima if necessary,
        e.g. if only a few input points (x, y) are given, local maxima can be
        determined by means of agglomeration (note: update MAX as maxi-
        mum value of the Hough space);
S := MAX / 2;            {example of in initial threshold in the Hough space}
repeat
        NUM := 0;                        {number of detected straight lines}
        S := S + S_inkr;
```

```
                    {parameter S_inkr for increasing the threshold step-by-step,
                                                     e.g. S_inkr = MAX/10}
        for Hx := 1 to Hx_max do
            for Hy := 1 to Hy_max do
                if Hough(Hx, Hy) ≥ S then begin
                    α := (Hy – 1) · α_step;          {α in degrees}
                    d := (Hx – 1) · d_step;
                    the straight line d := x · cos α + y · sin α is
                        detected in the xy-space, e.g. draw this
                        line in the input image;
                    NUM := NUM +1
                end {then}
    until NUM ≤ MAX_NUM
  {parameter MAX_NUM is the a-priori maximum number of possible straight lines}
```

This program implements a special method for controlling the number of detected straight lines. It is possible that the evaluation of detected straight lines also depends upon further criteria, e.g. that parallel straight lines should be avoided if these lines are very close in the image. Often the task consists of detecting straight line segments instead of straight lines. A possible approach is to register minimum and maximum x- and y-coordinates, i.e. four values, for each counter value at (Hx, Hy) and for all points (x, y) which contribute to that counter value.

(5) Ballard, D.H., Brown, C.M.: *Computer Vision.* Prentice-Hall, Englewood Cliffs, 1982.

Haralick, R.M., Shapiro, L.G.: *Computer Vision, Volume I.* Addison-Wesley, Reading 1992.

Leavers, V.F.: *Shape Detection in Computer Vision Using the Hough Transform.* Springer, Berlin, 1992.

7.3 SIGNAL ANALYSIS OPERATORS

The image processing literature contains extensive material about global filterings by means of Fourier, Walsh-, Hadamard, *DCT* (discrete cosine transformation) or other transformations, cp. for example

Gonzalez, R.C., Wintz, P.: *Digital Image Processing.* Addison-Wesley, Reading, USA, 1987.

Pratt, W.K.: *Digital Image Processing.* Wiley, New York, 1978.

Wahl, F.M.: *Digital Image Processing.* Artech House, Norwood, 1987.

This section deals with the two-dimensional Fourier and Walsh transformations. These transformations could also be performed on image windows instead of complete images. However the Fourier transformation is based on continuous basic functions. The Walsh transformation should be preferred to the Fourier transformation for small window sizes, e.g. for $n, m \leq 64$. The *DCT* is often used with small windows, e.g. 16×16, for the encoding of static images.

7.3.1 Fourier Transformation

(1) The discrete Fourier transformation maps a gray value image f into a complex-valued matrix **F** or a complex valued image F. The result represents an input image f as the composition of basic periodic functions with different frequencies. After this transformation the input image f is represented in the *frequency domain*. This transformation generates $M \times N$ complex image values $F(x, y) = a + i \cdot b$ for $M \times N$ gray values $f(x, y)$. The image F can be represented by two $M \times N$ gray value images F_1 and F_2, where F_1 stands for the real part a and F_2 stands for the imaginary part b.

The numbers a and b are real and can also have negative values. For representing F in the gray value scale, the range of F-values has to be shifted. However in general the result of the Fourier transformation is only an intermediate result, and a complex-valued matrix **F** is used for storing values in the frequency domain.

The values in the frequency domain can be considered as features of the given input image. They allow an interesting "new viewpoint" on the textures of the input image. High values of high spatial frequencies correspond to thin structures, good contrast, sharp edges etc., and high values of low frequencies correspond to smooth regions, low contrast, unsharp edges etc.

The Fourier transformation is of special interest for introducing some modifications in the frequency domain before transforming the image back to the spatial domain. This filtering approach is a topic of the next two Sections. However for this reason the Fourier transformation program is also interesting for complex-valued input matrices. The input is given by two gray value images f_1 and f_2, in this case standing for the real and the imaginary part.

Variant 1 ($VAR = 1$) assumes f as a (normal) gray value image, and Variant 2 ($VAR = 2$) assumes f as a complex-valued matrix.

Attributes::

Images:	$N \times N$ gray value images, $N \times N$ complex valued matrices, N is a power of two
Operator:	global, signal theoretical

Inputs:

Variant $VAR = 1$ or $VAR = 2$.

(2) Let f be a matrix of $M \times N$ real or complex numbers $\mathbf{f}(0, 0)$, $\mathbf{f}(0, 1)$,..., $\mathbf{f}(M-1, N-1)$. The Fourier transformation maps f onto a $M \times \mathrm{N}$ matrix $\mathbf{F}$ where

$$\mathbf{F}(u, v) = \frac{1}{M \cdot N} \sum_{x=0}^{M-1} \sum_{y=0}^{N-1} \mathbf{f}(x, y) \cdot \mathbf{exp}\left[-i2\pi\left(\frac{xu}{M} + \frac{yv}{N}\right)\right],$$

for $u = 0, 1,..., M-1$ and $v = 0, 1,..., N-1$. In this equation,

$$i = \sqrt{-1}$$

denotes the imaginary unit.

(3) The division by $M \cdot N$ in the equations in **(2)** can be integrated into the forward and the backward transformation. The division by $M \cdot N$ in the forward transformation as used in **(2)**, is only one possibility. A widespread approach is that a division by M is carried out in the forward transformation, and a division by N in the backward transformation.

The Fourier transformation is *separable*. It holds

$$\mathbf{F}(u, v) = \frac{1}{M} \sum_{x=0}^{N-1} \left[\frac{1}{N} \cdot \sum_{y=0}^{N-1} \mathbf{f}(x, y) \cdot \mathbf{exp}\left(\frac{-i2\pi yv}{N}\right)\right] \cdot \mathbf{exp}\left(\frac{-i2\pi xu}{M}\right)$$

$$= \frac{1}{N} \sum_{y=0}^{M-1} \left[\frac{1}{M} \cdot \sum_{y=0}^{N-1} \mathbf{f}(x, y) \cdot \mathbf{exp}\left(\frac{-i2\pi xu}{M}\right)\right] \cdot \mathbf{exp}\left(\frac{-i2\pi yv}{N}\right)$$

for $u = 0, 1,..., M-1$ and $v = 0, 1,..., N-1$. Therefore this transformation can be performed as follows for $M \times N$ matrices:

At first the one-dimensional transformation maps either all rows or all columns into transformed rows or columns respectively (cp. procedure *FFT* in Section 3.4.8). Then the one-dimensional transformation is applied to either all columns or to all rows. Altogether, the iterated application of the *FFT* vector procedure is sufficient for transforming matrices.

The *FFT* procedure of Section 3.4.8 is based on a value of N which is a power of two. Accordingly this property is assumed for the number M of columns and the number N of rows. For simplicity, let $N = M$. For $VAR = 2$ the input f is given in two $N \times N$ matrices. $\mathbf{f1}(0...N-1,\ 0...N-1)$ is the real part and $\mathbf{f2}(0...N-1, 0...N-1)$

is the imaginary part. The program computes the result F in these arrays **f1** and **f2**. This is a transformation in place and the original input values of f are replaced by the computed ones. For $VAR = 1$ the input f is stored in the $N \times N$ matrix **f1**, and the matrix **f2** is initialized with zero in all of its positions. The real-valued matrices **f1** and **f2** are combined into a complex-valued matrix **f = (f1, f2)**.

The program assumes N to be a power of two, and the Fourier transformation is performed in the order

$$\mathbf{F}(u, v) = \sum_{x=0}^{N-1} \left[\frac{1}{N} \cdot \sum_{y=0}^{N-1} f(x, y) \cdot \mathbf{exp}\left(\frac{-i2\pi yv}{N}\right) \right] \cdot \mathbf{exp}\left(\frac{-i2\pi yv}{N}\right).$$

Note that a division by N occurs only once.

It can happen that the matrices **f1** and **f2** have to be transposed if memory limitations allow only row-wise transformations. In such a case both matrices are transformed row-wise including a division by N. Then both matrices are transposed converting columns into rows, and finally all rows are transformed again (without a division by N). A row-wise access to the image data is sufficient for this approach.

If the Fourier transformation result should be correctly arranged, then a second transposition step can be performed. Since normally the Fourier transformation is followed by the inverse transformation, these additional transposition steps are not necessary. However if the Fourier transformation result is displayed in an image, then this should be taken account of.

In general, it will be possible to have **f1** and **f2** in direct access in the main memory. Then, a transposition is superfluous and the second iteration performs column-wise transformations. Of course in this case the *FFT* procedure of Section 3.4.8 has to be implemented for row-wise as well as for column-wise transformations. In the sequel the transpositions will not be considered.

A *complex number* $z = a + i \cdot b$ is also exactly characterized by its *amplitude*, i.e. by the absolute value

$$\sqrt{a^2 + b^2}$$

of z, and by its *phase*, i.e. the angle with a fixed coordinate axis, e.g. $\mathbf{arctan}(b/a)$. Thus the result of the Fourier transformation can also be represented by a gray value image h_1 of the amplitudes, and a gray value image h_2 of the phases. Absolute value and phase angle are non-negative values, i.e. the gray value scale has not to be shifted for displaying these images h_1 and h_2. The following program uses the representation by the real part **f1** and the imaginary part **f2**.

(4) Note that the index range of the matrices **f1** and **f2** begins at zero. For images it has been assumed that coordinates begin at one. Thus a shift by one is necessary if images are read into the matrix data structure.

```
if (VAR = 1) then begin
        read image f into the matrix f1, with value f(x, y) into position
                f1(x − 1, y − 1);
        initialize matrix f2 with the constant zero;
        end (then}
else    begin
        read the real part of f into the matrix f1;
        read the imaginary part of f into the matrix f2
        end {else};
for  y = 0  to  N − 1  do begin
        call the FFT procedure for row y of the matrix f = (f1, f2);
        divide all results by N and write the resultant line into row y of the
                matrix f
        end {for};
for  x = 0  to  N − 1  do begin
        call the FFT procedure for the column x of the matrix f = (f1, f2);
        write the resultant column into the column x of f
        end {for}
```

The resultant matrices **f1** and **f2** can be read into the gray value images F_1 and F_2. Note that the coordinates are shifted by 1 and that the gray value scale has to be adjusted, e.g. by a shift of $G/2$.

(5) Gonzalez, R.C., Wintz, P.: *Digital Image Processing*. Addison-Wesley, Reading, USA, 1987.
Pratt, W.K.: *Digital Image Processing*. Wiley, New York, 1978.
Wahl, F.M.: *Digital Image Processing*. Artech House, Norwood, 1987.

7.3.2 Inverse Fourier Transformation for Filtering

(1) The Fourier transformation is a one-to-one mapping. The *inverse Fourier transformation* allows an exact reconstruction of the original image. However in several practical applications this inverse transformation is used to implement filters for images by performing the following steps:

(a) Fourier transformation of the given image f into matrices **f1** and **f2**, cp. Section 7.3.1,

(b) opportune modifications of the spatial frequency content of f by calculating new values in the matrices **f1** and **f2**,
(c) inverse Fourier transformation of **f1** and **f2** into a gray value image g.

The image g is the result of a *Fourier filtering* applied to image f. The same result could also be obtained by a *convolution* of f in the spatial domain with a suitable convolution kernel h. However Fourier filtering has two advantages in comparison to a direct convolution:

(i) The fast *FFT* algorithm allows efficient computations but a similar algorithm can not be used for this direct convolution and
(ii) with the convolution kernel h in the image domain a direct frequency interpretation is often not straightforward but of course this is possible for the transformed function h in the frequency domain.

The *convolution theorem* in mathematical analysis says that a convolution of f with h is identical to the result of the following steps: let F and H be the results of the Fourier transformation of f and h, multiply F and H in the frequency domain point-by-point, and then transform this product into the spatial domain via the inverse Fourier transformation.

The program of this Section computes the inverse Fourier transformation. An experimental filter design can lead to important insights into the gray value structure of the given images. The availability of the Fourier transformation of Section 7.3.1, and of the inverse mapping as given here allow such a filter design; cp. also the non-centered or the centered representation in the frequency domain discussed in Section 7.3.3.

The design of the filter function H can follow certain general rules where the non-centered case is assumed:

- modifications of the values close to positions $(0, y)$, $(N-1, y)$, $(x, 0)$ or $(x, N-1)$ in the matrices **f1** and **f2** correspond to modifications of low spatial-frequency components of the input image f, and
- modifications of the values close to the center $(N/2, N/2)$ in the matrices **f1** and **f2** correspond to high spatial-frequency components of the input image f.

A *high-pass filter* can be implemented by attenuating the values of low-frequency components but without modifications of the high -frequency values. A high-pass filter can *enhance details* as texture, isolated points, thin curves, steep edges etc. A *low-pass filter* can be implemented by attenuating high-frequency values without changes in the low-frequency values. Such a low-pass filter reduces detail information in the image by introducing a *smoothing effect*.

Attributes:

Images: $N \times N$ complex valued matrices, N as assumed to be a power of two

Operator: global, signal theoretical

Suggested preprocessing:

modify values of matrices **f1** and **f2** in the sense of a certain task-depending filter function.

(2) The inverse Fourier transformation of a matrix **F** of $M \times N$ complex numbers $\mathbf{F}(0, 0), \mathbf{F}(0, 1), \ldots, \mathbf{F}(M-1, N-1)$ is defined as

$$\mathbf{f}(x, y) = \sum_{u=0}^{M-1} \sum_{v=0}^{N-1} \mathbf{F}(u, v) \cdot \mathbf{exp}\left[i2\pi\left(\frac{xu}{M} + \frac{yv}{N}\right)\right],$$

for $x = 0, 1, \ldots, M-1$ and $y = 0, 1, \ldots, N-1$. Altogether a division by $M \cdot N$ has to be performed in the forward and backward transformation for the value normalization. For example, a division by M can be performed in the forward transformation, and a division by N in the inverse mapping.

(3) The inverse Fourier transformation is also *separable*. It holds

$$\mathbf{f}(x, y) = \sum_{u=0}^{M-1} \left[\sum_{v=0}^{N-1} \mathbf{F}(u, v) \cdot \mathbf{exp}\left(\frac{i2\pi yv}{N}\right)\right] \cdot \mathbf{exp}\left(\frac{i2\pi xu}{M}\right)$$

$$= \sum_{v=0}^{N-1} \left[\sum_{u=0}^{M-1} \mathbf{F}(u, v) \cdot \mathbf{exp}\left(\frac{i2\pi xu}{M}\right)\right] \cdot \mathbf{exp}\left(\frac{i2\pi yv}{N}\right),$$

for $x = 0, 1, \ldots, M$ and $y = 0, 1, \ldots, N-1$. Assume that a division by N has to be included in this inverse transformation. The input **F** is assumed to be a $N \times N$ matrix of complex numbers, i.e. with $M = N$ and N is a power of two, cp. Section 3.4.8. The following program computes the inverse Fourier transformation according to the equation

$$\mathbf{f}(x, y) = \sum_{u=0}^{N-1} \left[\frac{1}{N} \cdot \sum_{v=0}^{N-1} \mathbf{F}(u, v) \cdot \mathbf{exp}\left(\frac{i2\pi yv}{N}\right)\right] \cdot \mathbf{exp}\left(\frac{i2\pi xu}{N}\right)$$

The *conjugate-complex number* z^* of a complex number $z = a + i \cdot b$ is defined as $z^* = a - i \cdot b$. It holds

$$\mathbf{f}^*(x, y) = \sum_{u=0}^{N-1} \left[\frac{1}{N} \cdot \sum_{v=0}^{N-1} \mathbf{F}^*(u, v) \cdot \mathbf{exp} \left(\frac{i2\pi yv}{N} \right)^* \right] \cdot \mathbf{exp} \left(\frac{i2\pi xu}{M} \right)^*$$

$$= \sum_{u=0}^{N-1} \left[\frac{1}{N} \cdot \sum_{v=0}^{N-1} \mathbf{F}^*(u, v) \cdot \mathbf{exp} \left(\frac{i2\pi yv}{N} \right) \right] \cdot \mathbf{exp} \left(\frac{i2\pi xu}{N} \right),$$

for $x, y = 0, 1, \ldots, N$. This gives the possibility of implementing the inverse transformation by a simple call of the (direct) Fourier transformation: at first map **F** onto its conjugate-complex values, and transform $\mathbf{F}^*$, obtaining the result $\mathbf{f}^*$. If $\mathbf{f}^*$ has only real values, then the inverse transformation stops. If $\mathbf{f}^*$ has to be represented as a gray value image then a simple absolute value computation should be sufficient in general. If a complex-valued matrix **f** has to be computed, then the matrix **f2** of **f** is multiplied by -1.

(4) Perform opportune task-depending filter operations in the Fourier transforms
$F = (F1, F2)$;
read the real part $F1$ of F into the matrix **f1**;
multiply the imaginary part $F2$ of F by -1, and store the result into the
matrix **f2**;
for $y = 0$ **to** $N - 1$ **do begin**
call the *FFT* procedure for row y of the matrix **f = (f1, f2)**;
divide the results by N and write the resultant array into row y of **f**
end {*for*};
for $x = 0$ **to** $N - 1$ **do begin**
call the *FFT* procedure for column x of the matrix **f = (f1, f2)**;
write the resultant array into column x of f
end{*for*}

Optional: Multiply all values in **f2** by -1, or compute the absolute value of **f** and map these values into the integer gray value scale $\{0, 1, \ldots, G - 1\}$.

If only row-wise transformations are possible then some *matrix transpositions* are necessary (cp. Section 7.3.1). In this case altogether an even number of transpositions has to be performed in the direct and in the inverse transformation in such a way that the resultant image f is correctly arranged.

(5) There exists a very broad literature about the design of Fourier filters, cp. for example

Gonzalez, R.C., Wintz, P.: *Digital Image Processing*. Addison-Wesley, Reading, USA, 1987.

Pratt, W.K.: *Digital Image Processing*. Wiley, New York, 1978.

Wahl, F.M.: *Digital Image Processing*. Artech House, Norwood, 1987.

7.3.3 Spectrum

(1) The *Fourier spectrum* |*F*| of an input image *f* is defined to be the absolute value of the Fourier transform *F*. The program delivers different representations of the Fourier spectrum as gray value images.

The Variant *VAR* = 1 computes the spectrum without additional modifications, the Variant *VAR* = 2 includes also a logarithmic transformation of the computed values. For example, *peaks* (i.e. isolated bright spots) in the spectrum correspond to strong spatial frequency components of the input image. Such peaks can be due to periodic interferences which should be reduced by filtering (cp. Section 7.3.2). The Variant 2 leads to an improved visibility of such peaks.

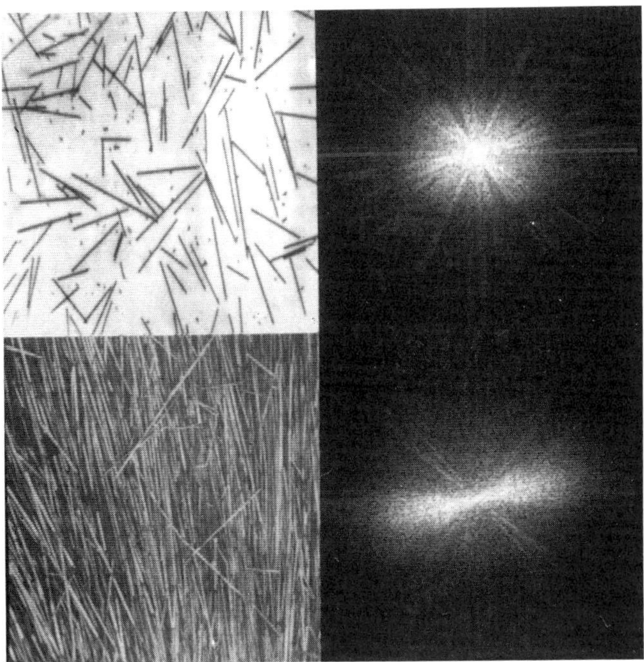

Figure 7.15: Two input images (top: *glass fibers* with about uniformly distributed orientation, below: straw with a dominant orientation) and their spectra which are centered and logarithmically enhanced for improved visibility.

The visual evaluation of a spectrum is simplified if the spectrum is shown in a centered position. The program computes a centered position with the input parameter value of *ZEN* = 1. *ZEN* = 0 produces the (normal) non-centered position, and in this case the low-frequency values are shown at the border of the spectrum image. In the centered case these low-frequency values appear close to the center($N/2$, $N/2$), and the high-frequency values are shown in the four corners of the spectrum image.

Figure 7.15 shows two centered ($ZEN = 1$) and logarithmically enhanced ($VAR = 2$) spectrum images. Certain "main spatial directions" in the input image appear in the spectrum with a 90° rotation.

Attributes:

Images: real valued (gray value images) or complex valued matrices
Operator: global, signal theoretical

Inputs:

Variant $VAR = 1$ without logarithmic transformation, and Variant $VAR = 2$ with this visual enhancement,
parameter input $ZEN = 1$ for centered representation, and $ZEN = 0$ for a non-centered output.

(2) Let f be the input image, and assume that f is transformed into $F = (F_1, F_2)$ via the Fourier transformation, cp. Section 7.3.1. Then $|F|$ is called to be the Fourier spectrum of f, i.e.

$$|F(x,y)| = \sqrt{F_1(x,y)^2 + F_2(x,y)^2},$$

for $1 \leq x, y \leq N$.

(3) The centered position of the Fourier spectrum can be achieved by multiplying all values $f(x, y)$ by $(-1)^{x+y}$ ("chessboard") before the Fourier transformation. This leads to a shift by $N/2$ in row and in column direction of the otherwise non-centered spectrum, without any further influence on the resultant values.

Normally only a few low-frequency values have high magnitude, and all the remaining values are close to zero. By representing the spectrum as a gray value image, this circumstance causes a very poor visibility of the weak spatial frequency components. To overcome this drawback the option of a *logarithmic transformation*

$$log(1 + |F(x, y)|)$$

is given.

(4) **if** ($ZEN = 0$) **then**
read f into the matrix **f1**, with $f(x, y)$ into **f1**$(x-1, y-1)$
else
read f into the matrix **f1**, with $(-1)^{x+y} \cdot f(x, y)$ into **f1**$(x-1, y-1)$;
initialize the matrix **f2** with value 0 in all of its positions;
for $y = 0$ **to** $N-1$ **do begin**
call the *FFT* procedure for row y of the matrix **f = (f1, f2)**;

divide the results by N and write the resultant array into row y of **f**
end {*for*};
for $x = 0$ **to** $N - 1$ **do begin**
call the *FFT* procedure for column x of the matrix **f** = (**f1, f2**);
write the resultant array into column x of **f**
end {*for*};
if (*VAR* = 1) **then**
write the absolute values $\sqrt{\mathbf{f1}(x,y)^2 + \mathbf{f1}(x,y)^2}$ into $h(x + 1, y + 1)$,
for $0 \le x, y \le N - 1$
else
write the logarithmically transformed values into h,
$h(x + 1, y + 1) := log\ (1 + \sqrt{\mathbf{f1}(x, y)^2 + \mathbf{f1}(x, y)^2}\ );$
for $0 \le x, y \le N - 1$
graphical display of the spectrum on a screen

The parameter setting *VAR* = 2 (logarithmically enhanced) and *ZEN* = 1 (centered) can be suggested in general, cp. Figure 7.14.

7.3.4 Walsh Transformation

(1) The Walsh transformation maps one-to-one a $N \times N$ gray value image f onto a $N \times N$ matrix **W** with real values (also negative values are possible!). The matrix **W** describes f as the result of the superposition of different *sequencies*. A sequency is a step-wave with a given periodicity. The image f is transformed into the *sequency domain*.

The program combines several processing options (Walsh transformation, Walsh filtering).

The matrix **W** can be represented as a gray value image, e.g. by a linear mapping of the sequency values into the gray value scale $\{0, 1, ..., G - 1\}$. This can be the final aim of the process, i.e. to visualize these sequency values, which can be considered as features of the input image.

By modification of the sequency values in **W** and a subsequent inverse Walsh transformation (cp. the analogous Fourier filtering approach, cp. Section 7.3.2), the input image f can be opportunely modified (filtered) in the sense of a given image processing task.

The Variant *VAR* = 1 computes only the Walsh transformation and outputs the matrix **W**. The Variant *VAR* = 2 allows a Walsh filtering, where the filter function has to be specified by the user. Each variant has several sub-variants, cp. **(4)**.

Attributes::

Images: $N \times N$-gray value images, N is assumed to be a power of two

Operator: global, signal theoretical

Inputs:

Variant $VAR = 1$ or $VAR = 2$.

(2) Let $N = 2^m$, and let f be a $N \times N$ matrix of real numbers $\mathbf{f}(0, 0), \mathbf{f}(0, 1), \ldots, \mathbf{f}(N-1, N-1)$. The Walsh transformation maps f onto a $N \times N$ matrix $\mathbf{W}$ of real numbers as specified by the equations

$$\mathbf{W}(u, v) = \frac{1}{N} \sum_{x=0}^{N-1} \sum_{y=0}^{N-1} \mathbf{f}(x, y) \prod_{i=0}^{m-1} (-1)^{B(m, i, x, y, u, v)} ,$$

for $u, v = 0, 1, \ldots, N-1$ and for

$$B(m, i, x, y, u, v) = b_i(x) b_{m-1-i}(u) + b_i(y) b_{m-1-i}(v) .$$

Here, $b_k(w)$ denotes the k-th bit in the binary representation of a non-negative integer w, cp. Section 3.4.9. The inverse Walsh transformation maps a matrix $\mathbf{W}$ of $N \times N$ real numbers $\mathbf{W}(0, 0), \mathbf{W}(0, 1), \ldots, \mathbf{W}(N-1, N-1)$ onto a matrix $\mathbf{f}$ of $N \times N$ real numbers

$$\mathbf{f}(x, y) = \frac{1}{N} \sum_{u=0}^{N-1} \sum_{v=0}^{N-1} \mathbf{W}(u, v) \prod_{i=0}^{m-1} (-1)^{B(m, i, u, v, x, y)} ,$$

for $x, y = 0, 1, \ldots, N-1$.

(3) The division by N has been equally distributed between the direct and the inverse transformation of **(2)**. If $\mathbf{f}$ has integer values, then $\mathbf{W}$ has also integer values before the division by N.

The Walsh transformation and the (identical!) inverse Walsh transformation are separable. The function $B(m, i, x, u)$ was defined in Section 3.4.9. It holds

$$\mathbf{W}(u, v) = \frac{1}{N} \sum_{x=0}^{N-1} \Big[\sum_{y=0}^{N-1} \mathbf{f}(x, y) \prod_{i=0}^{m-1} (-1)^{B(m, i, y, v)} \Big] \prod_{i=0}^{m-1} (-1)^{B(m, i, x, u)}$$

$$= \frac{1}{N} \sum_{y=0}^{N-1} \Big[\sum_{x=0}^{N-1} \mathbf{f}(x, y) \prod_{i=0}^{m-1} (-1)^{B(m, i, x, u)} \Big] \prod_{i=0}^{m-1} (-1)^{B(m, i, y, v)}$$

for $0 \le u, v \le N-1$. A transformation of a $N \times N$ matrix can be performed as follows: At first execute the one-dimensional transformation cp. procedure *FWT* in Section 3.4.9 either for all rows or for all columns. Then in a second phase execute the one-dimensional transformation either for all columns or for all rows on the resultant matrix of the first phase. Thus for transforming a matrix it is sufficient to perform repeatedly the *FWT* procedure for vectors.

The *FWT* procedure as given in Section 3.4.9 assumes that N is a power of two. For simplicity we also assume that it is $N = M$.

(4) In the program the matrix indices begin with zero. In comparison to the image coordinates, a shift by 1 is necessary.

input of image f into a matrix **f**, with $f(x, y)$ into $\mathbf{f}(x-1, y-1)$;
{Walsh transformation}

for $y = 0$ **to** $N-1$ **do begin**
call the *FWT* procedure for row y of matrix **f**;
store the resultant array into row y of **f**
end {*for*};
for $x = 0$ **to** $N-1$ **do begin**
call the *FWT* procedure for column x of matrix **f**;
store the resultant array into column x of **f**
end {*for*};
divide all elements of **f** by N and write the results back to **f**;
if (VAR = 1) **then begin**
{pictorial representation of Walsh transform}

subvariant 1: output of the absolute value of **f** elements, clips values smaller than 0 to 0, and clamps values greater than $G-1$ to $G-1$, or
subvariant 2: compute maximum and minimum value in **f** and scale all **f** into the range $0...G-1$ based on the computed max-min-values, or
subvariant 3: compute $\mathbf{log}_2(|\mathbf{f}| + 1)$ with proper scaling for the graphical output representation

end {*then*}
else begin
{filter operation}

subvariant 1: modify low-sequency components of the image f by changing opportunely chosen values of **f** close to $(0, y)$, $(N-1, y)$, $(x, 0)$ or $(x, N-1)$, or
subvariant 2: modify high-sequency components of the image f by changing opportunely chosen values of **f** close to the center $(N/2, N/2)$;

```
                                        {inverse Walsh transformation}
for  y = 0  to  N – 1  do begin
        call the FWT procedure for row y of matrix f;
        write the resultant array into row y of f
        end {for};
for  x = 0  to  N – 1  do begin
        call the FWT procedure for column x of matrix f;
        write the resultant array into column x of f
        end {for};
divide all elements of f by N and write the results back to f;
end {else}
```

(5) Gonzalez, R.C., Wintz, P.: *Digital Image Processing*. Addison-Wesley, Reading, USA, 1987.

GLOSSARY

The glossary lists terms currently used in the image processing field, with only a few exceptions, e.g. for characterizing tasks of digital image processing within the framework of computer vision. In a fast developing field like computer vision, the terms are often not yet consolidated. In the glossary a vocabulary is chosen that is uniformly used throughout the book. At the end of the term's explanation, numbers after the symbol ♦ point to relevant Chapters or Sections of the book.

agglomeration: iterative process for grouping single image points into *regions*[1], depending upon their image values or other local *features*, cp. *region growing*. ♦ 6.4, 6.5.3 to 6.5.7;

amplitude scaling: mapping changing the value range of a quantity (e.g. gray level) by means of a monotone transformation in the definition domain of this quantity. ♦ 5.1.1, 5.1.2, 5.1.3

analog/digital converter (AD-converter): hardware unit transforming analog images to a digital form (pictorial data) as an integral process of the transmission of images from a camera to a computer. An AD-converter generates millions of pixels per second. An AD-converter discretizes the analog images spatially (picture grid, or raster) and in amplitude (discrete gray value scale) ♦ 1.1.1, 1.1.2

artifact: artificial *picture object* resulting from processes during image acquisition or image processing not related to the image content which has to be analyzed. ♦ 6.1.7

background: union of all *image segments* which do not belong to picture objects. This assumes a binary image model such that segments can be classified either as background or as object segments. ♦ 5.4.1, 5.4.2, 6.3.2

bilevel image: image f with only two possible image values $f(x, y)$, e.g., 0 and $G - 1$. Special variants are *binary images* and *halftone images*. ♦ 6.1.7, 7.1.2

---

1 A word in Italics emphasizes that this word is an entry in the Glossary. The converse is not true (e.g. *gray level* or *image* are entries, but not in Italics throughout the Glossary).

binarization: transformation of a gray level image into a *binary* or a *half-tone image*. It maps specified gray levels onto the value 0, and the other gray levels onto the value 1 (in the binary case), or onto $G - 1$ (normally, if half-tone images are generated). A *binarization threshold* is the criterion used in the simplest gray level binarization approach. ♦ 5.3, 6.1.8

binarization threshold: value S, $0 \leq S < G - 1$, that defines a binarization of gray level images f into *binary* or *half-tone images b*, according to the general rule

$$b(x, y) = \begin{cases} 0 & , \text{if } f(x, y) \leq S \\ G - 1 & , \text{if } f(x, y) > S \end{cases}$$

♦ 5.3

binary image: image having only the gray values 0 or 1. ♦ 5.3, 6.1.7, 6.1.8, 7.1.1, 7.1.2

binomial filter: linear filter whose discrete impulse answer ($n \times n$ convolution kernel) is a product of a $1 \times n$ column vector and a $n \times 1$ row vector. Binomial coefficients

$$C(n-1, i) = \binom{n-1}{i} \quad \text{with} \quad i = 0, \ldots, n-1$$

are the components of both vectors. ♦ 6.1.3

blob: synonym for *component*.

boundary: one or several pixel thin set separating an image *segment* from another. This pixel set can be an *edge*, a border between two differently textured segments etc. Closed boundaries can be approximated by *contours*, e.g. by thinned boundaries. ♦ 6.2

box-filter: linear low-pass filter in which all coefficients of the *convolution kernel* are equal. ♦ 6.1.2

CCD camera (charge coupled device camera): camera type preferably used for image input in computer vision, measuring analog irradiances in a discrete sensor array. This array can be a single row (row camera), or a matrix (matrix camera). ♦ 1.1.1, 1.1.2

chain code: see main direction.

change detection: comparison between two images. If geometrically matched images (cp. *registering*) are given, then changes at image points p can be detected by certain point-to-point differences between the images. Generally, a change detection is based on measuring local deviations between two images at the same image point p. The simplest approach consists in a point-to-point image subtraction. ♦ 2.7, 5.4.3, 5.6.1

Dark Skin	Light Skin	Blue Sky	Foliage	Blue Flower	Bluish Green
Orange	Purplish Blue	Moderate Red	Purple	Yellow Green	Orange Yellow
Blue	Green	Red	Yellow	Magenta	Cyan
White	Neutral 8	Neutral 6.5	Neutral 5	Neutral 3.5	Black

Table G1: Placement of different colors on the ColorChecker card of Macbeth, cp. Figs. 1.8 and 1.9.

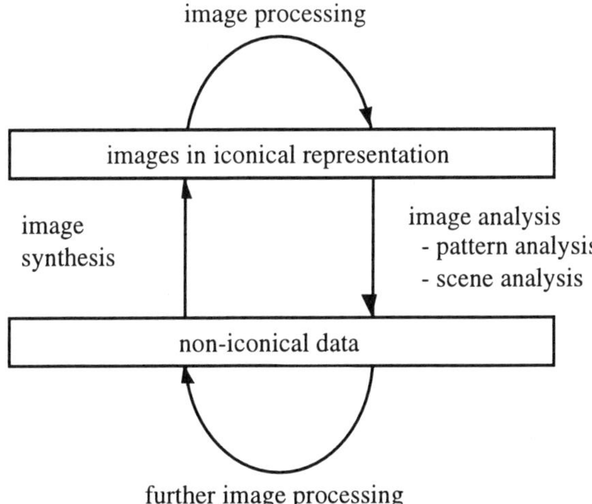

Figure G1: The subfields of computer vision: Image processing deals with computations of images from images. Iconic data structures are used in image processing operators.

charge coupled device camera: cp. CCD-camera.

color diagram: card or map of standard colors. It can be used to adjust or to calibrate a given color camera. A gray value representation (i.e. all color channels) of two color diagrams (and of a gray level diagram) is given in Figs. 1.8 and 1.9. There, the upper diagram is the KODAK gray level scale and the KODAK color diagram, available at photo-shops. Above the ColorChecker[2] of Macbeth (see Table G1) is shown. ♦ 1.1.3

component: topologically connected subset of a defined set of pixels. ♦ 1.1.1

2 ColorChecker is a registered trade mark by Macbeth, Division of Kollmorgen Corporation.

computer vision: field in computer science having the analysis of pictorial data as the main subject. Instead, computer graphics aims at the synthesis of pictorial data. *Image processing* and *image analysis* are the major subfields of computer vision, cp. Figure G1.

concavity: subset of the complementary set of a non-convex *region*, cp. Figure G2. The union of a region and its concavities defines the *convex hull* of this region. A point set is non-convex or concave, if there exist at least two points p, q in this set such that the straight line segment pq does not lie entirely in this point set. ♦ 6.4.2

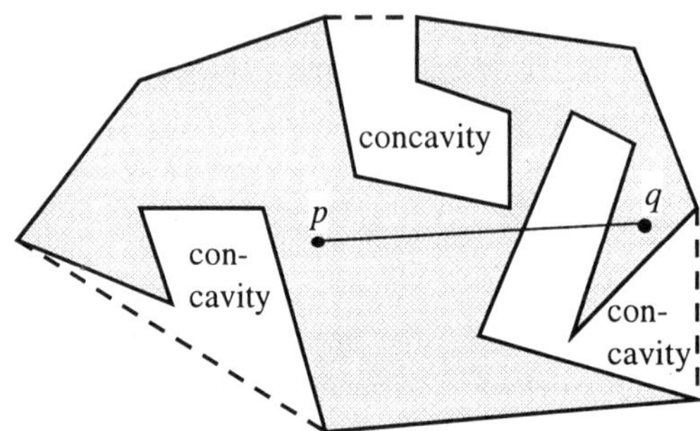

Figure G2: A non-convex region with three concavities. The straight line segment pq does not lie entirely in the region.

connectivity: point set property depending upon the used *neighborhood* definition (e.g. 4- or 8-neighborhood, cp. Figure G6). A set of image points **A** is connected, if, for any two points p, q of **A**, it exists a *path* from p to q lying entirely in **A**. ♦ 1.1.1

contour: simple closed curve which bounds a *region*. A contour is a *path* of image points. It can be obtained by *boundary* thinning. ♦ 7.2.1

contour detection: determination of *contour* locations in digital images. A parallel procedure typically computes first "candidates" of boundary or contour points (e.g., edge points), then contours are extracted on the base of such points by means of iterative procedures as, e.g., parallel relaxation. Serial contour detection can be based on various methodologies (e.g. dynamic programming, contour tracing) ♦ 6.2, 7.2.1

contour polygon: approximating polygon of a *contour*. A contour $p_1 p_2 \dots p_n p_1$ of grid points can be approximated by a polygon within certain limits of accuracy. Generally, such a contour polygon has much less vertices than the number of grid points on the given contour, thus simplifying further processing (e.g. computations of geometrical features). ♦ 1.1.1

contour tracing: serial procedure for *contour detection.* Starting with an initial point, a closed grid point path is constructed step by step. This path defines the contour of a *region.* ♦ 7.2.1

contrast: difference between the maximum and the minimum gray level of an image (global contrast), or of an image segment (local contrast). ♦ 5.1.2, 5.1.3

contrast enhancement: image processing task aiming at generating images with higher contrast. Increasing the image value difference between "dark and bright *pixels*" is possible, e.g., by enhancing object image values (higher values), or by reducing background image values. The enhancement factor can be constant in the entire image, or locally adaptive. ♦ 5.1, 6.3.1, 6.3.3, 6.3.4, 6.5.8, 6.5.9

convex hull: superset of a set of points defined by its convex closure. The convex hull $CH(\mathbf{A})$ is the intersection of all *convex sets* which contain the given set $\mathbf{A}$ of points. If $\mathbf{A}$ is a polygon, then $CH(\mathbf{A})$ is the union of $\mathbf{A}$ with all of its *concavities*, cp. Figure G2. Different definitions of the term "convex set" are possible in digital geometry. ♦ 6.4.2, 7.2.2

convex set: set of points $\mathbf{A}$ such that for any two points $p, q \in \mathbf{A}$ it holds that the straight line segment pq lies entirely inside $\mathbf{A}$. It follows $\mathbf{A} = CH(\mathbf{A})$ if $\mathbf{A}$ is convex, where $CH(\mathbf{A})$ denotes the *convex hull* of $\mathbf{A}$. There exist different approaches in digital geometry for defining digital straight line segments, and thus for defining digital convexity. ♦ 6.4.2

convolution: linear operator mapping two functions (images) into a new one. For an image f of size $(N+n) \times (N+n)$, and a *convolution kernel* g of size $n \times n$, with $n = 2k+1$, the convolution $f * g$ is an image of size $N \times N$, with

$$(f * g)(x, y) = \frac{1}{n^2} \sum_{i=-k}^{k} \sum_{j=-k}^{k} f(x-i, y-j) \cdot g(i, j)$$

for $1 \le x \le N$ and $1 \le y \le N$. ♦ 3.1.3, 6.1.1, 6.1.2, 6.1.3, 6.2.2, 6.2.4, 6.2.5, 6.6.2

convolution kernel: predefined function used in a *convolution* operation. A given convolution kernel g, cp. Figure G3, defines an image transformation T_g, with

$$T_g(f) = f * g \ .$$

The function g can also be defined in the *frequency* domain if the convolution is carried out via the *Fourier transformation.* ♦ 3.1.3, 7.3.2

convolution kernel, point-symmetric: special convolution kernel. A convolution kernel g of the size $m \times n$, where m and n are assumed to be odd, is point-symmetric if it has certain symmetry axes passing through the center point $((m+1)/2, (n+1)/2)$ of the kernel g, cp. Figure G3. ♦ 6.1.1, 6.1.2, 6.1.3

(a)

1	1	2	1	1
1	2	3	2	1
2	3	5	3	2
1	2	3	2	1
1	1	2	1	1

(b)

-4	-2	0	2	4
-4	-2	0	2	4
-4	-2	0	2	4
-4	-2	0	2	4
-4	-2	0	2	4

(c)

-4	-4	2	3	4
-7	-2	1	2	3
-5	-3	0	3	5
-3	-1	1	2	1
-2	1	2	4	3

Figure G3: The convolution kernel (a) has four axes of symmetry (horizontal, vertical, two diagonal), the convolution kernel (b) has one axis of symmetry (horizontal), and (c) has none.

convolution operator: image transformation T_g defined by a given *convolution kernel g*. ♦ 3.1.3

co-occurrence matrix: matrix $((a_{ij}))$, with $0 \leq i, j \leq G-1$, representing frequencies a_{ij} of gray value events ξ_{ij} which occur in a given (window of a) digital image f. Such a co-occurrence matrix is of size $G \times G$. For example, a gray value event ξ_{ij} (with parameter d) can be such that "there exists an image point p in the given window with $f(p) = i$, and in the given window there exists also at least one point q in distance d to p such that $f(q) = j$". Features of co-occurrence matrices can be used for defining *window functions*. ♦ 1.3.2

cursor position: interactively or automatically labeled screen position. A cursor can be used, e.g., for the visualization or for the interactive fixation of a current *reference point* position of a moving window. ♦ 1.2.2.

data, iconic: data array representing an image as an image value matrix. ♦ 1.1

density function: function in probability theory. A discrete random number ξ takes on a value out of a finite, or of an enumerable infinite set $\{a_1, a_2, a_3, \dots\}$. Let

$$p_i = P(\{\xi = a_i\})$$

be the probability that ξ takes value a_i. The function ϕ,

$$\phi(i) = p_i,$$

is the density function of the discrete random number ξ. For example, a gray value histogram is an estimation of the density function of the random number "image value". ♦ 1.3.2, 6.4.3

desktop publishing system: cp. *DTP-system.*

detail: fine image structure with high spatial frequencies near to the cutoff *frequency.* For example, a detail can be a thin line, a periodic pattern of high frequency, a fine texture, or stochastic noise. ♦ 6.1.7, 6.5.9

digitization: mapping of analog images (input signals) into digital images. This process consists of a spatial digitization (*raster* digitization) and of an amplitude digitization (*quantization*). ♦ 1.1

discretization: synonym of *digitization.*

discretization noise: random variation of digital image values arising in the image acquisition process (instability of the CCD-camera, changes in illumination etc.). This noise is not correlated with the image content. For example, an image *binarization* should be robust with respect to discretization noise (binarization with hysteresis, deletion of small artifacts etc.) ♦ 5.3.1, 6.1.7

distance, Euclidean: distance measure

$$d_2(p_1, p_2) = \sqrt{(x_1 - x_2)^2 + (y_1 - y_2)^2}\ .$$

between two points $p_1 = (x_1, y_1)$ and $p_2 = (x_2, y_2)$ of the plane. For different distance measures e.g. Manhattan and maximum metric see ♦ 1.1.1.

disturbance, distortion: alteration of *pictorial data* in amplitude or with respect to the geometrical mapping arising during image acquisition (e.g. single sensor elements of a CCD-array can measure irradiances with different I/O-curves, geometrical distortions can be due to optical distortions of the camera lens, or shading can influence the measured values), image transmission (e.g. certain interferences can occur during radio transmissions), or image digitization (e.g., the AD-converter is characterized by a non-linear behavior). Normally for a given image an interaction of several distortions must be assumed. Disturbance and distortion models help in guiding the design of image transformations for image enhancement. ♦ 5.1, 6.1, 6.3, 6.5, 6.6.2

distribution function: function in probability theory. For a discrete random variable ξ with *density function* ϕ,

$$\phi(i) = p_i = P(\{\xi = a_i\}),$$

it is also defined that

$$\Phi(i) = \sum_{a_j \le a_i} \phi(j)$$

is the density function of ξ. The sum histogram of *gray levels* is an estimation of the distribution function of the random variable "gray level". ♦ 1.3.2, 5.1.4, 6.1.6

DTP-system (*desktop publishing system*): computer work place, provided with a high-resolution screen and high-quality printer, allowing text layouts including graphics and (gray level or color) images. Several software functions are available for image processing by means of such a DTP-system. An object-oriented interface simplifies the user-system interaction. ♦ 2

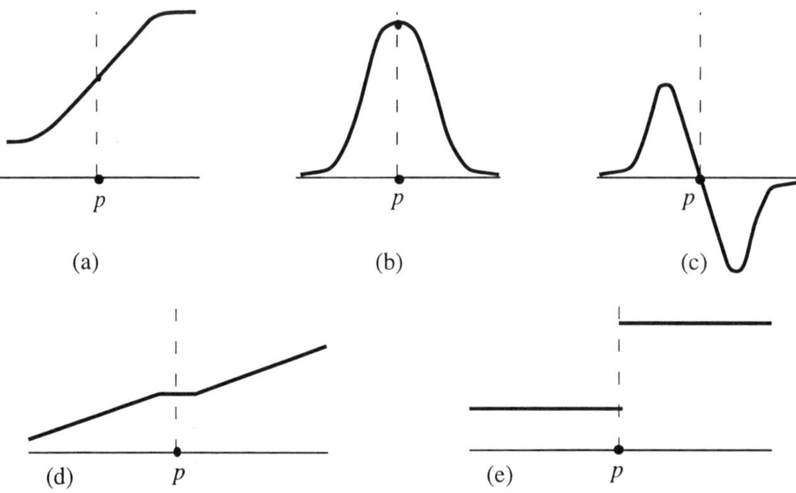

Figure G4: (a) Cut through a "linear edge" (assumed as continuous curve), (b) first derivative of this curve, and (c) second derivative (p is at the zero crossing position), all at the same edge position p. (d) An edge between two "slope segments" can be a locally constant gray value. (e) An ideal step edge is not a continuous curve.

edge: *boundary* characterized by strong gray value changes between two neighboring image segments. The ideal case of a linear edge between two gray level plateaus is shown in Figure G4 (a) and in Figure G5 (a). Such an edge type allows the design of edge detection operators inspired by derivation models (cp.. Figure G4 (b, c): maximum values of the first derivative, or zero-crossing of the second derivative). Further edge variants are illustrated in Figure G4 (d, e) and Figure G5 (b). In real gray value images edges are irregular in their gray level structure or affected by different disturbances. Image segments on both sides of an edge normally do not have uniform gray levels. Some disturbance models should also be assumed for these segments. Edges are very important features of an image. ♦ 1.1.2, 6.2

edge extraction: task in image processing directed on computing *edge* positions in digital images. The result can be visualized by an edge image. ♦ 6.2

edge operator: image transformation for edge extraction. ♦ 6.2

edge image: image visualizing the result of an edge operator. All edge pixels are labeled by certain image values. The edge slope and the edge amplitude can contribute to these image values. ♦ 6.2

edge slope: change in image values orthogonal to the edge direction. In Figure G5, an edge is assumed to exist between pixels $(p, f(p))$ and $(q, f(q))$, where f denotes an image, and the edge slope is equal to the slope of the line joining these two pixels (i.e. equal to **tan** α or **tan** β in these two examples). ♦ 6.2

edge slope, steeper: is associated with thinner edge zones, cp. Figure G5. An edge can be better localized if the edge slope has been enhanced before. ♦ 6.3.1, 6.3.4

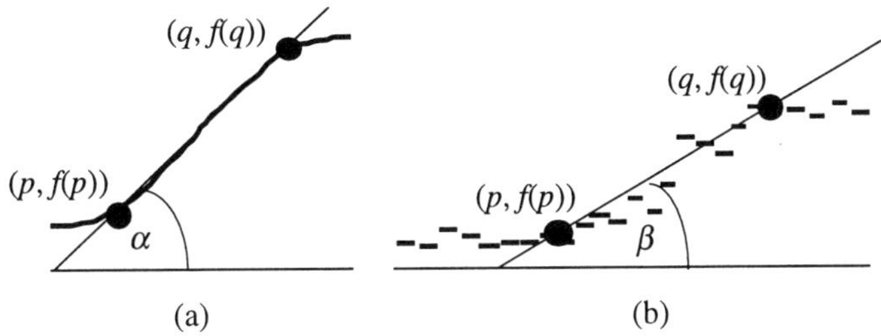

Figure G5: (a) A linear edge with slope **tan** α and (b) a non-linear edge with estimated slope **tan** β . Edge (a) has a steeper edge slope than edge (b).

false color: not corresponding to the coloring of the imaged objects. False colors can efficiently be assigned by using a color value look-up-table. ♦ 1.1.3, 5.6.3

feature: property of a pixel or of an image segment. Features can be continuos or discrete scalar values, as a gray level, the absolute value of the gradient, or the value of an isotropy measure. Furthermore, features can be binary predicates, as "the gray level is smaller than S", "the region is convex", or "the reference point is an isolated point". ♦ 1.1.2

FFT: cp. *Fourier transformation, fast*.

filter, 2-D: synonym of *spatial filter*.

filter coefficient: value of a *convolution kernel*'s element. A 2-D convolution kernel g consists of $m \times n$ filter coefficients $g(i, j)$. ♦ 1.3.3, 6.1.1

FIR-filter: cp. *linear filter, non-recursive*.

Fourier transformation: transformation of vectors or arrays of complex data into vectors or arrays of complex data representing the frequency components of the given data. ♦ 1.4.4, 3.4.8, 7.3.1, 7.3.2

Fourier transformation, fast (*FFT*): special algorithmic realization of the Fourier transformation. Instead of $\mathbf{O}(n^2)$ arithmetic operations required with the formula of the Fourier transformation, this algorithm performs only $\mathbf{O}(n \cdot log\ n)$ arithmetic operations if n complex numbers have to be transformed. ♦ 3.4.8, 7.3.1

frame buffer: special memory for storing pictorial data. For displaying an image on a screen, it is sufficient just to address the content of the corresponding frame buffer. Special image processing cards, e.g. for a PC or a work station, contain several frame buffers, and support efficient operations within (e.g. histogram calculation) or between (e.g. binary point-to-point operations) different frame buffers. ♦ 3.3

frequency: coordinate in the spatial frequency domain. A *Fourier transformation* maps data into this domain.

Gaussian low-pass: linear low-pass filter whose *filter coefficients* are values of a Gaussian function. These filter coefficients are given as a point-symmetric *convolution kernel*. ♦ 6.1.3, 6.2.4, 6.3.2

grading function: mapping of gray levels into gray levels. The diagram of such a function is a so-called grading curve, cp. Figure G7. ♦ 1.3.3, 5.1.2, 5.1.3, 5.1.4, 6.5.8

gradient: vector of derivatives of a function $u = f(x, y)$ into x- and y-direction. At a point (x, y) this is the vector

[**grad** *f*](*x*, *y*) = \B\BC(\F(∂f(x, y),∂x) , \F(∂f(x, y),∂y)) .

This vector is characterized by its orientation

$$\textbf{arc tan}\left(\frac{\frac{\partial f}{\partial y}}{\frac{\partial f}{\partial x}} \right)$$

and its length. Its slope is defined as ratio

$$\frac{\frac{\partial f}{\partial y}}{\frac{\partial f}{\partial x}} .$$

Often this ratio is called gradient of an image *f* in image processing. Local operators are used to calculate approximate derivatives of an image *f*. ♦ 1.1.2, 6.2.2

gray level: integer in the range 0, 1,..., *G*–1, with G ≥ 2, corresponding to a gray shade of an image point. The maximum gray level *G* – 1 corresponds to "white", and the minimum gray level 0 corresponds to "black". ♦ 1.1.2

gray level distribution: *distribution function* of the random number ξ of gray levels in one or several images. The random number ξ takes on values in the set {0, 1,..., *G* – 1}, and the general distribution function is defined by

$$P_\xi(\mathbf{B}) = P(\xi^{-1}(\mathbf{B})), \text{ with } \mathbf{B} \subseteq \{0, 1,..., G-1\},$$

where $\xi^{-1}(\mathbf{B})$ denotes the set of all image points (a subset of the raster **R**) in which ξ takes on a value out of the set **B**. However normally the special distribution function

$$P_\xi(u) = P(\text{"}\xi = u\text{"}), \text{ with } 0 \le u \le G-1,$$

is considered. The gray level histogram is an estimation of this function. ♦ 1.3.2

gray level dynamics: interval [*max*, *min*] in the range 0, 1,..., *G* – 1, where *max* and *min* are the largest and the smallest gray value occurring in an image, or in an image segment. ♦ 5.1

gray level image: scalar image *f* with image values *f*(*x*, *y*) in the set {0, 1, ..., *G* – 1} of gray levels. ♦ 1.1.

gray value agglomeration: an (iterative) image segmentation process controlled by a gray value similarity criterion. ♦ 6.1.6, 6.4, 6.5.7

gray value input-output function: synonym of *grading function.*

grid: regular tessellation of the *image plane* into squares (cp. *grid point*, *image cell*). Further regular tessellations could be into regular hexagons, or into equilateral triangles. However these tessellations are normally not used in image processing. ♦ 1.1.1

grid point: point with integer coordinates in the *image plane* which is used for defining a *pixel*. ♦ 1.1.1

halftone image: see *bilevel image*.

halftoning: transformation of a *gray value image* into a *halftone image* in which the visual impression of the gray level image should be preserved as well as possible. ♦ 2.2, 5.3.4, 6.1.8

high-pass filter: filter for attenuating low spatial frequency components, and for enhancing high spatial frequency components in an image. ♦ 6.3.2, 6.3.3, 6.3.4, 6.5.8, 6.5.9

histogram: graphical representation of frequencies of occurrence by means of a diagram. In computer vision this term is normally identified with the gray level histogram, used as an estimation of the global or local gray level distribution. ♦ 1.3.2, 5.1.4

homogeneity: attribute of a segment meaning that this segment is considered to be uniform in some sense. ♦ 6.1.4, 6.1.5, 6.5.9

hysteresis: preserving a state after the end of its cause. ♦ 5.3.1

IIR-filter: cp. *linear filter, recursive.*

image: two-dimensional arrangement of scalar or vectorial measurements representing pictorial information. An image f has a value $f(x,y)$ at image point (x,y). During a process of image acquisition, an analog image (*input signal*) is transformed into a discrete image (cp. *AD-converter*, *digitization*, *raster*). ♦ 1.1

image, analog: see *image*.

image, discrete: see *image*.

image, iconic: image in its direct pictorial representation (array). Non-iconic image representations can be obtained by special coding schemes (e.g. run-length coding, contour coding, etc.). ♦ 1.1

image analysis: subfield of computer vision oriented towards the interpretation of the content of pictorial data (cp. *image understanding*, *pattern analysis*).

image cell: basic area element in the *image plane* (one grid square) which is used for defining a *pixel*. ♦ 1.1.1

image coding: transformation of pictorial data into a different data representation which needs less memory than the iconic representation, or which has some other benefits (e.g. encryption of images). ♦ 7.2.1, 7.3

image coordinate: xy-coordinates in the *image plane*, xyi-coordinates of multichannel images with image values $f(x, y, i)$, or xyt-coordinates of image sequences with image values $f(x, y, t)$. ♦ 1.1.1, 1.1.3

image enhancement: image processing task aiming at improving the visibility or identification of image structures or details. For example, the interference or noise suppression can be a specific aim of image enhancement. ♦ 5.1, 5.4, 6.1, 6.3, 6.5, 6.6.2

image matrix: matrix representation of the $M \times N$ image *raster*. ♦ 3.2

image model: abstract specification of a class of real images by means of ideal (typical) features of image segments. Each segment is homogenous in the sense of a certain criterion of uniformity. In structural image models the gray value function of such segments is approximated by analytical functions with variables x,y. In statistical image models, homogenous segments are characterized by statistical features, e.g. by the average value, the variance, properties of the *co-occurrence matrix*, or prediction coefficients. ♦ 5.4.2, 5.5.2, 5.5.3, 6.4.3, 6.6.1, 6.6.2

image operator: synonym of *image transformation*. ♦ 1.4

image parsing: *segmentation* by dividing images into *regions*. ♦ 6.1.4, 6.1.5, 6.1.6, 6.4, 6.5.5, 6.5.7, 6.5.9

image plane: real xy-plane tessellated by a *grid*, having *grid points* or *image cells* as basic discrete elements (image points). A digital image is defined by values at such image points, where a neighborhood relation is assumed for defining topological connectedness, cp. Figure G6. ♦ 1.1.1

image point: *grid point* or integer address of an image *cell*.

image processing: discipline of computer vision dealing with transformations of pictorial data into pictorial data. The design of an image transformation follows special aims, e.g. in image segmentation, image enhancement, or image restoration. Image processing often plays the role of a preliminary early processing in view of image analysis. cp. Figure G1.

image processing, parallel: simultaneous performance of image processing by means of specialized hardware, as parallel processors, associative memory or transputers. However computers normally used in image processing are still serial processors. That means that such a computer can perform window operations only at one current image point at a time. Assuming a parallel processing scheme for such a serial machine means that the processing order of current image points does not influence the result of the process. ♦ 3.2.2, 3.2.3

image processing, sequential: unlike parallel image processing, operations are carried out on one pixel at a time, consecutively. Thus the operation's arguments are, in general, partly original and partly previously processed gray values. The resultant image depends upon the pixel processing order. ♦ 3.2.2, 3.2.3

image refresh memory: special memory containing the screen data. *Look-up-tables* are used for controlling the display mode. ♦ 5.6.3

image restoration: image processing task aiming at compensating or eliminating distortions or impairments due to the process of image acquisition or image transmission. An ideal compensation is only possible if the image acquisition or transmission process is mathematically describable. ♦ 6.1, 6.3, 6.5

image segment: connected set of *pixels*. Unlike a *region*, an image segment can also feature holes or cavities. ♦ 1.1.1

image segmentation: process of dividing an image into disjoint image segments. This can be achieved either by computing border lines (e.g., *contour tracing*), or by grouping of pixels (e.g., *region growing, agglomeration*) ♦ 5.5, 6.2, 6.4, 6.5.2, 6.5.7, 6.5.9

image transformation: mapping of pictorial data into pictorial data. An image transformation generates one or several resultant images out of one or several given images. For example, the Fourier transformation produces two data arrays (real and imaginary part) for a given gray value image which can be transformed into two gray value images. ♦ 1.4

image transformation, affine: mapping of pictorial data based on a linear coordinate transformation. Because of the discrete nature of the image points and of the finiteness of the *raster*, each transformation must be discussed separately. ♦ 4.3

image value: scalar (gray value) or vector (color value, value of a multi-channel image). The term $f(x,y)$ denotes the value of an image f at the image point (x,y). It can be defined on a continuous (analog image) or on a discrete (digital image) scale. ♦ 1.1.2, 1.1.3

image understanding: special discipline of *image analysis* dealing with the interpretation of images with respect to three-dimensional scene objects.

input signal: continuous input (*analog image*) of a computer vision system. ♦ 1.1

isotropy: invariance of an image *feature* with respect to angular orientation. In digital images, this invariance property is usually evaluated based on rotations which are multiples of 45°. ♦ 6.5.7, 6.5.9, 6.6.1, 6.6.2

linear best fit: data approximation by a linear function which is optimum according to a certain distance measure or optimization criterion. As examples of linear best fit operations, a *linear regression* of a random variable with respect to an other random variable is given by a straight line (the regression function G), and with respect to two other random variables by a plane. In image processing often it is desirable that gray values in an image *segment* are approximated by a linear function, i.e. by a plane. ♦ 5.4.2

linear filter, non-recursive (*FIR-filter*): filter whose result is a linear combination of the image values of an image segment. ♦ 3.1.3, 6.1.1, 6.1.2, 6.1.3, 6.2.4, 6.3.2, 6.6.2

linear filter, recursive (*IIR-filter*): filter whose result is a linear combination of the unprocessed gray values as well as of some resulting gray values of an image segment. ♦ 3.2.2, 3.2.3, 6.1.1, 6.2.5

look-up-table: representation of a functional relation $H(a_i) = b_i$, for the direct access. The value $H(a_i)$ is simply obtained by reading from the table the adequate value b_i which is defined by argument a_i. For example, a look-up-table can assign different color values $b_0, b_1,..., b_{G-1}$ to the gray levels 0, 1,..., $G-1$. These color values can be used for the visualization of pictorial data. The gray value $a_i = i$ is represented by the color value b_i on the screen. Without altering the image values, the visualization can be modified by loading different look-up-tables. Such look-up-tables are commonly used for the visualization of the image refresh memory (for the gray level scale, or for the value scales of the three color channels). ♦ 5.1.4, 5.6.3

low-pass filter: filter for attenuating high spatial frequency components, and for enhancing low spatial frequency components in an image. ♦ 6.1, 7.3.2

main direction: a possible direction of a step in the picture grid leading from the current point to one of its neighbors. For example, the main directions of the 8-neighborhood are $i\cdot(\pi/4)$, with $0 \le i \le 7$, encoded by the numbers 0, 1,..., 7. ♦ 1.1.1, 7.2.1

median: see *quantile*.

mathematical morphology: mathematical discipline of the set theory studying pattern transformations based on combinations of the basic operations of erosion and dilation. The definition of these basic operations includes the specification of a *structuring element*. Erosion and dilation were originally defined as set theoretical operations, i.e. for binary images. Extensions to gray level images are possible. ♦ 6.5.2

multi-level thresholding: point operator defined by several gray level thresholds in the gray level range 0, 1,..., $G-1$. A gray value $f(x, y)$ is mapped onto a selected gray value if it is comprised between two consecutive thresholds. Such a selected gray value is specified for each pair of two consecutive thresholds. ♦ 5.5

neighborhood operator: local operator in which the placed window $\mathbf{F}(p)$ consists only of p and of the neighbors of the current point p. The chosen neighborhood must be specified. ♦ 1.1.1, 1.4.3

neighbor of an image point: attribute specifying the relationship existing between discrete image points. This relationship is symmetric, non-transitive, and non-irreflexive. ♦ 1.1.1

noise suppression: image processing task aiming at reducing certain additive and/or multiplicative components which are not correlated with the pictorial content. ♦ 6.1, 6.5

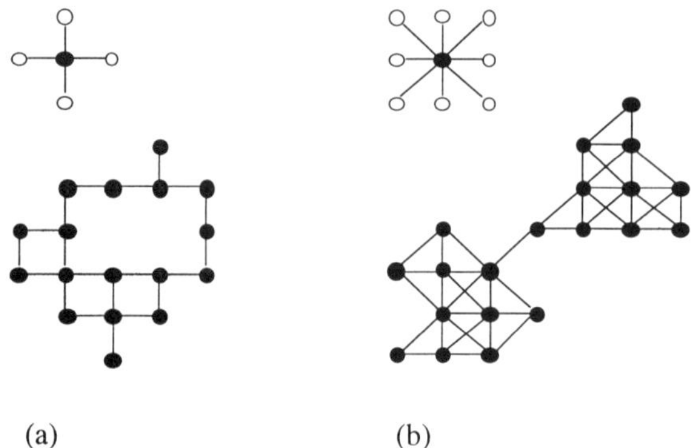
(a) (b)

Figure G6: (a) Representation of the 4-neighborhood and of a 4-connected set of image points; (b) representation of the 8-neighborhood and of an 8-connected set of image points which is not 4-connected.

object: image segment which is not a *background* segment. ♦ 2.5

operator, iconic: synonym of *image transformation.*

operator, local: operator defined by a window **F** such that each computed image value $h(p)$ of the resultant image h depends only upon the original and eventually upon some processed image values within the placed window $\mathbf{F}(p)$. ♦ 1.4.3, 6

operator kernel: synonym of *window function* .

path: sequence of image points $p_1, p_2, \ldots, p_n$ leading from p_1 to p_n, where p_i is a *neighbor* of p_{i+1}, with $1 \leq i < n$. ♦ 1.1.1

pattern: image or multi-dimensional data which is subject of a processing not associated with interpretation of these data as three-dimensional real world object (e.g. characters, line drawings, point clusters, also often microscope images or aerial views, if a three-dimensional interpretation is not relevant). ♦ 1.1.2

pattern analysis: special discipline of *image analysis* dealing with the interpretation of structures in *patterns.*

pictorial data: data representing an image. Different data structures can be used for pictorial data. Image processing applies iconic representations (arrays). Large memory requirements are typical for pictorial data. The coding of pictorial data aims at the reduction of the memory requirement. ♦ 3.2

pictorial signal, continuous: see *image, input signal.*

pictorial signal, discontinuous: see *image.*

picture element: see *pixel.*

picture object: real world object which is visualized by the given pictorial data. For example, in microscope images an picture object can be a nucleolus, a tissue region or an *artifact*. ♦ 2.5

pixel: picture element $(x, y, f(x, y))$, consisting of a *grid point* (x, y) and an image *value* $f(x, y)$ at this point (grid point model of the *image plane*), or of an image *cell* at position (x, y) and a constant value $f(x, y)$ within this cell (cellular model of the image plane). ♦ 1.1

point operator: image transformation where a resultant image value only depends upon a single *pixel* of the input image. ♦ 1.4.2, 5

poster effect: visual impression given by an image consisting of segments with constant gray value. ♦ 2.9, 5.5

quantile: value of i such that $\Phi(i) = q$, with $0 \leq q \leq 1$, where Φ is the *distribution function* of a discrete random variable i . More exactly, this value is denoted to be the q-quantile. For example, the 0.5-quantile is the median. ♦ 6.1.6.

quantization: selection of a finite number of discrete values out of a continuous value scale. ♦ 1.1.2

rank-order filter: local operator in which the *window function* is a function of the rank-ordered gray values within the current picture window $\mathbf{F}(f,p)$. The median filter, or the so-called L-filters, where the window function is a weighted sum of the rank-ordered window gray values, are examples of rank-order filters. ♦ 6.5

raster: set of image points (x,y) at which image values are assumed, normally a rectangular array of $M \times N$ image points. ♦ 1.1

raster scan order: scanning order for digital images, starting at the lowermost row $y = 1$ and ending at the uppermost row $y = N$, and left to right in each row, i.e. from $x = 1$ to $x = M$. ♦ 3.1.4, 3.3

reference point: selected grid point within a window. ♦ 1.2.1

region: simply connected set of image points, i.e. without holes or cavities. The connectivity definition follows from the choice of a neighborhood relation for image points, cp. Figure G6. ♦ 1.1.1

region growing: segmentation procedure based on a pixel grouping process. A larger segment is generated step by step, starting with a single pixel (seed), and merging a pixel with the current segment if it is a neighbor of a previously merged pixel, and if a predetermined uniformity criterion is still satisfied after merging this pixel (e.g. a special *texture* feature). ♦ 6.4

registering: geometrical fitting of several images showing the same picture object, e.g. aerial views of the same territory taken at different time, and with different viewing angles. ♦ 4.3

regression: approximation of statistical dependencies. Let ξ_1 and ξ_2 be two random variables. A function G is a regression of ξ_2 with respect to ξ_1, if $G(\xi_1)$ is an approximation of the statistical dependency of ξ_2 in relation to ξ_1. The random variable ξ_2 is a sum of two random variables,

$$\xi_2 \equiv G(\xi_1) + K(\xi_1, \xi_2),$$

where $K(\xi_1, \xi_2)$ is a difference function. Often the function $G(\xi_1) \equiv E(\{\xi_2 \mid \xi_1\})$ is used as regression of ξ_2 with respect to ξ_1. This approach minimizes the square deviation $E(\{K(\xi_1, \xi_2)^2\})$. ♦ 5.4.2

regression, linear: a *regression* using the linear function

$$G(\xi_1) = \alpha_{12} + \beta_{12} \cdot \xi_1$$

as regression function $G(\xi_1)$, where

$$\alpha_{12} = E(\xi_2) - \rho_{12} \cdot \frac{\sigma_1}{\sigma_2} \cdot E(\xi_1) \quad \text{and} \quad \beta_{12} = \rho_{12} \cdot \frac{\sigma_1}{\sigma_2}$$

are the regression coefficients,

$$\rho_{12} = (\sigma_1\sigma_2)^{-1}\, E\big(\{\xi_1 - E(\xi_1)\} \cdot \{\xi_2 - E(\xi_2)\}\big)$$

is the correlation coefficient, and σ_1, σ_2 are the standard deviations. The correlation coefficient ρ_{12} can be used as quality measure of a "best" linear approximation. ♦ 5.4.2

resolution: the number of image points per image, e.g. 512 × 512. Image *smoothing* allows a gradual change from higher resolutions to lower resolutions. ♦ 1.1.1

scene: real three-dimensional world as seen from a visual sensor. Scenes are analyzed in computer vision by studying a finite number of two-dimensional visual projections (*images*).

segmentation: see *image segmentation.*

sequency: coordinate in the spatial sequency domain. A *Walsh transformation* maps data into this domain.

shading: gray value attenuation at image boundaries due to the image acquisition process, e.g. typical for microscope images. ♦ 5.4.1, 5.4.2, 6.3.2

sharpening: increasing the visibility of image structures with high spatial frequencies (details or texture elements), or increasing the edge slope. ♦ 6.3, 6.5.1, 6.5.4, 6.5.7, 6.5.8, 6.5.9

signal processing, 2-*D*: discipline of handling the information exchange between technical systems (signals that are characterized by a representation and by its content), in which (two-dimensional) pictorial signals are considered. Coding and noise suppression are typical signal processing tasks. Analog image signals are represented on the two-dimensional real plane, and discrete image signals on a *grid.* ♦ 1.1, 1.4

smoothing: image transformation for enforcing an equalization of neighboring gray levels (reduction of high-frequency components). Gray value spikes ("outliers") are suppressed, noisy gray value variations are equalized, and texture patterns are attenuated. A secondary effect of smoothing is that steep edges are transformed into smooth edges, and *details* are deleted. Global 2-D filters or local operators can be used for smoothing. ♦ 6.1, 6.4, 6.5.1, 6.5.3, 6.5.4, 6.5.5, 6.5.6, 6.5.7, 6.5.9

solarization: special effect of photo developing, which can be simulated by means of a *point operator* defined by a grading function as shown in Figure G7. ♦ 5.1.3

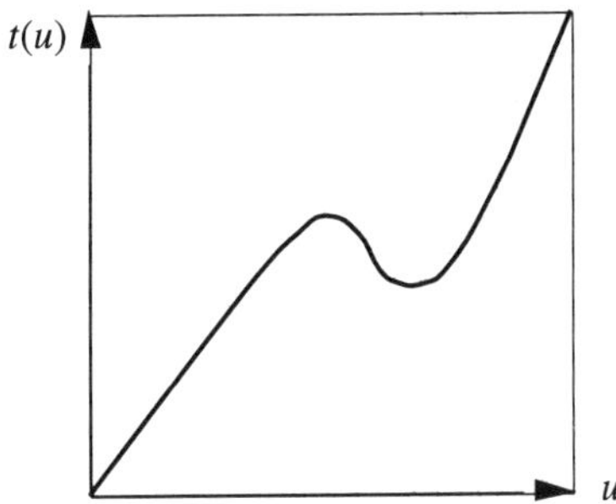

Figure G7: Grading curve of a transfer function t for achieving a solarization effect.

spatial domain: synonym of *image plane.*

spatial filter: image transformation in which the resultant value $h(x,y)$ at point (x,y) is computed in a certain spatial surrounding of the point (x,y) in the original image f. Generally, the value $h(x,y)$ is strongly influenced by the value $f(x,y)$. A linear spatial filter can be realized by a convolution with a certain convolution kernel g (filter function). ♦ 1.4.3, 1.4.4

spatial frequency domain: complex plane for representing the results of the *2D-Fourier transformation.* The inverse Fourier transformation maps data from the spatial frequency domain into the *spatial domain.* ♦ 7.3.1

spectrum: absolute values of the complex coefficients in a Fourier transformed image. ♦ 7.3.3

structuring element: window (in general, non-rectangular) for defining the area of influence in basic operations (erosion, dilation) of the *mathematical morphology*. The set of image points in the placed window has to be checked for intersection with object segments or background segments. Defining a structuring element requires to define also the position of its *reference point.* ♦ 1.2.1, 6.5.2

template matching: comparison of an picture window with a predefined pattern of image values. A "best match" minimizes the (weighted) deviations between the two templates. A cross-correlation is a possible way for carrying out a match process. ♦ 6.2.2, 6.6.1, 7.1.2

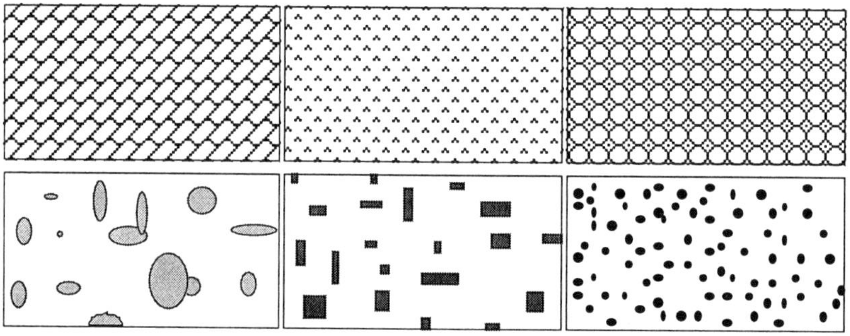

Figure G8: Examples of textures, (above) regular repetitions of geometrical texture elements,, and (below) random distributions of texture elements.

texture: structured variations of the surface radiance of real world objects with repetitive character, or synthetically generated gray value, or color value patterns. A texture can be characterized by elementary structural texture elements (e.g., brickwork, magnified views of textiles), or by random distributions of image values, single or in groups (e.g., grass, moss, leaves), cp. Figure G8. The description of a texture depends upon the resolution of the image. Figure 2.1 shows (synthetic) gray value textures. ♦ 2.1, 6.3.2, 6.3.3, 6.5.8

threshold: selected value dividing a value scale into two intervals, e.g. into values that are equal to or lower than the threshold, and into values that are higher than the threshold. ♦ 5.3, 5.5, 6.3.2

thresholding: image transformation defined by one or several thresholds. The classical threshold operator transforms *gray value images* into *bilevel images*. It is characterized by a threshold S, with $0 \le S \le G-1$. Gray values u are either mapped to $G-1$, if $u > S$, or to 0. ♦ 5.3, 5.5, 6.3.2

transfer function: synonym of *grading function*.

uniform distribution: special *distribution* of a discrete random number ξ on a set $\{p_1, p_2, ..., p_n\}$ of possible values (e.g., gray levels), with

$$P(\{\xi = p_i\}) = \frac{1}{n}, \quad \text{for } 1 \le i \le n.$$

A discrete uniformly distributed random number has the expected value

$$E(\xi) = \frac{1}{n}\sum_{i=1}^{n} p_i$$

and the variance

$$\sigma^2 = \frac{1}{n}\sum_{i=1}^{n}\left[p_i - \frac{1}{n}\left(\sum_{i=1}^{n} p_i\right)\right]^2,$$

where the addition between the values p_i has to be defined in an adequate sense (e.g., the common addition if the p_i's are gray values). ♦ 5.1.4, 5.2.2

window: rectangular set of grid points. More generally, a window could be any set of grid points forming a special point pattern. ♦ 1.2.1

window function: function defined for picture windows $\mathbf{F}(f, p)$. This function yields exactly one (new) image value, which must be stored at the position p at the resulting image. ♦ 1.3

window operator: synonym of *local operator*.

INDEX

Schimke: Handbook of Image Processing Operators – Disk

The disk contains programs which enable you to view the images directly in the 'raw' or PMG format, although not in colour. Programs on the disk are available in both 32bit and 16bit versions, and all source code is provided.

To run the DOS programs you will only need a VGA card. For higher resolutions, a VESA compatible BIOS is required. The image size is only limited by the graphics mode used.

The images can be viewed with nearly any graphics adapter card, since the authors have used the graphics mode 13H with a resolution of 320x200 pixels and 256 colours as default mode which is present with any VGA compatible graphics adapter card.

ISBN: 0 471 96705 X Price: £24.95 (plus VAT)

TO ORDER:

Schimke: Handbook of Image Processing Operators – Disk 0 471 96705 X

Please send me.....copies of **Schimke: Handbook of Image Processing Operators – Disk** at £24.95 (plus VAT) each.

Postage and handling free for cash order or payment by Credit Card

[ ] Remittance enclosed..............Allow approx. 14 days for delivery

[ ] Please charge this order to my credit card (All orders subject to credit approval)

Delete as necessary: AMERICAN EXPRESS, DINERS CLUB, BARCLAYCARD/VISA, ACCESS/MASTERCARD.

CARD NUMBER [..............................] Expiration date.................

VAT EC countries only – pleased complete as appropriate

[ ] I am registered for VAT; my VAT registration number is:..........................

[ ] Please send me an invoice for prepayment. A small postage and handling charge will be made

Software purchased for professional purposes is generally recognised as tax deductible

Name/Address: ....

....

....

....

Official Order No: ....

Signature: ....

If registered for VAT please quote your VAT number above. For non-registered customers, it may be necessary to add VAT to your order. Please contact our Customer Services Department on +44 1243 779777 (tel) or +44 1243 775878 (fax) or email customer@wiley.co.uk for further advice.

Affix
stamp
here

Customer Service Department
JOHN WILEY & SONS LIMITED
Distribution Centre
Shripney Road
Bognor Regis
West Sussex
PO22 9SA
England